THIRD EDITION

formerly *Electronic Design and Publishing Business Practices*
fully revised and updated

DIGITAL
DESIGN
BUSINESS
PRACTICES

FOR GRAPHIC
DESIGNERS AND
THEIR CLIENTS

including
DESKTOP PUBLISHERS
WEB DESIGNERS
PRINT DESIGNERS
ART DIRECTORS
WRITERS
ILLUSTRATORS
PHOTOGRAPHERS
PROJECT MANAGERS
PREPRESS HOUSES
COLOR SEPARATORS
PRINTERS
PROGRAMMERS
IT MANAGEMENT

This book is especially dedicated to the memory of **Michael Waitsman** who passed away as the Second Edition was being printed in 1995. It was his last design project. Michael was the spirit behind this effort. He brought the Chicago publishing community together in 1986 to pioneer desktop publishing. As a visionary, designer, photographer, writer, inventor, and one of the country's first interactive artists, Michael was a great influence on our profession. He inspired and mentored many professionals. It was Michael who saw the need for this book and helped me with the first two editions. He died on February 19, 1995 of a brain tumor at the age of 47. Some say the good die young. Michael's talents and integrity made him one of our industry's best. It is unfortunate we can't benefit from the many visionary and artistic gifts he would have brought to Internet design, as he left us by only helping to define its infancy. He saw this potential and believed that the Web would not only transform the graphic design industry, but bring it together in a new community of interest. His reach will be felt forever. To my partner and husband, I dedicate this third edition, for without him, I could never have created the first two or had the wherewithal to continue on without him to do this third.—Liane Sebastian, February 19, 2001.

Published by **Allworth Press**, an imprint of Allworth Communications, Inc., 10 East 23rd Street, New York, NY 10010. Visit www.allworth.com/dd Distributor to the trade in the United States: Consortium Book Sales & Distribution, Inc. 287 East 6th Street, Suite 365, Saint Paul, MN 55101.

Library of Congress Catalog-in-Publication Data:

Sebastian, Liane.
   Digital design business practices : for graphic designers and their clients / by Liane Sebastian.-- 3rd ed.
      p. cm.
   Rev. ed. of: Electronic design and publishing. 2nd ed., rev. and expanded. c1995.
   Includes bibliographical references and index.
   ISBN 1–58115-086-5
   1. Printing industry--United States--Management--Handbooks, manuals, etc. 2. Electronic publishing--United States--Management--Handbooks, manuals, etc. 3. Web site development industry--United States--Management--Handbooks, manuals, etc. I. Sebastian, Liane. Electronic design and publishing. II. Title.

   Z244.6.U5 S43 2001
   686.2'2544536--dc21

                                                      2001022587

Printed in Canada
Book cover by Liane Sebastian
Book interior design by Michael Waitsman and Liane Sebastian
MichaeLight Communications, Evanston IL, **www.michaelight.com**

This is a revised and updated edition of a book previously titled *Electronic Design and Publishing Business Practices*.

ACKNOWLEDGMENTS

CONTRIBUTORS

Communication Arts Magazine

Graphic Artists Guild

Graphic Design USA Magazine

HOW magazine

www.creativepro.com

www.imagingmagzine.com

www.lunareclopse.net

www.zone zero.com

CONFERENCE AND INTERVIEW SOURCES

AIGA Conferences

International Design Conferences in Aspen

GraphX and GraphExpo Conferences

Seybold and CONCEPPTS Conferences

Print Outlook Conferences

CONTENT EDITORS

Betsy Shepherd, Michael Waitsman,
Van Tanner, Eda Warren, Susan Wascher-Kumar

COUNSEL

William T. McGrath, Davis Mannix & McGrath

SPECIAL CONTRIBUTORS

Stephen P. Aranoff, Emily Ruth Cohen, Tad Crawford,
Gary G. Field, Robert L. Fitzpatrick, Caryn R. Leland,
Jon Leland, Maria Piscopo, Ellen Shapiro, Nancy E. Wolff.

TABLE OF CONTENTS

PLANNING

DEVELOPMENT

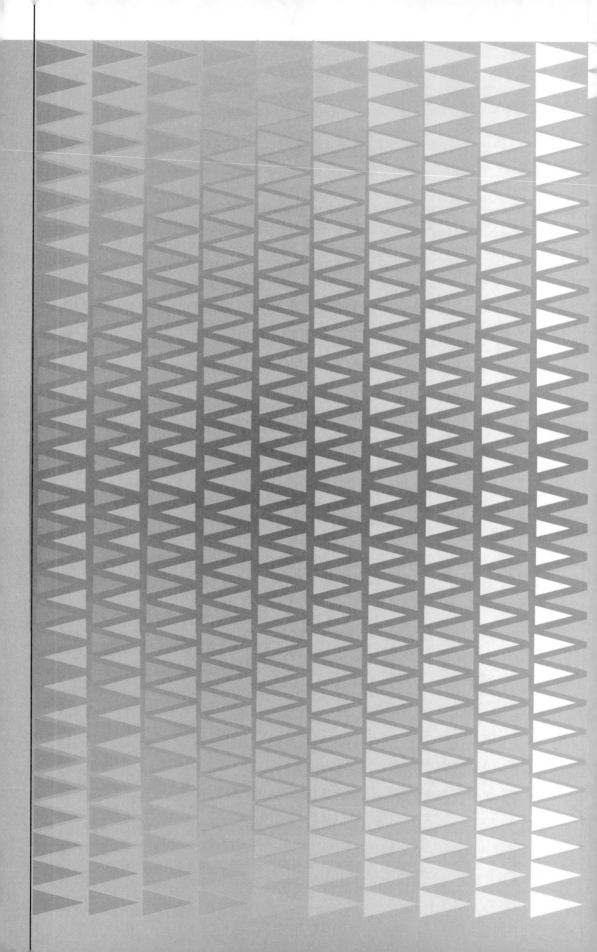

PREFACE THE PUBLISHING EXPERIENCE

Digital Design Business Practices is a guide to procedure in the quickly changing publishing industry. It is not law, though several sections are based on laws that apply to the industry. It is not a book of *musts,* but rather of suggestions to help avoid the most common problems that arise in publishing projects. Following the suggestions of this book will help the graphic design professional be more profitable, saner, and able to best grow with the profession. It will help clients achieve their marketing goals through successful relationships.

Communication is a challenge in any industry, but more so in the communications field. Digital publishers are continually moving into unknown territory. The purpose of this book is to help find the best practices and working methods. Adapt it. Use it as a springboard for thought: Agree with it, disagree with it, challenge it. But most importantly, let it inspire and improve procedures.

Although this book is written by one person, it has been reviewed by professionals throughout the industry. Several organizations contributed to the origins of the text, and each recognize the great variety of the publishing experience. Other viewpoints and opinions are quoted from clients, designers, writers, photographers, prepress professionals, printers, suppliers, and other related disciplines. If you have suggestions, would like to participate, or are curious, please see *www.allworth.com/Catalog/ GD044/dd.htm* for updates, new resources, and a forum for sharing experiences and ideas.

We hope that all in the industry find these *Practices* to be instrumental in planning publishing projects—again and again.

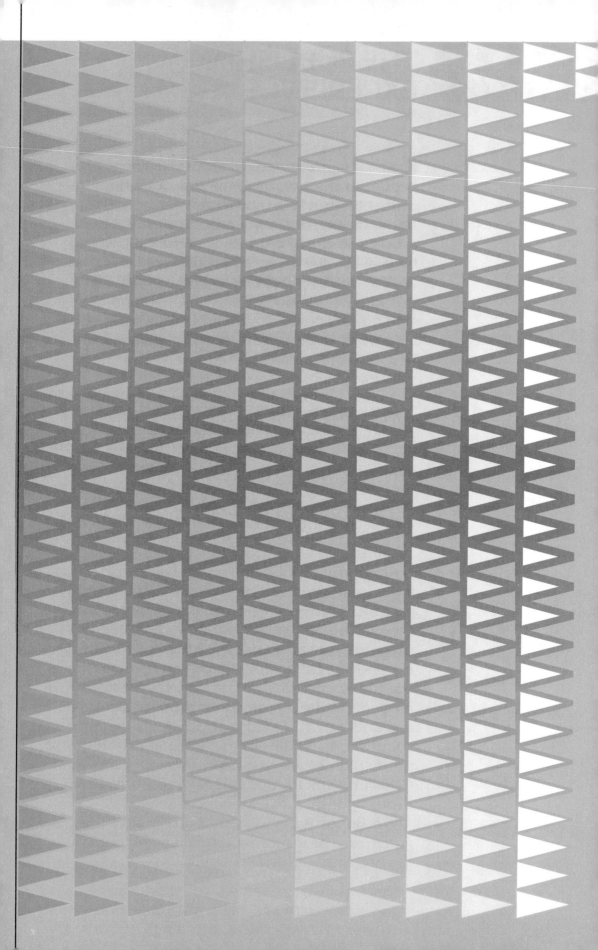

INDUSTRY TRUISMS

No matter how the technology changes, people are people. We bring psychology to every form of communication and publishing. Often unpredictable, technology inspires new applications for motivations and fulfillment of desires. Graphic designers who embrace new technological tools find more ways to interact, and the patterns of business adapt to the new arena with tendencies and expectations that confound many efforts.

Defining the industry is like defining scenery before seen: constantly moving and changing away from predictable descriptions. This book tries to capture truisms in the graphic arts that are timeless, inherent parts of the process. It only looks back at relevant traditions—trying instead to look at what is moving forward and what will stay. How clients, designers, and suppliers behave encompasses variations on themes such as communication, project management, balance of viewpoints and skills, ways to make money, and pitfalls that lead to losing it. The methods may vary, but motivations and techniques usually follow a variation on one of these themes.

By using advice in this book, many pitfalls and dangers will be avoided. In fact, varying from these practices can easily lead to trouble. *Digital Design Business Practices* hopes to be the book that captures the best and worst of serving clients while minding the store. It may not have all the answers to every challenge, but it identifies them and provides as many helpful suggestions as possible.

Digital design thrives on collaboration and establishes a momentum that defines the success of projects. This book provides methods for keeping projects on track and profitable. Roles and responsibilities are woven throughout and reading of others' experiences will enrich any publisher's background.

CLEAR ROLES FOR NAVIGATION

All communication goals remain consistent through time. However, the way publications reach audiences change and evolve. Books, magazines, newsletters, catalogs, brochures, booklets, stationery, posters, flyers, menus, advertisements, and Web sites are created to inform, sell, or motivate. The reasons may stay the same, but the methods evolve, bringing new issues and new considerations that reveal themselves as never ending challenges in growing a profitable publication venture.

Beginning with movable type, advancing with offset lithography, and now with computers, publication technology takes another great leap in capabilities through the Internet. People change to keep pace with the technology, and the Internet has provided a vast array of changes in the way that professionals communicate, work together, and seek information. The job of the graphic designer is critical in successfully communicating any client message; never more so than interactively. Managing changes in methods, processes, and expectations are part of having a momentum with vitality and growth.

The roles of professionals transform along with the process itself. Although the end product is substantially the same, responsibilities, roles, and ownerships are being redefined. As the chain becomes increasingly interlinked, questions on these topics arise more often. Perhaps the biggest changes are that the creator of a publication is empowered to do more and the process happens faster. Communications have become immediate.

In the past, responsibilities were clearly defined by sequential phases as a project went

This book is designed to cover print, presentation, and interactive graphic design produced through the use of personal computers. If it is to be as useful as possible, it may be referred to in legal proceedings. Therefore, several of the reviewers are attorneys with expertise in publishing. Nevertheless, a lot of effort has been put into writing this in plain English rather than legalese. Caution needs to be taken in utilizing this document legally. For it to have any influence in a legal dispute, there must be proven prior knowledge of these practices by all parties involved. References should be made in the reader's own professional agreements that refer to *Digital Design Business Practices* to ensure that everyone is aware of responsibilities. Copies of this book can be shared by all those participating in a project to facilitate communication.

from the hands of one expert to another. As phases are compressed, changed, or eliminated, professional roles overlap or are reassigned. As activities and capabilities become more complex, responsibilities become more confusing.

It is helpful to have a comprehensive document the entire industry can rely on to clarify responsibilities. Ideally, this document covers each step of the process, from the originating client's need for a publication, through the many creative, production, financial, and delivery stages.

Many organizations have contributed previously published specialty trade practices for inclusion into this comprehensive guide.

While this book deals in part with various legal subjects and contains general information about various aspects of the law as it applies to digital design and publishing, the author is not a lawyer and this book is not intended to be taken as legal advice. Resolution of legal issues for specific situations depends on the evaluation of precise factual circumstances. For this, always consult an attorney.

Many new participants to the publishing process are unaware of traditional trade practices. It has become even more important to define responsibilities in the computerized environment due to shorter schedules, quickly changing technology, and evolving skills. This document functions best when reviewed at the critical phases of a project. It can be a reminder of key considerations and gives advice on how to achieve the most desired results.

It's far better for there to be rapport and cooperation between all parties throughout a project than to have misunderstandings as the project progresses. This guide can help in building such rapport.

THE INDUSTRY'S MATURING PROCESS | THIRD EDITION

The honeymoon is over and a deeper fascination sets in. The marriage between computers and publishing has evolved into the everyday challenge of living together and running a business. The fascination of new media, once magical and all-consuming, has matured to a symbiotic functionalism. The dust is settling and the roles, responsibilities, and expectations are getting sorted out—polished by experience. Health requires balance: between the initial romantic fascination of a new technology, to the patient understanding and nurturing of a new culture.

The challenges of the market, however, never become less. There's no room for complacency or getting too comfortable if someone wants to have a viable business that can compete in a quickly shifting climate.

Entering into digital publishing's third decade means it's important to collect the lessons earned from experience. These can take an active role in both individual and industry potential. *Digital Design Business Practices* was begun during the turbulent late eighties and first published in 1992, just as the dust was shaking off the boots from a hike through unknown desktop terrain. That path traveled is old now. It was preparation for another redefining hike: the Internet. Cross-media capabilities have become the terrain of the designer in a new arena of possibilities and complexities. The trends, momentum, new challenges, and a new visual language propel business activity. Technical decisions are commonplace and a backdrop for a global world of communication.

More difficult than the challenges of technology

TALENTS FOR HANDLING:

► CREATIVE CHALLENGES
> resourceful
> insightful
>perceptive
> organized
> flexible
> open minded
> diligent
> masterful
> developed range
> masterful skill set
> diplomatic

► WORKFLOW CHALLENGES
> capable of handling intangibilities
> insightful
> appropriate
> charming and tactful
> willing to experiment
> perceptive
> flexible
> managerial
> able to integrate disparate concerns

are the challenges of people—defining roles, turf, control, skills, and responsibilities. Human needs become more obvious with the technological tools. There are new ways to address them.

However, creative workflow is less tangible in a digital environment. It is harder to measure than in traditional production. All professionals need to become more technologically savvy. Success balances talents and technical skills. The use of digital tools has become so pervasive, the notion of craft is continually reexamined. Though the technological side of the business craves its virtuosos, the most meaningful projects are always the blend of concept, positioning, and execution.

The first edition of *Electronic Design and Publishing: Business Practices* was formulated with the help of ADEPT (The Association for the Development of Electronic Publishing Technique)—a comprehensive industry user group begun in 1986. Other industry associations also contributed to the framework of the content. Compiled for several years, the first edition was published in 1992.

In 1995, the second edition expanded workflow concerns. Quotations were added to express more industry voices. Most came from conferences, interviews, and key publications. All have a ring of timeless wisdom.

Now, the third edition (with the new title of *Digital Design Business Practices*) adds Internet design and a greater corporate point-of-view. Future editions and the Web site will continue to mirror industry changes, define what remains the same through time, and have an outreach to every facet of digital publishing.

► TECHNOLOGICAL CHALLENGES
> patient
> scientifically focused
> sharp analytical skills
> dedication to constant study
> able to experiment with techniques
> interface experimentally
> handle teaching
> keep craft-orientation
> curiosity

► HUMAN CHALLENGES
> able to integrate technological tools
> develop process definition
> keep questioning
> plan and project
> fascination for and love of change
> able to expand capabilities
> possess continuous financing
> able to look at problems in new ways
> understanding of business practices.

STRUCTURE OF CONTENT OVERVIEW

The introductory section of each chapter gives an overview for all readers. This segment emphasizes development, relationships, communication, efficient process management, and ensures successful graphic projects. Every reader will benefit from reading this introductory material, as it gives context and each topic.

Relevant organizations, magazines, and online resources are listed at the beginning of chapters oriented to specific industry concentrations.

EXPLAINING IN MORE DEPTH

The use of divider headlines help to make finding specific information easier. Also, please see the index and the glossary, set up to quickly find explanations and links to pages.

The content is taken from twenty-five years of experience as a designer and art director—from working freelance to owning a design firm to running a corporate department. Industry practices from each segment were used to reflect content presented.

Experiences, advice, and explaining further points are given in examples and sidebars throughout the text.

SPECIFIC PERSPECTIVES

Each segment is written first as a general overview and then breaks into the three major roles: client group, creative group, production group (see next page).

Although many readers will zero in on the segments of greatest interest (i.e. the one that the reader fits into), it can be helpful to read the segments for the other groups. This can expand and improve knowledge of process considerations, understanding, and hence, communication.

Research and interviews with other professionals round out the core text, offering further insights, examples, and advise. Each segment brings several voices by having quoted material throughoutas sidebars and at the end of each chapter.

CLIENT GROUP

The first point of view presented is for the client who is the buyer of creative services. These segments explore

► decisions to be made

► key points in the process

► tips for collaborating with creative professionals

► advice for keeping projects on track

► guide for asking the right questions

It covers what each client needs to think about when supervising the development of and being responsible for a publishing project.

CREATIVE GROUP

Scaled for the art director and designer who both deal with the client group and the production group. Each segment covers tips and examples for

► managing projects

► efficient processes

► client relations

► handling presentations

► establishing effective communications

► working collaborations

► presenting and negotiating fee structure

► delineating responsibilities

► establishing profitability

► defining ownerships

► giving advise for common pitfalls

PRODUCTION GROUP

Directed to the craft of finishing projects, these segments cover both the print and the interactive media sides of the business. Suppliers of software and equipment are included. Each segment covers

► working in a collaborative environment

► keeping buyers happy

► building accurate expectations

► handling projects as they change and evolve

► developing service orientation

► expanding businesses for future vitality

PLANNING

DEFINING ROLES

Trying to categorize the various players on the field of publishing is like trying to guess who will catch a ball before it is thrown. Roles are unclear because traditional definitions have been transformed by electronic ones, and are continually evolving.

EXPANSIVE GRAPHIC COMMUNICATION

Publishing includes the blended media of print, presentation, and online delivery of literary, audio, and visual information. How these work together to represent companies and organizations continues to be refined through branding, global exposure, and more direct ways to reach others. Communication continues to grow more visual. Knowing how to manage publication projects can make a difference between growth and stagnation.

All publishing activities, independent of size or the number of people involved, are handled by three major groups:

► **client group**—initiates publishing projects

► **creative group**—creates the projects

► **production group**—includes tech and print that produce the projects.

EVOLVING CAPABILITIES

Digital publishing changes traditional methods in ways that should be understood to attain proficiency. Digital processes:

► **Reduce the number of disciplines** and may simplify the process as each individual "wears more hats." Part of this is driven by faster methods. Part is due to some processes becoming obsolete. It is important to remember the phases needed and not to skip steps for quality assurance.

► **Create more complex skills**. Each professional has an unlimited amount of information to keep up with, new skills to develop, background to search for, and processes to evolve. Everyone is overwhelmed, so understanding responsibilities outside of specific project pressures will make working on them much smoother.

► **Compress project phases** and demand faster schedules. Although this is attractive to those needing published materials quickly, such compression can allow more errors, for there are fewer checkpoints and fewer eyes reviewing each stage. Mistakes cost extra money, so an ap-

"The publishing process proceeds most smoothly when those working together keep three guiding C's in mind:

• Communication
• Common sense
• Courtesy between professionals."

—Matthew Doherty
Designer
Evanston, IL

proval and proofing process is the best protection against oversights.

ROLES AND RESPONSIBILITIES

Each project necessitates a different configuration of staffing. Though the possibilities of collaboration are infinite, the roles and responsibilities are not. These may be divided in a myriad of ways, but there are decisions, phases, processes, and activities that must be done.

Each group has defined functions, though some functions may overlap, such as a Webmaster collaborating inside the client group, or a page composer working for the printer. The talent resources available can take several forms:

As software and hardware evolve, their capabilities become more powerful. With each new release, there is more to learn. Success will reward those who can learn the fastest, not necessarily those who learn with the most depth. And because it is impossible for any one person to know everything in the process, learning from one another will enhance both individual potential and the process itself. It can shortcut experience.

► **Publishing team**—some roles involve more than one person, such as large projects or marketing campaigns, shared by several professionals

► **One professional**—can perform more than one role (such as a designer who handles prepress or programs for Web sites)

► **Two professionals**—do the whole process for small projects, collaborating for conception, creation, and production.

No matter how many people are involved in the process, there are always three distinct portions and roles to any communication endeavor.

Crucial role intersections happen between the originating client and the creative group, and between the creative group and the production group.

CLIENT GROUP

The client group is the organization or person needing a publication and so originates the project. This group begins the process, ultimately pays the bills, and receives the intended benefit of the project or system.

► **Client** is the decision-maker (and invoice-payer). This could be a business owner, a marketing and sales professional, or a publisher. Sometimes this role is served by a committee comprised of organization officers, directors, or marketing advisors. *The client group is responsible for deciding all parameters, content, deadlines, and intended results.*

► **Project director** interacts with the creative group most frequently and provides information, sets up meetings, checks quality, and facilitates processes. *The project director is responsible to carry out system or project implementation until finished.*

CREATIVE GROUP

The creative group is a design firm, an in-house department, a consortium of creative professionals, or a single freelancer that works on a project, campaign, or graphic system.

IDENTIFYING ROLES

Though most professionals will fit more than one role, breaking the process activities into definable modular chunks enables each to identify an individual mix appropriate to the project challenge.

► **Art director** (or the creative director) manages the project and supervises the quality of the entire design and production process. *The art director is responsible for creativity, originality of concepts, and execution of projects.*

► **Project manager** keeps track of all schedules, budgets, and communications. Easily accessible to oversee projects on a day-to-day basis, *the project manager is responsible for keeping projects moving on time and on budget.*

► **Researcher** finds all relevant facts, background information, handles any market research, and accesses library, news, or online resources. *The researcher is responsible for facts and authenticity of source materials.*

► **Designer** originates ideas and concepts, coordinates resources, and facilitates the production process. *The designer is responsible for providing creative ideas to assignments and fulfilling project parameters.*

► **Writer** generates the text for the written part of the publication. The writer pulls together all research and expresses the clients' message and parameters of the project in works or text. *The writer is responsible for correctness of content, style, accuracy, and originality.*

► **Editor** evaluates the writing for content and style, ensures correct versions, checks proofreading, adapts text to the design if necessary, and reviews proofs for typographical errors. *The editor is responsible for developing content, supervising research, editing, and style.*

► **Webmaster** determines how the Internet fits within an organization's overall strategy and supervises the site development. *The Webmaster is responsible for functionality and implementing plans for creation and maintenance.*

► **Proofreader** checks through the manuscript (sometimes called copy or text) for typographical and grammatical errors. Sometimes a copyeditor is utilized for consistency of style and word usage. *The proofreader is responsible for errors and bad punctuation, but not responsible for content.*

FINDING THE PERFECT FREELANCER

Resources for finding freelancers:

> personal referrals— word of mouth

> direct-mail promotions

> temp agencies

> art/photo reps

> technical trade schools

> local trade, design, and advertising publications

> networking at design and advertising associations

> asking other freelancers.

—Maria Piscopo
Creative Services Consultant
Costa Mesa, CA
http://MPiscopo.com

- ▶ **Illustrator** or **photographer** generates original visual images in accordance with the design concept. *Each image creator is responsible for originality and for fulfilling project parameters.*

- ▶ **Fabricator** or **prototype maker** constructs models or mock-ups of campaigns, components, print formats, packaging, point-of-purchase, or special products. *The fabricator is responsible for formulating a model that production plans and fees can be based upon.*

- ▶ **Page composer** (also called a desktop publisher) creates the document files, handles typography, formats and assembles all the elements into page form, and works with the production group on execution. Traditionally, this person would have been several people, such as typesetter, keyliner, and production artist. These functions are combined in the digital environment. *The page composer is responsible for the quality of file construction, meeting deadlines, and other project parameters.* The roles between the page composer and designer can vary greatly.

PRODUCTION GROUP

The production group provides the finishing aspects of the project. For print, high-resolution services have a linear workflow (print group). For Internet or database publishing, the technical professionals (tech group) program according to specified design and content formats. Publishing, whether ink on paper or pixels over a network, is a custom-manufacturing process.

IDENTIFYING ROLES

Technological advances will continue to streamline the finishing process, but these professionals will still be needed:

- ▶ **Buyer** purchases finishing capabilities. This person can be any of several professionals, so one term covers them all. For example, the project director, project manager, designer, or page composer can be the customer of the imaging center. The client, designer, page composer, or service bureau can be the customer of the printer. The client, project manager, or designer can be the client of the programmer. *The buyer is responsible for purchasing finished product, according to prototype specifications and parameters.*

- ▶ **Programmer** works with the art director, designer, and Webmaster (for Internet) on realizing the publication creation, updating, and maintenance. *The programmer is responsible for* technical functioning and currency. The developments of the Web and advances in prepress continually place more production power and capabilities in the hands of those creating the projects. The roles between programmer and designer can vary greatly.

► **Imaging center** (prepress provider, service bureau, or color separation) outputs camera-ready paper or plate-ready film for imposition. Sometimes the printer has capabilities in-house for multicolor preparation and plate-ready output. *The imaging center is responsible for the creation of final output to project specifications.*

> Most service bureaus grew out of the typesetting industry and have a strong typographical and black-and-white imaging background. Most trade shops evolved from the color separation industry and have a strong background in color. Today they are blending together as each takes on the other's capabilities; the composite term is imaging centers.

  ▷ **Production coordinator** is the contact person within the imaging center who interacts with the buyer. This person receives the files and makes sure they are properly prepared for imaging. *The production coordinator is responsible for communicating status to buyer.*

> Many imaging centers assign the production coordinator a preflight department that can put incoming projects through a series of tests to verify completeness before prepress work begins.

  ▷ **Prepress specialist** readies document files for output, matching the specifications of the buyer for the printer. Traditionally, these professionals are scanner operators, camera operators, and strippers. These functions now fall together and are controlled by digitally skilled professionals. *The prepress specialist is responsible for correctness of output per specifications.*

► **Printer** advises on output, receives film from the imaging center, creates proofs, handles all printing through delivery. *The printer is responsible for providing finished pieces on time, according to budget and specifications.*

  ▷ **Press operator** runs the presses and balances the color to match the proof. *The press operator is responsible for monitoring quality control throughout the press run.*

  ▷ **Binder** completes the manufacturing process by taking sheets or rolls from the press and converting (through cutting and folding) them into the final form, such as books, magazines, stationery, etc. *The binder is responsible for accuracy and matching instructions.*

  ▷ **Shipper** takes skids or boxes of the printed publication from the dock of the printer and delivers them to destination(s) selected by the client. *The shipper is responsible for timely and accurate delivery.*

THE AUTHORITY

Within this chain of professionals (each of whom may subcontract portions of their work), responsibility can shift from shoulder to shoulder, but always remains under the art director's purview, and subject to the client's final approval.

"Good work has rarely come out of an adversarial relationship. It's important to abandon relationships that are a no-win. A designer needs to be generous with clients, to accept responsibility for a good relationship, but also to be demanding that the client come forward and meet them halfway.

"The majority of the work we do is difficult, not because our clients are incapable of being good clients. They may be educated or uneducated and that's not what really matters as much as whether they enjoy the process and trust us. Those with whom we have a cooperative working relationship can help get through any problems. All of our clients are trouble because they're all parts of organizations with time schedules and budgets—design is a really difficult business. Clients never have enough time, never have enough money, their products are always being introduced before they're ready, they're always rescheduling, they never get products to you for the scheduled photo shoots or they change the name of the product after the photos are shot, but the direct mail is already committed so the deadline doesn't change. All clients have these problems.

"If the client cares, there are ways to resolve problems. Even our best clients have horrible thoughts about us sometimes—it's a love/hate relationship, and they always think they're paying too much. For what they're paying, they think you should be able to answer all their problems. I understand this because I feel this way about my own staff! I'm very sympathetic to the client."

—Nancye Green
Designer
Donovan and Green
New York, NY

"Years ago I thought it was important for design to be recognized as a profession. But I have come to the conclusion that the only thing a designer can do is develop respect within his or her own sphere of influence."

—John Massey
Director of Design
Container Corporation
(from 1964-1983)

"On a daily basis, there is no telling what a designer may be asked to design, produce, or provide as a service to his or her client. One must be prepared to be asked anything or be ready for anything within the next hour or the next three phone calls."

—Bruce Scott
Scott Graphic Design
Grover Beach, CA

"If you're a designer (note the intended omission of the word graphic) who's making a serious attempt to keep pace with the emerging technologies of interactive media and the Web, you must be changing your business cards as frequently as your underwear.

"Are you an information architect? User interface designer? Web designer? An interactive designer? How about a producer? Any and all of these are appropriate titles for designers working with new media."

—Shawn Freeman
focus2
Dallas, TX

"The challenge for the designer is to create the opportunity for good design. Before you can begin to put pencil to paper or mouse to pad, you have to evaluate, educate, and take control of the circumstances in order to create an opportunity that is conducive to good design! This means that designers have to approach every project from a number of angles. You have to be the businessman that puts the project into financial perspective for the client. You have to be the research department that evaluates new processes and techniques. You have to be the educator who explains the new processes and mediums available. You have to be a CFO to figure out the client's budget and proportion of moneys to allow for good design. And if you do all of these things well, then you've earned the right to be the designer that knows how to deliver a great design."

—John Brady
Designer
John Brady Design
Pittsburgh, PA

"The Internet of tomorrow will show the role of designers increasing in importance due mainly to higher speed connections. People want 'eye candy,' and a site's hit count will be greatly affected by its appearance; content and information will become secondary factors... visitors need to be 'wowed.'"

—Michael Johnson
Inacom Corporation
Omaha, NE

"The designer's role is to map out and oversee the entire image (overall) of the Web site. Designers provide boards and files to the programmer, who is responsible for executing the designer/client's image."

—Mika Toyoura
MZ Design, CA

"As advertising focuses on smaller and more defined markets, understanding how people think and feel makes for better design results. It is also important that a designer realize they are part of an ensemble, not a solo act. Everyone has a role to play—and if the designer is part facilitator, mediator, and communicator, he or she will have a much easier time in the whole process. In the never ending battle of which is more important—the written word or visual impact—a designer who is able to partner with a writer maximizes the impact of both."

—Barbara Krouse
Amber Design Associates
Hackettstown, NJ

"Designer/creative director(s) offer tasteful, functional, informative input. They do not serve as programmers—although they may assist. Internet has created a huge increase in the importance of design. I see a reversion to basic principles emerging from the overkill of technical gadgetry, special effects, etc. This is where honest design skills can be greatly utilized in simplifying communication and getting results. Simplicity of design and function serve the user and provider with the best measure results— design will make or break the effort in most of these instances."

—Jim Brown
Action Ltd.
Nebraska

"Designers currently do not develop entire sites, but design artwork for portions of sites. In a limited sense, the designers also serve as programmers and developers; however, programming is also provided by freelance. Designers are underutilized in Web development. Anybody can get up and running in a day. That is not to say it will be aesthetic. The same can be said about what happened in traditional publishing—anyone can produce a brochure with the proper software, but I believe standards are lower than ten to fifteen years ago. That said, the Internet has highlighted the importance of the designer because of the all-pervasive bad or mediocre design on the Web."

—Mike McCartney
MM Design Inc.
Oklahoma

"I view the Web and Web design as very significant, but I feel that most agencies and studios are pushing designers to be programmers and programmers to be designers—the result is often lackluster. Let the art directors and designers work with the programmers. In a communications/electronics overloaded world, the true essence of graphic design is becoming lost in a 'forced-to-learn' sea of never ending software programs and upgrades."

—Dennis Voss
Crosby Marketing
Communications
Annapolis, MD

"The Web has reinvented how design is conceptualized, as well as recreated the designers' role in the production and creation of a Web site.

"Designing for print and designing for the Web require distinct approaches and strategies for how the information is presented and accessed. The designer must be willing to both subvert the limitations of this technology and exploit its capabilities."

—Christine Klaehn
Cory McPherson Nash,
Watertown, MA

"As the first generation of designers who are versed in the use of computers, we are blessed [and cursed] with its double edged blade. On one edge we can now provide creative-to-press services to our clients and employers. On the opposite edge we must now be competent in all aspects of design. This means we can no longer specialize in particular functions, nor can we be generalists with a smattering of knowledge; we need to be able to perform all functions with a high degree of skill."

—Stephen Buchheit
Rose Advertising
Incorporated
Willoughby, OH

"Is there a size to ethics? Big projects or small, designers battle with ethics. How many graphic designers have cringed as they created a beautiful ad touting falsely how wonderful something is—when they know it is not. Truth in advertising is ethical. 'New and improved,' 'Better than Ever,' 'Faster,' 'Leaner,' 'Best Food in Town' lies are not ethical. Do graphic designers have any say in that? No. Can't handle lying on a daily basis? Then try graphic design in an editorial department."

—Joyce Halliday
Halliday Design Studios
Harvey Cedars, NJ

"Multiple 'media' as seen today is in the process of blurring. There will not be a design just for print, or packaging, or the Web, but the seed of the designer's conceived thought for a particular project will hold all of the options at the ready. Designers will need to be cohesively versed in all areas of design and the life expression of their design communication target."

—Lynniel Q. Van Benschoten
MHO Networks
Denver, CO

"Whether we like to think this or not, the role of the graphic designer has always been to sell something. Graphic designers don't 'design' a product per se; they design with the marketing of that product in mind. Don't get me wrong: We do somewhat shape the culture, being different by nature. But, I think it's the client's ever changing product or service that reflects the trends of society. The challenge we face as designers is staying aware of our clients' target audience and the trends in that marketplace. We must expand on these trends with experience. The ultimate goal of quality creative is to sell the consumer."

—Kim Allen Farnham
Pear Creative Group
Endicott, NY

"Perhaps the best word to describe the role of the designer in today's world is storyteller. Designers are the best storytellers. We're able to articulate the most intricate details of an event or issue. And at its essence, storytelling is strategic. You have to be able to understand an issue fully in order to tell its story."

—Noreen Morioka
Adams/Morioka
Los Angeles, CA

"One might argue that the client is paying for the designer's services so why shouldn't he or she do exactly what the client says? What the client is paying for is a person who has the ability to discern what images will attract and inform the viewer, then configure them in a way which makes it unique to that particular company. The client is NOT paying for a hired hand to push buttons or recycle effects."

—Matt Ohnemus
LoConte Goldman Design
Boston, MA

FOUNDATIONS 2

New tools inspire people to be more effective. They also add new venues to learn and absorb. In response, professional skills must keep pace. Cultural changes, such as new ways of thinking and new working environments, come more slowly than the technological changes that instigate them. However, some changes that inspire new methods can be fast and conscious, such as transforming from faxes to e-mails or from meetings to videoconference calls. The unconscious changes, such as those that transform methods, processes, and whole industries, can also happen with surprising speed, leaving most members breathless in their effort to keep up. These changes sneak up on many who find themselves suddenly out-of-date.

PROFESSION IN TRANSITION

Digital publishing and communication is causing all professionals to continually redefine how they work, who will do the work, and how responsibilities will be divided. Major changes transforming publishing include how:

► **Production grows more automated** and old skills become obsolete. With the demise of typesetting, keylining, and stripping, the new craft of digital imaging has taken over. Analog methods and processes are turning into fine art as digital methods and processes are now mainstream. The older crafts become boutiques and younger professionals aren't learning these skills. The bad news is that not learning them is limiting. The good news is that there is more time to practice other skills.

► **Technology is becoming easier to learn,** placing capabilities in the hands of more workers and enabling more to publish than before. Anyone can have their own personal publications (cards and newsletters) and communications (Web sites and e-mails, chats, etc.). High school students can do presentations and displays that look as good as most professional ones used to in the past. Some of these individuals will become professionals in company or organizational publications. Some will become clients. Some will become designers or page composers.

ORGANIZATIONS

www.designfortheworld.org
Design for the World

www.unesco.org
United Nations Education Scientific and Cultural Organization

WEB RESOURCES

www.penrose-press.com/idd/org/graph.html
International Directory of Design

▶ **Roles between professionals** blend and grow less definable. Responsibilities become blurred as capabilities expand. Specialties are redefined. Who does what becomes more of an individual process question. Most professionals have a skill set that encompasses several roles. Understanding these is part of the planning process.

▶ **An influx of design college graduates** are lured by technology. The industry continues to expand as communication expands. The Internet is particularly seductive for aspiring publishers. Higher education is transformed as art schools need computer labs and must find the funding to keep them attractive to students.

▶ **Design students invest in learning software** versus learning hand skills. This leads to an incoming workforce without the ability to draw or have fast alternatives to present concepts.

▶ **The material investment is higher** both as an entry point into the field for beginning designers, and to sustain systems continually through upgrades. Cost of doing business for a creative group is much higher than in pre-electronic days. Most groups spend a significant percentage of income on technology. (See magazine resources, chapter 10, page 101, for surveys on how firms allot operating capital.)

▶ **Perception of fees**—many managements express a belief (or a desire) that design and production should cost less due to faster technology. However, most fees have remained relatively unaffected by technology as, hopefully, system costs are offset by speed. (See Fees, chapter 5, page 50.) This overhead, is a greater percentage of a creative group's overall budget than in the past.

▶ **Images are cheap and available**, affecting how clients and designers use stock images versus commissioned ones. Proliferation of copyright free images encourages greater use, popular applications for collage and altering images, and the shrinking of budgets allotted for spot commissions. Illustrators and photographers are affected by the competition of lower priced, albeit less original, work.

▶ **Designers grow in recognition** for marketing differentiation and visual skills. Status of design is both more valued on the strategic end and devalued on the hands-on end as capabilities for novices grow. There is more design, both bad and good. It is a challenge of competencies and visibilities to create a successful design enterprise.

LEVELS OF DESIGN: WEB SITES

Appropriateness has never been more important than on the Web. Sites have various purposes and design must support the nature and goals:

> NEWS—often clients want to convey a large amount of information quickly and make the site very dense. The key is speed—easy on and off to get sought-after information. Design must focus on clarity and organizing content. It must be fast and simple.

> PORTALS AND BROWSERS—need to be easy to use and enticing. They give information a context, and are the beginning of a community of interest. Design needs to clearly provide categories of choices and express the coalition of the content.

>INFORMATIONAL—offers more depth to reward the viewer, who may arrive from a search. Site design has an immediacy and a denseness of topics.

> EDUCATIONAL—often for intranets, educational sites instruct the viewer in a subject. Many e-magazines fit in this category. Design offers more visual support to content.

> ORGANIZATIONAL—membership-based groups that give some portal functions to members via database directories, and other membership services. Design needs to be simple and fast, but also must be distinctive and express the personality of the group.

> SALES—for commercial purposes, online sales, and services for sale of products. These sites can be extensions of existing retailers. Many are virtual entities. Design needs to dynamically entice the viewer to shop. Building a visual atmosphere challenges campaigns and cross-media efforts.

> ARTISTIC—sites that are statements in themselves, utilize the Internet for its visual and organizational potential. They make statements that could not be made in any other media.

Design addresses each of these kinds of sites differently, as the interactive nature is unique in each.

With all sites, navigation and brand identity are key design concerns.

► **Cottage industries are spawned** by career-shifters who find the ease of use and portability of digital publishing attractive. It is particularly appealing to flexible work schedules and home-based practices. New lifestyles are encouraged based on variable hours and remote communication connections.

► **Work methods become decentralized**—with communication links for e-mail, Web presentations, and Web publishing, professional communicators are geographically free, unless the client base demands physical proximity. Many functions like writing, programming, research, compilations, building databases, assembling documents, illustrating, printing, and electronic distribution all mean that those who create the communications can work anywhere.

► **Cybercommunities are built by each professional**—much time in the work plan must be devoted to the involvement in a community of interest with colleagues, clients, and suppliers via e-mail, subscriptions, memberships, and participation. Many educational and confer-

LEVELS OF DESIGN: PRINT

Designers accomplish marketing goals using their talent, training, evolved skills, and depth of experience. Any publishing project requires certain levels of creativity, planning, and execution. From the simplest to the most complex, the different levels of design services are:

> KNOWLEDGE AND USE OF TYPOGRAPHY—the simplest projects organize information and require formatting text, utilizing templates, and producing documents. Training is required in software and in the rules of typography.

>FORMAT, COMPOSITION, AND READABILITY—projects become more involved when the designer arranges and composes the words rather than simply placing them on the page—lead the reader through information with the effective use of typography.

> PAGINATION AND SEQUENCING OF INFORMATION—content is planned through multiple pages and what should be on each page is determined. Sensitivity is needed to make information easy to read with composition and typographic crafting.

> OVERALL VISUAL SCHEME—projects grow in challenge when they need more distinction and impact by adding visual devices such as illustrations, borders, or photographs.

> GRAPHIC THEME—higher level design expresses the content visually rather than having the reader rely on words alone—design builds a visual language and recognizability, becoming the foundation for branding.

> EXPANSIVE PROGRAMS OR CAMPAIGNS—the most challenging projects develop flexible themes that give visual cohesion and recognizability to communications—in other words, an overall image that can be applied to a marketing vision as expressed through many applications.

The first three categories involve layout, the arrangement of elements. The last three involve the conceptual underpinning which conveys information through visual language. Marketing purpose and client goals dictate the appropriate level of design.

ence activities and seminars can be replaced more economically by the Web. Having connections to international resources can be part of the value each professional brings to project challenges.

COMMUNICATION PURPOSE

Capabilities vary as widely as the quality and effectiveness of the projects. What *doesn't* change are the reasons to communicate which address the marketing goals of the client group. Each project fulfills one or more of these functions:

► **shares information**.

► **sells a product or service**.

► **builds an identity**.

These goals can be achieved in more ways today than in the past. As always, they require strong vision and the appropriate use of talents, skills, and resources. With a variety of meals on a menu, choosing the most nutritious one makes health sense. With a variety of media and methods available, choosing the most appropriate and targeted ones makes business sense. Utilizing technology to its greatest potential means addressing user needs with as much attention as technical ones. Also, defining roles and responsibilities is as important as choosing projects and developing content. The more roles collaborate, the more successful the results.

A recurring theme throughout this book is communication, which can never be practiced enough. How to make the greatest use of time by knowing what information to communicate to whom and at what point in the process affects project success as much as the quality of ideas and positioning in the marketplace. To keep a fast business pace, no one has enough time to perfect the fine art of communication. But trying to, and finding suggestions throughout these pages, should help streamline the process.

CLIENT GROUP

MAJOR DECISIONS

set project foundation

The pre-digital publishing process hummed along like a well-oiled machine. The writers wrote, the designers designed, the typesetters keystroked, the keyliners pasted, and the printers made film and plates and ran the presses. It was linear and craft-based. The client was in a better position to remain detached once the initial direction and design was chosen. Once begun, the production process for a magazine, book, brochure, advertisement, etc. was not very flexible.

As technology has made roles more ambiguous and the process less linear, the client needs to show more leadership and knowledge to best manage the entire process and integrate it into business and communication plans. As projects can be executed more quickly, having a strong idea and well-targeted approach allows a focused and efficient development that can respond to opportunities. This is particularly true with Web sites which redefine all business operations and need to be flexible for a variety of content and purposes.

FOCUSING ON THE RIGHT STRUCTURE

Decisions the client needs to control in the publishing process to build a foundation for strategic and project planning:

- ► **goals and parameters** of communication needs. (See chapter 4, Estimates and Proposals, page 38.)

- ► **match the challenge to the best creative source**—make sure that staff or subcontractors are knowledgeable and have the resources and experience to best get the job done. Review portfolios and check credentials.

- ► **balance tools to the assignment**—develop the ability to make decisions without being hands-on by understanding the issues and capabilities of the media and processes.

- ► **develop own method**—adapt new processes and approaches through research, reading, experts, and experiments.
- ► **define process expectations** —such as in deadline and cost. Are the goals possible with the resources available?
- ► **define scope and simple plans**—scale expectations of technology. It's easier to build a campaign or a site in segments than it is to emerge from the gate full-speed.
- ► **arrange process for ongoing work**—series that build, or segments that need to be regularly updated. Knowing how each project fits within an organization's total communication efforts provides context for design approaches and can encourage branding.
- ► **utilize resources**—stay current and evolve plans.
- ► **listen to professional sources**—learn new ways to solve challenges and build a repertoire of solutions.

CREATIVE GROUP

The design industry keeps growing—with techno-vitamin shots like desktop publishing in the eighties and the Internet in the nineties. The industry also becomes more complex as roles are added to the creative group, with few becoming obsolete. There is always more for the professional to know— about new media opportunities, software developments, new techniques, and additional production capabilities. There are always more ways to enhance creative ventures through a talented use of technology. Every creative group must balance business practices and technological savvy with the creative offerings. Good business is part of good work. Every creative group strives to produce the most creative work for the budgets available. Those who place the positioning, strategy, and constituent point of view as paramount are the ones who end up the most successful and vital.

EXPANDING VOCABULARY

As the creative professional must balance conceptual and artistic skills with technological skills, each must specialize to a degree, for no one can know everything in the range of communication pursuits. Teams can cover the spectrum of industry needs, requiring that project management become as important as any other ingredient for success. The farther away from idea generation the work grows, the more technical expertise can be shared, and, thus, managed.

To fully take advantage of interactive media, each creative professional must understand its visual language opportunities, and how their portion supports the other components.

The publishing artist must:

▶ **conceptualize for integrated media** campaigns or projects with ideas that are elastic and streamlined, yet still recognizable. Variations must be possible upon the design structure.

    ▷ *Grasp the differences between media.* Ideas can be translated from one media to another, but must conform to each medium's vernacular and translation, which can't be done without working experience.

    ▷ *Know what skills apply* and expand with what is most current. Operating on the "bleeding edge" is not profitable. There are those always willing to try the experimental and push the very edges of technology through their time and investment, but this should not be undertaken as part of client work.

    ▷ *Have presentation and production skills* to support concepts. Be able to explain the process to decision-makers to build understanding and collaboration. Don't depend on the work to speak for itself, as most conceptual work can be pushed in stylistic directions or combined with other ideas.

    ▷ *Apply talents* to best address client requirements. Happy and repeat clients are the only way to make money. Clients who recommend to others are gold. All creative groups thrive on referrals. Plus, joining forces with the client to fulfill their purpose gives work its greatest challenge and depth.

▶ **understand strengths** and try not to be everything to everybody. No one artist can be proficient in all visual media or aspects of the design business.

▶ **know the capabilities** of the tools to support concepts versus those for mere technique-enhancements. Choose appropriate tools or resources.

▶ **define appropriateness**—identify and develop the level of design to fit the assignment. (See pages 14 and 15.)

> It is important to recognize the distinction between the computer as a design tool and the computer as a production tool. The computer makes most production tasks faster than traditional methods. But as a design tool, the computer does not compress design time; the tools might enhance the execution of ideas, but generally (except for drawing type and patterns), it takes the same or more time than traditional methods. Planning helps ensure the best use of computers as design and presentation tools.

PRODUCTION GROUP

Technical expertise comes in definable levels. But technical expertise alone turns the products of skill into commodities. Those professionals in the production side of publishing know that service skills, business connections, and business mastery are needed to remain competitive and keep clients happy. Hardware and software are easy to buy—especially as systems grow cheaper and more capable. But they can't substitute for human experience and judgment. Printing is forced into scientific scrutiny by a technological fervor to print the most color for the least amount. The world of color grows. But printing remains a craft. It is custom manufacturing, for every project is unique.

SERVICE FOCUS VERSUS PRODUCT FOCUS

Every production-oriented business must be a mix of those who focus on the customer and those who focus on the technology. The communication between those two perspectives is a mark of business health. But often, those immersed in the technical can lose touch with the client and overall goals. Any tech side of the business needs to keep focused on service more than the next techno development. Simultaneously, it needs to keep everyone on the team conversant in capabilities to build the right buyer expectations. Technology can be seductive, so it requires skills to keep the bigger picture of balancing business decisions and capabilities to meet buyer needs.

Managing in the spectrum of technological possibilities requires that professionals and their organizations:

► **have clear business definitions**: markets, their niches, and best areas of expertise. New business areas are worth exploring, but not at the expense of current ones.

► **have management vision** that can choose appropriate technology for business goals, anticipating new developments on the horizon and the worth of investing.

► **emphasize customer service** and responsiveness in addition to technical expertise.

► **maintain the ability to change quickly** in response to customer needs and new developments. Watch customer trends (see chapter 3, page 23) carefully and adapt services continually—finding technology to support those needs.

► **compress the processes** to turn work around faster and less expensively. Doing more with less is a trend that will continue. Offering suggestions and savings to clients can earn their loyalty.

► **add new business services** that complement and enhance current offerings with client ease as mandatory. By doing for clients more efficiently than they can do for themselves, the skills in handling the tools can become more important than the tools themselves.

► **retain a visionary technician** who can keep capabilities on the forefront and be a research and development support for clients. Staying on the "bleeding edge" can be expensive without the perspective of someone who can make informed and solid technological business decisions and see opportunities to maximize investments.

"Electronic publishing has clearly facilitated our editorial and business operations. The flip side is that it has reduced inter-personal interaction. It still takes people to do layout, design, and write text. What the PC has done is facilitate getting the message from the brain to the page. Has it improved the editorial product? No. The quality of our various magazines is still in the minds of the creators."

—Gilbert Grosvenor
President and Chairman
National Geographic Society
Washington, DC

"Bosses need to understand that no machine can substitute for professional judgment and experience. Communication is a highly sophisticated and specialized activity, demanding skills that take time to learn and hone. Just because someone can write a letter does not mean they can become a skilled newsletter editor overnight, no matter what the software makers claimed."

—Jan White
Designer, Author, Consultant
Westport, CT

"Basically we have two kinds of projects. We need designers to help us sell generic services and to help us sell specific services. Generic services include things like our accounting firm's financial capability in tax practices or in telecommunications consulting. Specific services include things like providing an audit for a particular company, which we call an 'engagement.' The design projects that sell generic services are likely to be brochures, newsletters, or an identity. When we are going after a specific project, designers are more likely to be called in to design a presentation format, to help design charts, or to do a lot of nitty-gritty stuff that's going to clarify our proposal to get that engagement. In that way, we'll use design to really help with our communication process, with new prospects. As far as I'm concerned, that's where the rubber hits the road. For instance, when you have a new business presentation, and it looks like hell, you give it to a designer, who looks at it and says, 'Okay, we need sections and organization, and let's do this and that.' He sets it up on his desktop publishing system and all of a sudden it becomes something very unique and appealing. When we are competing for a significant piece of business, we will do whatever it takes to position our firm successfully against the competition."

—Robert Moulthrop
Senior Marketing Director
KPGM Peat Marwick
New York, NY

"Graphic design is a silly title that has little or no meaning these days. Just about anyone can design a newsletter. Everyone can get fonts, everyone can get charcoal drawings with a software plug-in filter applied onto a photograph. The things that used to differentiate us from other professions and businesses have turned into everyday commodity products. In a sense the computer is a great equalizer. Why should someone hire us to design a newsletter when they can buy this off the shelf? This is a difficult question; the answer, I believe, is that we need to provide added value and experience that cannot be bought 'off the shelf.'"

— Clement Mok
Designer
Sapient Corporation
San Francisco, CA

"Clients want revisions overnight and graphic designers have become production artist, typesetter, photo retoucher, client contract, baby-sitter, copywriter and business psychologist."

— Archie Boston
Archie Boston Graphic Design
Los Angeles, CA

"Almost everyone, up through the highest ranks of professionals, will feel increased pressure to specialize, or at least to package himself or herself as a marketable portfolio of skills. Executives and what used to be called managers, will have undergone probably the most radical rethinking of their roles.
    "More and more of the population will be caught up in the defining activity of the age: scrambling. Scrambling for footing on a shifting corporate landscape—cynics will call it a freelance economy—where market forces have supplanted older, more comfortable employment arrangements. Scrambling to upgrade their software, their learning, their financial reserves. Scrambling even to carve out moments of tranquillity under a banner blazoned FIGHT STRESS, a banner flapping like a Tibetan prayer flag in the gales of change."

— Walter Kiechel III
Executive Editor
Fortune magazine
New York, NY

STRATEGY 3

There is no substitute for good planning. Time spent on positioning and organizing projects establishes the momentum and potential success. It makes the greatest use of investment dollars.

FACTORS TO BEGIN RELATIONSHIPS

Before starting any communication effort, these are the best components to compile for any project or campaign:

► **goals of communication** plans are established.

► **well-targeted definition** and focus on audience needs. What do they think about current products or services? How should they react as a result of marketing message?

► **rough budget determined** through an approved proposal. (See Estimates and Proposals, chapter 4, page 38.)

► **decision-makers determined** and influencers considered.

► **client champion** chosen who shepherds decision-making process. If a committee, there needs to be a decisive leader and agenda. The leader of the communications effort makes the best client when committed to the goals, the process, and a willingness to do what needs to be done. The leader must

  ▷ *act as a consensus maker*—keep discussions and decisions moving. Keeping agendas defined and prioritized keeps the focus progressive.

  ▷ *support project director* with decision-making autonomy for the development process. This is for larger projects when the client has to delegate the development to a creative team. The client must first make sure that the goals are defined and understood.

  ▷ *fight opposition to important new ideas*—this takes belief in the new approach, as the most unusual concepts lead to the most successful marketing. They are also the most risky.

► **rough project plan**—sketch out the scope, positioning, staffing, and creative requirements that can be used for the proposal or as a basis for beginning financial agreements.

A certain amount of description is necessary in an effective plan. Often clients can be so anxious to start projects they often don't take time to understand what they are paying for. Description also demonstrates consistency to the entire team.

ORGANIZATIONS

www.writerswrite.com/org.htm
The American Communication Association (ACA)

www.ama.org
American Marketing Association

www.theabc.org
Association for Business Communication

www.marketing.org
Business Marketing Association

www.dba.org.UK
Design Business Association

www.the-dma.org
The Direct Marketing Association (DMA)

www.iabc.com/homepage.htm
International Association of Business Communicators

www.icahdq.org
The International Communication Association

www.ipma.org
International Publishing Management Association

www.natcom.org
National Communication Association (NCA)

www.nmoa.org
The National Mail Order Association

www.nsaspeaker.org
National Speakers Association (NSA)

VISION AND THOROUGHNESS

How well a project develops depends on forward thinking. For an extreme example, animation firms plan the frames of a film down to the second, continually adding details to the rough skeleton of the production plan. How the project is defined sets the tone and character of the development momentum.

Beginning well ensures a good finish. Any possible later disputes can be minimized by an initial well-crafted plan. There is the 80/20 rule that 20 percent of the time spent in planning saves 80 percent in efficiency.

RESPONSIBILITIES OF EACH TEAM MEMBER

Planning defines responsibility and accountability. As the process changes and roles evolve, questions of who is responsible for what will continue to transform. But there are certain roles that will always be prevalent in projects, no matter the level of technology or who is doing what task.

Each participant, no matter what their role in the publishing process, should always be prepared to accept responsibility to:

► **prevent potential difficulties**

  ▷ plan an *efficient procedural calendar*—the more detailed, the more surprises can be detected early.

  ▷ *communicate concerns* and keep team members informed about progress.

In the continuum of the project, a careful time frame feeds activity without the pressure of compromise forced by too little time. It seems all projects escalate to a crescendo of activity at the end. But again, as always, knowledge and planning form the best insurance policy to underwrite success

---

WATCH AND LEARN: ANALYZING TRENDS

"Careful businessmen keep tabs on competitors' actions, analyze behavior patterns and watch general economic trends. It is important that your business's strategy is flexible enough to adapt. Trends on the Internet come and go, and having a competitive advantage may only last a few months. Where do you look for trends that keep you ahead of the pack?

>YOUR CUSTOMERS—Start here. Your customers' actions may lead you to adjust how your site works, where you focus marketing dollars, or something as drastic as what you sell. Beta test new features on your most valuable customers before spending money on change.

> YOUR COMPETITION—Check out their Web sites. Search for articles and information on them. Do your homework to find out how you can set yourself apart and give you a focal point in your marketing efforts.

>THE BIG GUYS (AOL, Yahoo, Amazon, Lycos, Excite, etc.)—These companies are your competition as well. They send your customers to your competition.

> BOOKS—This source may not provide up-to-the-minute trends, but it can provide important information on what businesses are doing.

> MAGAZINES—Know the trade magazines of your industry but check some of the major news or portal sites first. They may provide some more general trends that can be helpful."

—Adam Strong
StrongVisuals, Springfield, IL
www.StrongVisuals.com and www.lunareclipse.net

> ▷ *define expectations* of what part each participant must contribute.

▶ **control quality**

> ▷ explain through *clear instructions and examples* whenever possible.

> ▷ *monitor progress* and excellence with collaborators.

> ▷ *present phases to decision-makers.*

▶ **facilitate the work flow**:

> ▷ *handle the approval process*—make sure each step is complete before the next step begins.

> ▷ *deal with unexpected results*—allow time for experimenting if connecting to other systems or if dealing with new processes.

> ▷ *control the budget*—throughout the process to ensure that there are no surprises at the end.

Check this list for each coworker, for if these factors are not met, there will not be a smooth project.

CLIENT GROUP

No project should start without compiling the components needed to form first a proposal (page 38) and later a project plan. (For print, page 62, or for Web, page 70.) First be sure to define the:

▶ **goals and scope** —communicate to the entire team

▶ **assigned budget** to all general parameters, which will require cost estimates. (See Estimates and Proposals, page 38.)

▶ **deadlines** to highlight and arrange key dates.

▶ **decision-makers agree** on general parameters to base fees and proposal.

PLANNING POINTS: BUILD A SKELETON PLAN

As the originator of all publishing projects, the client group is responsible for the strategic thinking to launch design directions and decisions. The best backdrop builds a rough plan containing:

▶ **time frames**—final delivery and the rough calendar to establish the project pace:

> ▷ *place critical dates* and paths by working with the creative group figuring backward from the date needed.

> ▷ *get feedback on dates* from team-members. Discuss how to streamline process if deadline is tight.

Corporate identity includes a system of visual elements to represent an organization. It includes the company's symbol or logotypes, its name, colors, and typographic style. To ensure consistency throughout an identity system, a graphic designer creates a standards manual that shows how these visual elements should be applied to stationery, signs, brochures, packaging, advertising, Web sites, giveaways, and other visual items.

A standards manual is like an easy-to-read blueprint that can be followed by those responsible for supervising the production of communications.

MAJOR DECISIONS

set project foundation
plan rough project skeleton
assemble category outline
solicit proposals

Being a good client involves two overall attributes as the project develops: knowing what questions to ask and trusting instincts when making choices. Decisiveness is only possible for those who can think quickly through a problem. Educated hunches develop and sharpen from being in touch with market needs rather than personal tastes. Professionalism involves knowing the difference between personal and marketing judgments.

▷ *have meeting to launch project* so all team members meet each other, can ask questions, and give input. (See pages 64 and 72 for preparation and participation.)

▷ *fill in calendar* and share with the team. The more detail, the greater the early warning when reality deviates. Plans are only maps to be adjusted when circumstances intervene. A calendar evolves throughout the process, but maintains the critical end date.

► **schedules**—add to the calendar and include each group:

▷ *assign the various tasks or segments to individuals* giving the plan substance and tangibility:

▷ *make schedules realistic*—build in catch-up time.

▷ *utilize e-mail* to keep teams informed of progress.

▷ *have time compiling system* to keep all workflow and databases up to date.

▷ *mechanism to report early* on time frames in jeopardy.

► **monitoring of the elements**—requires special skills with defined needs:

▷ *track critical dates* in project process. Check status before due date to see if there are any concerns instead of when the due date comes.

▷ *modify plan as needed* and communicate these changes both to other decision-makers and to the creative group.

▷ *check the budget periodically*—don't wait for reports to be offered even though specified in the plan. Key points are when selecting creative concepts, after finished artwork is completed and at the end. Further points should be filled in by each group's concerns.

▷ *have everyone's participation*—do not have decision-making meetings unless the creative group is represented and all decision-makers are present. If not, time is wasted later when an influence comes in with parameters no one knew about. Prevent this in the beginning and it will save time, money, and tempers.

► **corporate identity or brand strategy.** Whether following an existing identity or originating a new one:

▷ *identify targets*—design needs to have reason that is reflected in each component. Know what the design and the project must accomplish.

▷ *don't demand a design to convey more than three or four messages*—identity depends on instant recognition and imparting a memorable message. With short attention spans, give communication elements a hierarchy and simplify wherever possible.

▷ *target perceptions toward the brand*—identify values and attributes that constituents can recognize. Build on that trust.

▷ *understand brand essence*—the key attributes, emotions, styles, and personality. Utilize these qualities as building blocks of communication.

▷ *communicate key values to target market*—keep their point of view paramount in defining what to make consistent and what to change.

▷ *conduct design audit* (see sidebar, page 28) to bring all elements of strategy and visual requirements together. Evaluate possible directions through the filter of key objectives.

▷ *set up for flexibility and elasticity*—define the lowest technical requirements of a design—often, it is the fax machine).

BENEFITS OF A GOOD PLAN

Ultimately, it is the client group who inspires a strong project plan because they initiate the project and set its pace. Systematic and careful planning can

▶ **be a discussion point** and help define criteria when selecting the creative group, especially in the project launch meeting.

▶ **administer each of the planning points** (above) through follow-ups and enforcements. Especially watch for key dates.

▶ **recap goals** when prefacing meetings and express anything that may affect them.

▶ **don't be afraid to try new solutions**—unless a change is tried, old solutions will yield old results. Communication is most effective when it is new, builds, and sparks recognition.

▶ **reach a consensus** from decision-makers on scope and content, especially when dealing with committees, teams, or tight budgets:

▷ *gather all the opinions* needed and be careful not to leave anyone out.

▷ *don't expect department managers to understand technology;* use examples to explain intentions.

HELPING EMPLOYEES
FEEL OWNERSHIP

"How to involve everyone arriving at a successful strategic plan:

> INVOLVE YOUR DEPARTMENT managers and other key employees in the planning process.

> MAKE SPECIFIC ASSIGNMENTS to managers and employees as a part of the plan. Hold regular meetings to account for assignments.

>REVIEW STRATEGIC PLAN REGULARLY with staff. Congratulate yourselves on accomplishments and refocus efforts on shortcomings.

>CREATE OPPORTUNITIES TO HAVE INFORMAL DISCUSSIONS. Often, the feedback provided is more accurate, and more valuable, than the results of any management meeting.

> BE FLEXIBLE. As circumstances change—markets, finances, technology, personnel—don't be afraid to change your strategic plan."

—Marty McGhie
CFO and General Manager
Creative Color
Salt Lake City, UT

> *assemble all influencers* (if possible) at one time for a single discussion. (Send an advance e-mail on the purpose so participants can prepare.) Try to have one meeting which gives everyone the opportunity to give input. Conference calls can function for such a discussion with a designated moderator and concrete agenda. All participants should be given the opportunity to ask each other questions.

> *hear directly from each person* if a single discussion is not possible. Don't depend on intermediary messages.

> *designate clear deadlines* for receiving input, especially if it isn't possible to get it directly. Be clear in e-mail communications if a project is moving quickly. For example, all Internet projects must move quickly or they aren't keeping an audience.

> *define ongoing reporting schedule* for reviews and re-definitions. Analyze new data gathered, what has worked well, what needs to be changed, and integrate those changes into the schedule. Make sure all team members understand responsibilities.

CREATIVE GROUP

Project planning skills set up the potential for profitability and repeat business. If projects aren't planned effectively, the creative group can't fulfill parameters and keep satisfied clients.

ADD TO SKELETON PLAN

Project planning forms the blueprint for the project process. Details can be established and steps defined by the nature of the challenge. (See Print Design, page 100, and Web Design, page 126.) As part of planning, the creative group needs to

► **communicate consistently** throughout the process as the most important ingredient to success, both when working with the client group and the production group.

► **set up an inspiring environment** with client input and designer collaboration. Nothing inspires a designer more than satisfying a discriminating client.

► **build proposal**—use the foundation and strategy to build a plan (see chapter 4, page 38)—with scope, budget, and time frame.

► **understand approval process** and follow up on key dates.

"The true key to a successful design business—and gaining big-name clients—is using the concept of 'strategic brand design.' Strategic brand design adds value to your design business; the benefits are significant. Here's what strategic brand design can do for you:

> Separate you from much of your competition because you're working from a position of strength.

> Help produce clear design strategies by keeping design focused on achieving results and meeting the client's needs.

> Build your reputation as a problem-solver who gets the job done—the first time, every time.

> Help attract and keep outstanding staff who believe they're on a professional and winning team.

> And best of all, target clients who will pay a premium for better quality work.

"Clients are desperate for strategy. They need reassurance that you know what you're doing. With millions of dollars at stake, not to mention their jobs, they want results."

—Steve Coleman
Creative Director
Elton Ward Design
Australia

- ► **explain to client expectations** for quality, timing, and what happens at each phase. Use examples when possible.
- ► **follow the art director who manages** the project process and has a responsibility to
  - ▷ *gather all strategic and planning information* to solicit quotations from suppliers and create project proposal.
  - ▷ understand the *requirements of production* and what procedures are most appropriate. Have an overview knowledge and experience of all processes involved. If not, develop resources to compensate.
  - ▷ *work with the team on implementing the plan*, following the calendar to meet deadlines.
  - ▷ *monitor decision points* as the work progresses. Make sure the client approves key phases before proceeding.
  - ▷ *advance each next step* depending on "who has the ball" and keep the process moving. Project momentum must be fueled.
  - ▷ make all *project judgments* as an advocate for the client's audience. This requires guiding all collaborators towards professional judgments versus personal ones. It keeps all participants focused.
  - ▷ *report any deviations* from plan as early in the process as possible—include recommended solutions.

THE DESIGN AUDIT

"A design audit is an examination and analysis of issues that affect a packaging or corporate identity design project. With an audit, you have a road map for your design choices—recommendations that will ensure the design achieves maximum results.

A design audit is divided into three key phases.

1. DISCOVERY includes investigation and research. Here's a hit list of important components:
   > compile existing market research summaries.
   - conduct field visits
   - meet separately with marketing and sales teams.
   > conduct competitive comparison:
   - evaluate graphics
   - list current communication cues
   - build a matrix of brand and category attributes and requirements

> study consumer behavior:
   - needs, wants, and perceptions
   - where, when, how, and why they buy
   - likes and dislikes
> build a profile of the brand's strengths, weaknesses, and opportunities.
> establish hierarchy of communication objectives.
> identify solutions that bring results beyond client expectations.
> establish a ratings system to benchmark design alternatives against objectives.

2. BRAINSTORMING involves pulling the project apart to look for key objectives and opportunities.

3. RECOMMENDATIONS are your determination of what needs to be done and how best to do it."

—Steve Coleman
Creative Director
Elton Ward Design
Australia

SKILLS OF THE DESIGNER

Design becomes more strategic as the modes of communication expand. Integrating design into the full marketing health of an organization means the visual consistency and blending of media. To best service clients, designers need to be able to

► **think cross-media**. Ideas need to be flexible and recognizable. (See chapter 8, page 74; chapter 11, page 142; and chapter 19, page 236.)

► **adapt integrated thinking** even into a single project.

► **work collaboratively** with clients on parameters through understanding the market, address client requests, and create any new solutions appropriate.

► **recommend ways to repurpose content** into other projects, allowing the client to efficiently build on design investment.

► **consider branding issues** and how the plan ties in with the client's overall direction and potential. Be proactive with suggestions and approaches.

PRODUCTION GROUP: TECH

The technical portions of a project can be accurately scheduled and monitored, as opposed to the creative portions which are more intangible. Provide quotations to rough specifications so a buyer can use them in preliminary planning.

COLLABORATION STRATEGIES

Technical professionals work closely with the creative group to advise, build pages, bridge systems, convert data, set up processes, and execute goals. Use an approved quotation (see chapter 4, page 38) to start setting up work flow possibilities:

► **specify the form for submitted data**—define who does what process in converting it for publication. Determine

▻ *status of current setup*—if new systems are needed or conversions necessary.

▻ *any patterns from experience*—utilize solutions from similar projects to speed efficiency.

► **separate variables** that are for onetime development from those that are ongoing concerns. Anticipate elements that change and set up structure to handle.

► **create flowchart**—match dates to team assignments.

AVOID POTENTIAL CONFLICT

"All disputes within the graphic arts industry seem to fall within five areas of potential conflict. Always ask these questions when proceeding from one phase to the next:

1. Do we have the TRAINING and skills for this project, or do we need to factor in experimental time?

2. Does our PLANNING provide for an organized approval process, define decision-making setup, project management, and assign appropriate responsibilities?

3. Are we COMMUNICATING our expectations and budgetary considerations to everyone in the process?

4. Are all the right professionals PARTICIPATING together early enough in the process to receive input for planning, both inside and outside the organization?

5. Are the groups COOPERATING with one another? Are the participants open to suggestions and learning? Do they know what questions to ask?

"These are ingredients for beginning a successful engagement. Use this list initially to foresee what might derail plans, and, later, to discover what went wrong if something does."

— Barbara Golden
President
Computing Solutions
Chicago, IL

► **define reporting** and monitoring mechanisms, for accessible communication and to utilize the Internet to advantage.

PRODUCTION GROUP: PRINT

Digital tools enable smooth workflow strategies. The production side needs to remain as a service business, applying this fluidity to also make the interface with the creative group advantageous.

SIMPLE PROCEDURES

Having a concise order form and procedural information available can help customers plan efficiently. Most buyers specify the way project components need to be provided in the estimate. But giving the buyer guidance on defining factors, options, and, later, on file preparation, will help the project utilize the tools to potential.

Providing accurate time frames to the buyer is critical for planning. Usually budgets are worked out in advance and deadlines are mandatory. There is the least deadline pressure when the work is scheduled to realistically absorb deviations. Giving input during the planning process only happens with an educated client. It is the best way to make use of expensive finishing resources by knowing what is required and expected.

FIVE WAYS TO DO BETTER BUSINESS IN A DIGITAL AGE

"1. OPEN ALL CHANNELS OF COMMUNICATION. If customers use e-mail, you need to use e-mail. If they like fax, use fax too. To do better business and make more customers happy, we need to be able to respond to customers through whichever medium they prefer, regardless of how we may feel personally about any particular 'channel.'

2. CREATE THE COLOR ADVANTAGE. Those who learn to use these tools to enhance their messages can create a colorful competitive advantage for themselves.

3. BUILD A NETWORK OF RESOURCES. The 'virtual company' and 'out-sourcing' are buzzwords de jour for a reason. Digital technology requires a wider range of expertise than ever before. We used to get by with the plumber, the electrician, and the copier repairman. Now, we need network administrators, Web site programmers, presentation and design consultants, and more. Doing business in the digital age is most effective when you engage a community of collaborators. Think of digital communications as a 'team sport.'

4. LEVERAGE THE NET. From e-mailing or posting proofs on your Web site for client approval to using newsgroups, use online alliances and associations as low-cost marketing tools.

5. PUT PEOPLE BEFORE TECHNOLOGY. Don't get caught using technology for technology's sake. Like any other form of communication, digital communications must touch the intended audience. The more we learn to depend on technological innovation, the more we need to connect with our own and each other's humanity. Otherwise, there will be no lasting success."

—Jon Leland
President and Creative
Director
ComBridges
San Anselmo, CA
www.combridges.com

"There is a tendency for companies to go directly to a design firm for graphics and for design firms to go directly to companies before an overall marketing communications plan has been developed. This results in wasted time and money for both the design firms and the companies. Design flows from corporate objectives, and until the objectives are established, a design cannot be effectively developed. Design should be a part of a company's overall communications plan. Thus, design firms need to strategically align themselves with those who develop the communications plan to ensure they will be considered for the design work. On the other side of the coin, companies should be sure to develop a corporate communications program first, before requesting design services."

—George Sepetys
Comark Group
Detroit, MI

"If clients don't plan properly, they may end up with a beautiful piece, a style they like, but without the appropriate content. They may find that they are riding a beautiful camel when what they really need is a horse."

—Jim Madden
President
Rider Dickerson, Inc.
Chicago, IL

"As a professional consultant, the graphic designer can determine the feasibility of a project by incorporating his or her knowledge of the technical resources available. Often clients choose to develop projects and *then* bring in the designer. This can be an inefficient use of the designer's capabilities, since many decisions may already have been made that the designer should have been consulted on. The result can be unnecessary delays, additional costs, and inadequate design conclusions. The sooner the designer is called in to consult on a project, the easier it is for the designer to help steer the project to the best graphic solution."

—The Graphic Artists Guild
Pricing and Ethical Guidelines

"Designers need to spend a lot of time up front getting clients to talk about their processes, issues, concerns and problems. The more you get clients to talk, the more they start to answer the questions themselves. That forces them to see the logic of good design on their own. If they have ownership over that solution, they're more likely to accept it.

"Ultimately, you can never have too much creative input. The more the client talks and the more you can connect what they're saying to business objectives—and also understand where that's going to take you from a design perspective—the more valuable the experience."

—Mark Oldach
Creative Director
Arthur Andersen
Chicago, IL

"It's one thing to understand a market; It's another thing to conform to it. The purpose of design is to differentiate a product, based on its visual component, from what exists so that people have a distinct reason to act. If you take market research and say, 'This is what's working, so we should be doing this,' you're misreading its purpose.

"Many designers believe market research interferes with the creative process and leads to 'cookie-cutter' design solutions. But if done correctly, market research enhances the creative process by letting the designer know exactly who she is designing for and by providing a clear strategy for effective design solutions. Market research can give designers powerful ammunition for defending their visual strategies and, consequently, lead to fewer revisions and client hassles. Design firms can be as creative in how they obtain their market-research data as they are in how they approach their design solutions."

—Laurel Harper
Partner
Harper & Associates
Bedford, KY

"Good research results in focused work. But only if the same objectives, the function of the product, is understood by both the researchers and the creatives before the work begins.

"What is the core thought? What are we trying to say? To ask people to do? When the core thought is expressed in a single, declarative sentence, it's actually manageable to create a commercial or a billboard which really does leave the viewer with the message you want them to be left with. Brevity is bravery!"

—Todd Waterbury
Creative Director
Wieden & Kennedy
New York, NY

"The best people in the business use research to generate ideas, not to judge them. They use it at the beginning of the whole advertising process to find out *what* to say. When it's used to determine HOW to say it, great ideas suffer horribly."

—Luke Sullivan
Designer
Fallon McElligott

"People who receive the same message five times in the same medium are less influenced than people who receive similar messages in five different mediums. For instance, one is less likely to be influenced by seeing the same ad five times in the same newspaper than if one sees a newspaper ad, a Web banner, a company truck, and hears a radio commercial and a word-of-mouth endorsement. The cross-referenced messages seem to validate each other, resulting in greater overall impact and persuasiveness. Today, companies need a range of identity expressions to convey a range of qualities and to come alive with the power of new media."

—Philip Marshall
Durbrow
Vice Chairman
Frankfurt Balkind Partners
San Francisco, CA

"The role of identity is to find core beliefs that are powerfully motivating and differentiating, and to convey this character with clarity and conviction. The real power of identity comes from within. It comes from being and doing what you believe in. It does not come from being what others want you to be."

—Philip Marshall Durbrow
Vice Chairman
Frankfurt Balkind Partners
San Francisco, CA

"Designers need the self-discipline to submerge their voice beneath that of the brand. You have to speak the language of the target group. You have to immerse yourself in their lifestyle. The challenge is to make a real connection; otherwise, it's false and the consumer is too savvy to tolerate a fake."

—Marc Gobé
dga worldwide
New York, NY
www.dga.com

"Brands these days have much more to do with lifestyle than product lines and are much more about image than graphic design. The strength of the brand can be calculated by analysis of the four major factors:

> DIFFERENTIATION—is the brand unique from others of its kind?

> RELEVANCE—is the brand current with the times?

> ESTEEM—does the brand deliver consistently?

> REPUTATION—how well do consumers know the brand?

"All brands start by speaking to the needs and aspirations of an audience. The aspiration is the brand identity: that's a projection of how the brand wishes to be perceived by its target audience as opposed to the brand image, which is the way the brand is, in fact, currently perceived."

(Reprinted with permission from *CA Magazine*, June 1998)

—DK Holland
Director
The Pushpin Group
New York, NY
www.pushpininc.com

"Our clients entrust us with the equity they've built in their businesses. Building upon that equity requires us to take a holistic, integrated and strategic approach to design. Achieving success brings double dividends: by helping to increase the worth of our clients' businesses, we enhance our own value to them."

—Scott Curtis
Esser Design
Phoenix, AZ

"Great leaders stay focused on a mission without getting hung up on convention. Even creative people need a scope for their creativity. The best bosses and leaders are those who can articulate a project's mission clearly, but give their staff the creative latitude to determine the best way to meet the goal. What you plan to accomplish should be clear. How you do it can be highly innovative."

—Lorraine Monroe
Center for Educational
Innovation
New York , NY

"'Authorship' is the buzzword for intellectual graphic designers looking for ways to broaden the scope and increase the relevance of their cultural contributions. Authorship distinguishes commercial art from visual communication.

"Today there are two kinds of graphic designer. One is primarily production oriented, the other primarily idea oriented. Although the two are not mutually exclusive, one by-product of the digital revolution is a clearer distinction between those with skill and those with imagination.

"The computer has forced more responsibility onto designers to work that was previously assigned to middlemen. The designer can be a major player in a total production.

"One of the key benefits of the computer is the potential it offers for those with vision to turn ideas into products. The machine offers platforms for increasing integrated, rather than fragmented, creative activity. New media provide tools for desktop manufacturing authorship of a 'creative product.'

"Authorship is about originating concepts, bringing them to fruition, and marketing them. It is also about making things for consumption that will benefit the maker and the consumer. It is about taking responsibility for the quality and efficacy of the product. In short, it is a redistribution of a graphic designer's talents and energies in a product-oriented rather than service-oriented arena."

—Steven Heller
Art Director, Author, Cochair
School of Visual Arts
MFA/Design Program
New York, NY

GRAPHIC DESIGN AS A STRATEGIC BUSINESS TOOL

"In today's message-saturated environment, communication programs that produce positive results must stand out in order to get noticed. People have become graphic design sophisticates.

"Strategically guided graphic design positions an organization to set off a very desirable chain reaction: Positive impressions create higher perceived value which boosts sales. The final links in the chain tug nicely on the bottom line because the first links are forged into place with a results-oriented plan —a communication strategy.

"Without management endorsement of the communication strategy a company will soon find that it has not one, but several communication programs producing a variety of inconsistent messages. On the other hand, a management-backed communication strategy prevents separate agendas, reduces duplication, and aligns all messages with strategic objectives. In the most successful design projects, the CEO believes strongly in the importance of design and takes an active advisory role throughout the process."

—*The Graphic Design Handbook for Business*
American Institute of Graphic Arts
Chicago, IL

"Now our working milieu is in the world of 'RoboDesign.' Everything is based on speed: faxes, Fed Ex, modems, and the computer. Client design expectations have become marketing driven and more sophisticated, and they have a working knowledge of computer capabilities. Because of this, we still use the basic process of design, yet we get there much quicker and see things develop a lot faster. People want instant gratification—and clients are no different. It is a reflection of wanting to take care of the right here and now deadline crunch, rather than investing and taking care of the long term effects of such design decisions. This can create an atmosphere of trade-offs—price versus quality versus convenience."

—Keith Bright
Designer
Bright & Associates
Venice, CA

"There's one thing that can be said with certainty about all good creative managers: They've got great job security. With the proliferation of technology, the advent of the Web, and the continuing need to manage people and their talents in an ever-evolving creative environment, there's little chance that companies won't find a need for creative agers who can handle it all."

—Bob Schonfisch
Director of Creative Services
Sega of America
San Francisco, CA
www.creativepro.com

"The importance of communication in business is by now beyond dispute. Perhaps it always has been; but never before has communication been so widely acknowledged as one of the most significant functions of corporate life. There is almost universal agreement that the ability to communicate is essential both to the success of careers within a corporation and to the success of the corporation itself."

—Ralph Caplan
Consultant and Writer
New York, NY

"Good ideas often materialize as the client and designer work together in the planning phases. Everyone involved with the project should freely express their needs and goals. The more a designer understands the intent of the project the better the end result. If the designer is brought in too late, there is not enough integration between content and visual expression. He is merely spreading design frosting on a cake of information, which is an unrewarding way to spend promotional dollars."

—Michael Waitsman
Designer
Synthesis Concepts, Inc.
Chicago, IL

"Every point of contact is a contract with the consumer. Whether it's a TV commercial or a package design or a point of sale piece or a speech made by a vice president, it's also a contact with the consumer and everything from advertising and design to public relations and promotions to event planning and naming to public speaking has to be derived from the same strategy and under the same roof."

—Bill Oberlander
Kirshenbaum Bond & Partners
Kansas City, MO

"Multiple 'media,' as seen today, is in the process of blurring. There will not be a design just for print, or packaging, or the Web, but the seed of the designer's conceived thought for a particular project will hold all of the options at the ready. Designers will need to be cohesively versed in all areas of design and the life expression of their design communication target."

—Lynniel Q. Van Benschoten
MHO Networks
Denver, CO

"To consumers, design is the brand, whether it's a package, a product or a graphic image. And brands, like people, have a heart and soul. My job is finding out how to create a design with a heart and soul—that makes an emotional connection on a personal level at the cluttered point-of-purchase.

"If we don't give consumers the right choices, in the right package, in the right display, at the right place, they will go someplace else."

—Dean Lindsay
Principal and Cofounder
Kornick Lindsay
Chicago, IL

"Business in the global marketplace now must be more knowledgeable about culture and its importance in our global society. They no longer can enter a market with an assumed know-how, but must be more open-minded to learning and understanding. In our domestic territory, we also must be aware of the diversity of our people, in terms of generations, spirituality, ethnicity, economics, and education."

—Dominic Pangbörn
Pangbörn Design, Ltd.
Detroit, MI

"Business communication needs to be built on sound strategy and a position based on differentiation. Design's role is to visualize the strategy. And to visualize it, you need to understand it. To help create it. To provide context and clarity.

"Information without context has little value. Design can help turn information into knowledge. By integrating electronic media, with all of the other components of a marketing or corporate communications program, around a single compelling, unique brand position, it can make it an ever stronger more powerful tool.

"All this change requires us to help clients architect the information. So the right message hits the right person at the right time to build meaningful relationships with their customers. You can't separate creativity from strategy and technology. They work as one.

"Design isn't design until it's strategic. And strategy will get you nowhere if no one notices it. Technology and new techniques can help you do both."

—Guy Gangi
Principal
Mobium Creative Group
Chicago, IL

ESTIMATES AND PROPOSALS

Never begin a communication project without first knowing the economics. To launch a project most effectively, a sequence of preparations will set the right momentum in motion.

TWO KEY DECISIONS

No matter what relationships are established or new, the majority of possible trouble in projects will be alleviated if several factors are defined well in the beginning:

▶ **overall needs**—purpose and objectives for the publishing venture. Any committee meetings, business plans, and marketing strategies are roughly outlined (see chapter 3, page 22). Any market research has been gathered and a communication context defined by market needs.

▶ **budget parameters** and guesstimate of resources. This can be based on experience or rough estimates gathered from services during strategic planning. Possible scope is established.

REQUEST FOR PLANNING

Before any project begins the parameters of the communication challenge and a financial arrangement should be made tangible. The purpose of the agreement is to establish price and to predict cost, define expectations, and confirm the scope. Although these three words might be used interchangeably, each has its own meaning:

▶ An **estimate** is a nonbinding preliminary projection of project costs. It establishes the range of resources.

▶ A **quotation** is a firm statement of price for specified work to be performed. It can be subject to credit approval.

▶ A **proposal** gives a complete overview of the project:

▷ *outlines the intent*—such as market and purpose.

▷ *specifies parameters,* such as direction, elements, and timing.

▷ *includes quotations* for design and production fees.

▷ sometimes has a *schedule* and *work flow plan.* (See chapter 3, page 22.)

When asked for a spontaneous quotation, remember that nonbinding agreements can later be hard to deviate from. There is no such thing as a "ball-park." Once a figure is named, it sets the tone and is remembered. Also, it is very easy to underbid when an estimate is given spontaneously if provided in an initial discussion.

Estimates, quotations, and proposals are usually prepared by the suppliers of communication services without charge. However, fees may be appropriate for extensive proposals or research. If so, this should be agreed upon in advance.

It is wise to receive some estimates initially in planning to see if resources are appropriate, and then request a proposal to provide more detail.

Proposals vary from very formal and detailed to just a letter and an outline. In some cases—for small projects or for long-established business relationships—less formal means, such as verbal quotations or simple hourly billings, are preferable. At the other end of the spectrum are business arrangements based on elaborate proposals and contracts. (In general, it's easiest to resolve any problems when the agreement is in writing, but appropriateness should be the main guide for how projects are structured.)

CONSIDERATIONS FOR GRAPHIC PROJECT PREPARATION

1. APPROPRIATENESS:

> identify the purpose, change, and goal for graphic project or campaign.

> describe the market perceptions and the kind of appeal the communication needs to make.

> have the right number of projects and see if some can't be combined. Fit them into an overall communication strategy.

> define if project is freestanding or part of an established marketing momentum.

> identify overall style and mood to achieve.

> describe delivery mechanisms (direct mail, brochures, online, on-site, etc.).

2. BUDGET

> allocate budgets wisely and adequately to the purpose of the project.

> find ways to share elements between projects to maximize identity, branding, and investment.

> simplify the decision-making process as much as possible. The more decision-makers, the longer and more expensive the process.

> build a contingency percentage into the budget for alterations.

3. QUALITY

> determine the level of design needed: high visual impact or just visually appealing (see sidebar on next page).

> question formats and determine the best for the project.

> identify priorities in the communication needs: what has the most emphasis.

4. DEADLINES:

> establish target date of project completion.

> set adequate deadlines to provide the appropriate quality and budget.

> determine if content can be assembled at one time or if it must be in pieces.

5. STRATEGY

> determine if there are any organizational changes on the horizon that might affect the project.

> factor the involvement of the decision-makers and how much project detail they need.

> anticipate the communication skills of the team members and make assignments to compensate.

> schedule briefings for all crucial team members.

CLIENT GROUP

MAJOR
DECISIONS

. set project foundation

. **plan rough project skeleton**

. **assemble category outline**

. **solicit proposals**

As the client needs the communication project, it is their responsibility to find the best resource to meet their challenge—whether an in-house group or outside firm.

FIRST PROJECT PHASE: PREPARATION

This comprises a good portion of the client group's work on the project. Beginning with requesting estimates to formulate the budget, the client follows up with receiving and evaluating proposals.

► **When several firms compete**—evaluate through portfolios, proposals, fees, and rapport. Discussion can indicate the creative group's understanding and enthusiasm.

► **When working with an in-house department or continuing source**, proposals can be more abbreviated due to a relationship history. However, this process should not be so streamlined it does not include essentials. Never assume an understanding and if based on a previous project, determine what needs to be preserved, changed, or evolved.

"A GOOD CLIENT

> Is decisive and has the ability to inspire and articulate.

> Is thoughtful and plans well.

> Has the ability to see the bigger picture.

> Knows what creative service they need for what activities.

> Is sensitive to and recognizes their audiences.

> Imagines media solutions to their marketing challenges.

> Has general marketing knowledge and sensitivity.

> Links the right people together.

> Provides background information.

> Secures sufficient funding.

> Can delegate but still be accessible.

> Can handle the politics of their organization—anticipate and navigate through political changes.

> Questions and challenges the design team to excellence.

> Is not afraid of taking risks.

> Can express reasons for decisions.

> Understands budget allotments.

> Takes the time to understand the basic procedures.

> Has a strong follow-up or distribution plan.

GET WHAT IS NEEDED

One of the most important ingredients to the proposal—probably the first aspect prospective clients look at—are the fees. Be confident that fees are within industry ranges through the association and publication surveys.

The more specific the client can be about the kind of proposal needed, the more time will be saved during both the preparation and review processes. When getting several bids, give each supplier the exact same parameters. There are no standard ways that proposals are prepared: they are as unique as the firms that compose them. But the clearer the definition, the easier it is to compare proposals if all utilize the same marketing criteria. Complete advance preparation for obtaining proposals includes (see chapter 3, page 22 for developing these strategies):

- ► **purpose of project.**
- ► **components and scope.**
- ► **decision-making process.**
- ► **rough budget numbers.**
- ► **timeframe.**
- ► **business and communication plans.**
- ► **marketing considerations**

"A BAD CLIENT

> Knows exactly what he wants and leaves no room for exploration.

> Has made up his mind what he is going to like before a presentation.

> If it's not his idea, it is not a good idea. Uses designers as though they are pairs of hands.

> Asks everyone else's opinion and agrees with the last person asked.

> Hires the best design team he can find and then won't listen to them.

> Believes that 'cheap, fast, and good' can all coexist.

> Chooses design talent based only on price.

> Asks for proposals from more than three firms (demonstrating insecurity).

> Asks for design sketches with proposals so he 'knows what he is buying.' [See chapter 6, Speculation, page 56.]

> Has conflicts and poor rapport with his company and the design team.

> Chooses colors for projects based on what he wears.

> Stops and starts projects in spurts, changing decisions each time.

> Says 'yes, yes, yes' and then when the work is done says 'no, no, no.'

> Accepts any solution and doesn't really care.

> Doesn't get necessary approvals before doing final artwork."

—Michael Waitsman
Designer
Synthesis Concepts, Inc.
Chicago, IL

INGREDIENTS FOR BEGINNING

Anyone is ill advised to start a project without adding these components to the planning skeleton. With these elements, initial discussion with communication suppliers will be fruitful and yield well-targeted approaches.

Once a proposal is approved, the client reviews and:

► **obtains necessary organizational support** to begin the project, no matter how long it takes. If starting before all input and data are gathered, rework, which will cost more time and money than beginning later, is sure to be the result.

► **creates decision-making plan** for various phases and approval processes. If working with committees, be realistic in how much time it will take to reach everyone and incorporate each in workflow.

► **establishes key dates** to achieve the deadlines—first work back from the date the project is needed, and fill in intermediate checkpoints of design, content, final artwork, proofs, tests, and review processes. Project specifications, if tight in the beginning, offer less room for the creative exploration needed to evolve design development.

► **assigns the project** and gives the creative group the go-ahead to proceed, often secured with an advance payment, purchase order, contract, or other appropriate document.

► **identifies proprietary material** and any trade secrets made available to the creative group. This is sometimes covered by a nondisclosure agreement.

► **negotiates ownership** of project materials and possible future use for other purposes. (See chapter 29, Materials Ownership, page 330.)

CREATIVE GROUP

To begin any project with a client (whether the client is a department of the same organization or an outside firm), a clear agreement of scope, fees, and time frame sets the tone. The creative group may bid on a project with an estimate, quotation, or proposal, depending on the preferences of the client. Creative groups must be flexible in their business dealings to adapt to the way clients do business, but still cover essential arrangements both parties can agree upon. Each agreement is unique and subject to change with business fluctuations. Any agreement not accepted by the client within thirty days may need to be revised with procedures or fees.

ORGANIZATIONAL FORMS

How the proposal is assembled and presented will depend both on the construction of the creative group and their relationship to the client. There are three kinds that have different forms of management and satisfy clients in different ways. Although every group is unique, each will resemble one of these forms.

► **Creative boutique**—usually small, this group is organized around a single talent whose name is on the door. Most talents need to supplement their creative skills with employees to coordinate projects, manage the business, and facilitate client communications. These groups are usually specialized and their proposals will reflect the unique approach of the lead talent.

► **Department**—groups often encompass many disciplines and work within larger organizations. Able to offer a range of services and consolidated systems, proposals may be spreadsheet line items in departmental accounting. Partnerships work well along with designers who require both minimal supervision and the ability to communicate well with other team members.

► **Design team**—assembled for a specific project or goal, team members are chosen for talent and expertise more than place of origin. Different disciplines can be combined together to allow a variety of viewpoints. Within an organization, such teams can draw from as much diversity as is

When soliciting quotations from suppliers such as printers or programmers, be very explicit in how the variables are defined so that time will be saved when comparing fees. If suppliers are left to define the project themselves, resolving the differences between the way the project is quoted will take time. It will probably require going back to each supplier with redefined parameters and more questions.

BE PREPARED BEFORE YOU NEGOTIATE

"Preparation may be the most important, yet most overlooked, element of negotiation, according to negotiations trainer and trial lawyer John Dolan. Dolan recommends the following steps as you prepare for any negotiation.

1. Know what you want and don't want. Write a paragraph describing both in detail.

2. Write a similar description of what your counterpart wants and doesn't want. This can help reveal common interests that lead to creative details.

3. Know what concessions you're willing to make. Consider what's essential and what you can surrender.

4. Know your alternatives. Bargaining power comes from knowing your options if the deal falls through.

5. Know your counterpart and your subject matter. Research. Learn as much as you can about the personality of your negotiating partner.

6. Rehearse. Practice helps in negotiations. Attending a few swap meets and flea markets can sharpen your negotiation skills."

—Jack Neff
Business Writer
Batavia, OH

available. Proposals usually have sources to combine, often causing somewhat complicated billing arrangements. These need to be defined so that every team member is clear on project parameters.

How each kind of creative group operates will depend mostly on the style of the art director. If the design strategy needs a strong visual personality, the creative boutique may be the best. If the project is comprehensive, multidisciplinary, and involves a lot of decision-makers, the art director needs to be a good manager and organizer heading the department. If the client needs innovation, speed, or the greatest distinctiveness, the design team can only benefit from an art director who offers charismatic leadership. Proposals will define the approach and the nature of the team to assemble.

SETTING UP BUSINESS RAPPORT

The relationship between the client and the creative group is more important than any single project. Rapport leads to trust, which creates the best of projects or campaigns. To launch any publishing project, the creative group:

► **listens carefully to all parameters**—is prepared with questions and appropriate research. Clients generally know the market better than the creative group, so it is essential to inspire the client to articulate about the purpose and market.

► **creates a proposal** fueled by initial discussion to define project goals and scope. (See chapter 3, Strategy, page 22.) This may be as simple as a memo of understand, or a more detailed creative brief, or a full proposal that can act as the basis for the project plan. (See chapter 7, page 62, and chapter 8, page 70.)

► **diplomatically explains payment policies**—clearly arranges finances and is flexible to conform to requirements. Be sure to have scope, fees, and ownership rights to design and images spelled out. Many chapters will focus on each of these aspects.

► **shows examples** of projects with similar scope, where appropriate, to learn from responses. Examples from the creative group's portfolio are the best, as they will increase the client's comfort in the project uncertainties.

► **explains the process** and discusses expectations, provides an outline of the process, and establishes deadlines in a calendar format.

"Before requesting printing bids, you should compile information about your product in a request for proposal (RFP) package for printer candidates. Try to make your package as comprehensive as possible so all bidders receive the same information and none need to ask any further questions. Be sure to include:

• A SPECIFICATION SHEET listing all prepress, ink, paper, folding, binding, finishing and distribution requirements

• A PROJECT OUTLINE showing a production and distribution schedule

• YOUR LIST OF GOALS and priorities

• A SAMPLE, mock-up or schematic of the product

• A CONTACT LIST of involved personnel with respective phone, fax and e-mail addresses

• A REQUEST SHEET asking for samples, references, and pricing."

—Steven W. Frye
Print Buyer
Frye & Associates
Sun Valley, ID

- **explores resources** and obtains bids from other creatives on the team.
- **explains rights to client** of how images are priced and how they are contracted for use (see chapter 29, page 335).
- **signs nondisclosure agreement**, when appropriate and
  - ▷ *uses to communicate needs* to both the creative and production groups.
  - ▷ *binds both the creative and production groups* to content, materials, and process confidentiality.
- **solidifies fees**, deadlines, and payment arrangements before beginning the project. Use this agreement to launch further project planning and act as an outline for production. (See Print Planning, page 62, and Web Site Planning, page 70.)

Project planning needs initial structure to be approved and formulated before developing design concepts and creative applications. The faster a project must be done, the more important the planning and correct setup must be.

PRODUCTION GROUP

The tech or print groups enter into the financial discussion during the planning processes but often don't work on the project until later in the development. They rarely provide proposals. Rather, they bid on work to very specific parameters by giving an estimate or a quotation.

THEORY VERSUS REALITY

Often provided for planning early in the project, parameters are usually very rough, usually roughed out by the art director. Estimates may need to be revised if

- **thirty days have passed** and it has not been accepted by the buyer (see role definitions, page 5).
- **specifications change** or new information requires a different approach. Most projects undergo transformations during their development that affect the production fees. Most often, the estimate needs to be refined into a quotation before the project begins. A quotation
  - ▷ *is based on the exact materials, costs, and job requirements specified.* Sometimes it matches a comprehensive presentation that shows the intent of the design.
  - ▷ *should indicate appropriate taxes* (such as on film and materials), deliveries, and any other extras that will be charged.

Example of a commonly overlooked expectation: Many designers charge extra for editorial changes, and the client is surprised at the fees after the work is done. To avoid misunderstandings, the creative group should clarify extra fees in advance and perhaps recommend a contingency for alterations that may or may not be needed.

A quotation should give the buyer a clear idea of the future final invoice, all totals added.

▷ *should be in writing* and approved before starting work. Review of budget with the buyer is a good idea where possible to be sure that everyone has the same definitions of requirements and the same expectations for results.

► **the design changes**—always check throughout the design process, as small parameter changes can sneak into a project, altering it slowly, but changing the nature of its specification. Suddenly, it might be a different project. Often these developmental changes affect costs. If these costs aren't adequately tracked and communicated to the buyer, the tech group may not be able to collect for any extra work.

TECH PROPOSALS

Proposals are common and a way of life for most tech groups—requirements range in detail from technical segments of a site to a complete site structuring. Some processes may be difficult to assign a time frame, but most components are definable enough, that if awarded the project, a well-crafted schedule can be composed.

Project requirements will dictate the form of the proposal and should include:

► **submittal formats**—origin of data and the format that it will be provided in to the tech group.

► **translation or conversion of data** if different from specified formats. Any data that needs to be recreated or reconstructed would have to be quoted separately.

► **key dates** for segment development: receipt of materials and how it relates to the schedule of the whole. If materials provided are received late, then the deadline should be renegotiated.

► **scope of content** and how materials will be developed. Role of tech group is explained through assignments, usually in collaboration with the creative group. Define the elements and phases as the beginning of the project plan and as a way of building accurate client expectations.

"Workable agreements must be made between vendor and client to establish a true partnership. We must agree on such items as:

> What services do I expect from my vendors?

> How will I pay my bills?

> What alterations will I pay for?

> When should the vendor stop work on a faulty file?

> What approvals are necessary before starting again?

> Who is responsible for trapping?

> Who is responsible for copyrights?

> Who has ownership of the resultant data on the disk?

"The list of individual items should continue until both parties are satisfied that an agreement has been met. Only through establishing a foundation to a relationship can there be profitability on both sides. "

—James Hicks
Director of Print Services
Valentine Radford Communication
Kansas City, KS

"The single most important ingredient in estimating is prior experience. Design estimating is more difficult than manufacturing estimates because there are more undefined elements. Defining them is part of the design process. Design firm estimating will be dependent upon previous experience with similar work. Experience helps you understand the time needed, the problem potential, the quality factor, and the price history of what clients have paid."

—W. Daniel Wefler
Publisher
Wefler & Associates, Inc.
Evanston, IL

"We saw five designers' presentations and asked for three proposals. We were looking for a designer who projected excitement, visually and verbally. We get excited about what we do, and we like to see that kind of excitement from a designer. Then, we asked to see some numbers. As a matter of fact, the designer we chose was a little bit higher than the others. So it wasn't straight dollars; it was the rapport established in the initial meeting that made us comfortable."

—John and Bill Schwartz
Owners
Schwartz Brothers Restaurants
Seattle, WA

"I want a proposal to show me that a designer understands the project objectives; what the problem is that this particular project is going to solve. It can be structured in any way that truly accomplishes this.

"The purpose of a proposal is to make sure there are no surprises! You may want to consider it as a kind of discussion draft. If it weren't that, it would be a contract. A designer must reassure me that they understand what I said in our initial meeting. They must also make sure that I am clear about the specific services they are going to provide."

—Robert Moulthrop
Marketing Director
KPGM Peat Marwick
New York, NY

"Without up-front money, you may inadvertently be working on spec with payment promised only upon acceptance. Like other professionals (such as architects and lawyers), you're hired based on experience. This means that you're entitled to be paid regardless of whether your work is accepted or approved. (This is provided, of course, that your services follow the client's initial creative direction and is of the same quality and creativity you were initially hired for.)"

—Emily Ruth Cohen
Consultant to
Creative Professionals
North Plainfield, NJ
www.emilycohen.com

"Designers and clients must be responsible for communication and for understanding each other. The smallest and most insignificant detail can often be the one which causes the machine to break down. When it comes to working agreements, you need to work to agree. Be clear and complete."

—Richardson or Richardson
Hopper Paper Company

"Budgetary constraints should be viewed as basic problematic parameters and not loathed as obstacles to a project's full potential. This is difficult, but circumstances force a designer to be more resourceful and to make each specification work more, and hence worth more. With each aspect of a design scrutinized to be most effective, less expensive, materials and processes should look better. Designers must continue to innovate. The default option to use what has worked before or to arbitrarily revive precedent should be avoided. Clients and consumers will be using new gauges in criticizing items as appropriate or inappropriate. They will be inclined to call their new perceptions 'practicality' or 'pragmatism' when it is simply the next shift in aesthetic taste. We should remember not to cater to these changing perceptions, but as designers, change the perceptions themselves."

—Terence Leong
Context International
New York, NY

"Neither estimates nor budgets are predictive. They will not tell you what is going to happen. Their principal value to the client is to set a limit on costs. The principal value to the designer is to provide a guide to help them keep costs under control. Even the most carefully constructed estimate will not help you if you allow a project to get out of control."

—W. Daniel Wefler
Publisher
Wefler & Associates, Inc.
Evanston, IL

"Once you complete your research and fully evaluate the unique needs of each client and project, you can develop an effective payment schedule that includes several progress payments.

"Progress payments are based on a percentage or portion of your estimated costs. As mentioned earlier, each payment should be due at a specified, defined project phase and encompass defined deliverables and responsibilities.

"Once you have received payment, follow through with a thank-you note or phone call to show your appreciation."

—Emily Ruth Cohen
Consultant to
Creative Professionals
North Plainfield, NJ
www.emilycohen.com

"During the negotiation process, ask the client for credit references (three names is standard), then call the references to confirm credit history. The references should include, if available, a contact within a related industry like a photographer, copywriter, or illustrator.

"Then run a credit check on your client through a company like Dun & Bradstreet. Keep in mind that a credit report can't predict either your client's continued dependability, reliability, or their ethics. The report simply provides a useful credit history to the client."

—Emily Ruth Cohen
Consultant to
Creative Professionals
North Plainfield, NJ
www.emilycohen.com

"Act confident and assertive and clients will respond to your actions. Then, when you are successful, these elusive feelings will kick in. Cindy Brenneman, my workshop producer, always said, 'Unless you ask, the answer is no,' and I saw this demonstrated on almost every job."

—Maria Piscopo
Creative Services Consultant
Costa Mesa, CA
http://MPiscopo.com

"Set forth precisely what usage rights are being sold or transferred. Thus, for example, if the rights to first publish the work in a magazine are desired, consider using this language: 'First publication print magazine rights only. All other rights and uses, including and without limitation, all electronic or online rights, are reserved to the Artist.' If greater rights are sought by a client and the artist wants to sell them, be sure to negotiate a separate fee as compensation for the additional use."

—Caryn R. Leland
Licensing Art and Design
Intellectual Property Attorney
New York, NY
www.lelandlaw@erols.com

"[Creative clients] know what they want to achieve. They're able to think of who they're trying to target without personal preference. For example, a client who wants to launch a project for a teenager must be able to view the product as a teenager, not as a 60-year-old-man. Good clients are able to separate themselves from their projects and think objectively of what their products are supposed to do."

—Supon Phornirunlit
Supon Design Group
Washington, DC

Your proposal should state that it is limited to the one project and that if any additional work is required or if the project expands, there will be additional charges. A revised proposal would be appropriate, and you should, of course, have the client sign off on that proposal. It is also a good idea to set forth an estimate of the number of hours to be spent on any single project so the client will have some idea of the hourly charges to be racked up while work is being continuously revised, changed, and redone."

—Leonard D. DuBoff
Author
Portland, OR

Everyone wants to know exactly how much communications cost. The craft of proposing fees takes place before the projects are designed, so experience and researching industry fee studies are the best guides.

FORMS OF PAYMENT

Fees anticipate the costs for which all work will be performed and billed. Various forms of payment for creative services include:

► **hourly rate**—typical of in-house departments and utilized in pricing and managing project development.

► **per-project rate**—determines a not-to-exceed amount. This is probably the most common form of charging for the design firms.

► **monthly retainer**—particularly applicable for large ongoing projects and can pace payment with the process development. Many ad agencies work on retainer.

Communicating the financial aspects of a project at the beginning helps to determine the scope of the work and can avoid potential misunderstanding. It leaves the publishing team free to concentrate on the work and sets up a good collaboration. Usually fees are a part of a proposal (see chapter 4, page 38). But sometimes it is simply a budget provided. Some begin only with a purpose, a deadline, and a budget. Larger or complex projects are wise to require a proposal, no matter the creative source. Working with a new source also necessitates a proposal to build accurate expectations. Regardless of the form of agreement, clarity launches the project in the right direction.

CLIENT GROUP

There are only two ways to have a project come in on a predictable budget: know the components in advance (hard to do when a creative solution is needed), or set a ceiling on resources based on research, experience, and agreement with creative source.

POSITIONING INGREDIENTS

In planning and preparing for a project, the client defines parameters that are used in the proposal (see chapter 4, page 38). It is the responsibility of the client to provide:

► **goals**—what the project needs to achieve.

► **scope**—size, range of market, and positioning.

The *Graphic Artists Guild Pricing and Ethical Guidelines* is a good source for finding out what suppliers charge. For information: Graphic Artists Guild, National Office, 11 W. 20th Street, 8th Floor, New York, New York, 10011, 212/463-7730. www.gag.org

There are no standard fees in the graphic arts because no two projects and no two suppliers are ever the same. This increases the importance of clear agreements.

Three magazines do annual fee surveys and salary surveys of their readerships, which comprise a significant percentage, and cross-section, of the design industry:

Communication Arts
www.commarts.com

Graphic Design USA
www.gdusa.com

HOW Magazine
www.howdesign.com

MAJOR DECISIONS:
- set project foundation
- plan rough project skeleton
- assemble category outline
- solicit proposals
- choose creative group
- **finalize project plan**

Two newsletters could be eight pages, 8-1/2"x11," and two colors. One is on Strathmore Grandee paper and in a quantity of 2000 for a small non-profit association. The other could be on a house gloss text stock, print in a quantity of 50,000 for a major corporation. The two would cost vastly different amounts to design and produce.

► **intent**—the audience the publication should reach.

► **direction**—guidance on appropriate content and style.

► **background**—history, industry information, etc.

► **budget**—resources available and payment considerations.

► **time frame**—projected completion date.

These factors all affect the fees. The client approves costs before any work begins. Defining fees should never be skipped under heavy deadline demands. Similarly, if the project is being done just for expenses (pro bono), still describe what the project is worth so the client appreciates the gift. Money is one of the most critical parameters in the process. If a client commissions work from a creative source without first solidifying fees, this is an indication of misunderstandings to come.

CREATIVE GROUP

Most creative groups emphasize creativity over project management. Because the project begins with agreeing to a budget range, this determines the scope and approach. The art director, leading the project, should be equally skilled at the creative and management sides to meet financial requirements.

FEE VARIATIONS

Fees have many components and each organization has their own method for arranging, explaining, and presenting them.

If a potential client balks at your payment plan, unless they have their own company policies, this is an early-warning sign of trouble. Clients with good intentions are not offended by your asking them to sign an agreement or providing an advance payment. If there is trouble with these terms, it is best to know right away instead of later after the work is done and you are trying to collect.

► **In-house departments**—generally conform to company policy when charging other departments. Variations include:

▷ *bidding and charging*—as if other departments were outside businesses.

▷ *managing budgets* and allocating through a strategic understanding of each other's resources.

▷ *doing favors* where one department handles a pro bono project for another. Score is kept where each department trades services.

A client may prefer a retainer method of billing with design firms they use regularly to plan their costs over a long period of time. This can enable smooth budgeting within a defined timeframe. It also gives efficiencies between projects and insurance for cohesive marketing.

► **Outside vendors**—such as design firms, ad agencies, and freelancers, operate on agreements secured by a proposal or a quotation. (See Estimates and Proposals, page 38.) Possibilities for fee arrangements can be (see above):

▷ *per-project basis*—estimate time and materials.

▷ *payment plan*—such as one-third in advance , one-third with design approval, and one-third upon completion.

▷ *retainer*—a monthly billing arrangement guaranteeing a consistent payment matching the economics to progress. This stabilizes cash flow for large projects, series, or regular updatings.

BASIS OF FEES

One reason publishing has no standard fees is that every project is a different mix of factors. Also, every creative group is different, but fees generally reflect:

► **scope**—the wider the market appeal, the more design costs. Scope determines time and materials needed.

► **size and quantity**—variables come from the scope and affect design and production costs:

   ▷ *number of design solutions presented*

   ▷ *quality of the presentation materials*

   ▷ *number of components*

   ▷ *quantities of printing of number of Web pages*

► **uses to which the materials will be put**—the more benefit through extensive use, the greater the compensation.

► **complexity of parameters** and difficulty of assignment—the level of creativity needed dictates to the experience and talent applied. Also, narrow parameters can sometimes create a more difficult assignment.

► **nature of client organization**—a small company or not-for-profit group requires a lower fee structure. A large multinational corporation has projects with larger scales and corresponding fees.

► **project and account management**:

   ▷ *decision-making processes*—the more decision-makers, the longer and more expensive the project.

   ▷ *presentation processes*—how finished-looking the materials need to be depends on the decision-makers' abilities to visualize. Committees can be more demanding.

   ▷ *documentation*—reports to fit client requirements.

► **technological capabilities**—unusual equipment or special effects are required that entail purchasing or training.

PREPARING THE PROPOSAL

The art director provides a project description (usually as a proposal; see chapter 4, Estimates and Proposals, page 38), which may include:

► **time frame**—broken into key segment dates as a planning tool to measure progress and deviation. Try to extend deadline if budget is tight.

► **estimate of outside costs**—received according to rough parameters, to be solidified later.

► **itemization of fees**—suitable descriptions requested by

Because the project has not yet been designed, both you and the designer are discussing something that does not yet exist. There needs to be a certain amount of contingency built in to allow for a variety of approaches. Some firms automatically add a percentage for contingency, others give a range of fees, still others prefer to work on a not-to-exceed basis. In any event, the client should be informed about what level of design different fees will purchase. The designer may want to show samples of different scales of projects.

It is very dangerous to choose the lowest bidder. The firm may not be experienced enough to know what could happen. They may hound the client with additional charges on items that should have been made explicit in the beginning but weren't. The design source that is best at accurately estimating the project will probably be the easiest to work with during its changes and execution.

There are several cases where design firms may provide lower fees than they normally charge: for start-up organizations, nonprofit causes, markets they have not worked in before, or personal connections. Designers may do this for personal reasons, or to build a portfolio.

When buying print or programming services, the outside service can be more expensive than the design fee. When this occurs, it is better for the creative group to arrange to have the client billed directly and charge a supervision fee (see chapter 25).

the client and to show sufficient intent.

► **payment arrangement** that fits the business practices of both client and creative groups. This may require discussion and negotiation before an agreement is formed. Most successful negotiations involve refining the project scope.

Before proceeding, the basic plan and essential structure should be established.

PRODUCTION GROUP

The creative group (or sometimes the client group directly) requests a quotation as part of their planning process. (See Print Planning, page 62, and Web Site Planning, page 70.) It is not wise for a buyer to embark on a project without investigating the production expenses. Similarly, it is also not wise for a printer or a programmer to embark on a project without financial arrangements, if they want their projects to be profitable.

TANGIBLE PARAMETERS

Unlike the creative group, the production group quotes on specific parameters—roughed out at the beginning for the budget and refined when the design approach is selected. Charges for work may be based on the following:

► **job quotation**—includes all materials and overhead expenses.

► **hourly rate**—for in-house labor covering single or variable rates for staff time.

► **unit price**—achieved by dividing quantity by fees and averaging complexity throughout the project.

SET UP TO RECEIVE

The process of providing an initial quotation, refining it with final parameters, and receiving approval fits most every project. To best work with the buyer, concisely communicate various terms and payment processes. Further:

► **secure approval in writing**—estimates should be signed by the buyer before work is performed.

► **bill as quoted** when project is completed. Adjust the budget only with approvals and list deviations as line items.

► **report changes and alterations** if in additional to the quotation. (See chapter 26: Project Changes, page 310.) These are impossible to anticipate in advance, so handling them is an important part of trust and fairness between groups.

► **invoicing** is considered accepted unless the buyer contests within 15 days of receipt. If budgets have been reported prior, there should be no surprises.

"The most general principle for determining the price of artwork is that the price should be in relationship to the value of the intended *use* the buyer will make of the art. This means the more extensive the use, the greater the compensation to the artist. Some inexperienced art buyers are shocked by such a concept. They assume that they are buying a *product* at one flat price, with which they can do whatever they wish upon payment. But artists normally sell only certain *rights* to the use of their creative work. The greater such rights, the greater the compensation required."

—Graphic Artists Guild Handbook of Pricing & Ethical Guidelines

"There is no orderly market for design services. There is no formal mechanism to collect and transmit information about design prices. No stock market quotations or price indexes. Design is bought and sold in thousands of negotiated transactions that are only loosely related. Buyers can seek competitive proposals and get a feel for pricing. Designers must work a little harder seeking to find out what their competitors charge for the same work."

—W. Daniel Wefler
Publisher
Wefler & Associates, Inc.
Evanston, IL

"Anyone researching design fees will quickly learn that although there are several published guides for calculating hourly rates, very few sources for hard data on pricing exist. Since there can never be an absolute set of national fee standards, due to federal trade restrictions, the best sources of information continue to be informal AIGA peer group networks and the *Graphic Artists Guild Handbook of Pricing & Ethical Guidelines.*"

—Juanita Dugdale
212 Associates
New York, NY

"Excellence is not necessarily a function of budget. It's a function of attitude. When budgets are low, that's the perfect time to propose something a little bit different. And I think it's *always* important for designers to look for the opportunity to suggest an approach that goes beyond the usual. . . . If you don't have a huge budget, you have to leverage, and leverage comes from creativity on the part of the designer."

—Edwin Simon
President
Pelican Group, Inc.
Hartford, CT

"As dot-coms turn ever more corporate, a community of rebellious Web designers are building studios, commanding high fees and firing clients."

—Edmund Lee
Journalist
The Industry Standard
New York, NY

"The only way to sell quality in a budget-conscious world is to show the power of an idea. It stands on its own without help from trendiness. Creating an image with impact means having to examine the roots of our needs and perceptions. Powerful design performs well, but also hits an emotional chord. Ultimately, quality design will make us all successful. Satisfying business needs is why we're here—giving them the best is what makes us valuable even when budgets are tight."

—Mike Scricco
Keiler & Company
Farmington, CT

"I believe the biggest challenges facing the design industry involve working with greatly reduced client budgets, maintaining your own profitability as a design firm, and still delivering high quality work. The industry has come a long way in the past eight to ten years. Fortunately, it is finally being recognized for its tremendous worth to business. Although business in general has become increasingly more competitive in the 90s and clients are pressuring us even more in terms of price, we can only be a detriment to ourselves by making price, as opposed to quality and value, the primary issue as the basis for being awarded projects."

—Mary F. Pisarkiewicz
Designer
Pisarkiewicz & Company
New York, NY

SPECULATION

Obtaining creative design is critical to an organization's growth and success. As the world becomes more visual, design grows as a strategic investment. Design costs are paid for *before* benefits are reaped, which can make clients nervous. They may also be nervous about selecting the best concepts. Some think these risks can be avoided by asking for initial ideas as part of the proposal and awarding the project to the group with the best idea.

Speculative (or "spec") design work is produced by the creative group for a client's review in expectation of receiving the project. Without a design fee, this is not recognized as a fair business practice. In architecture, public art, or advertising, there may be speculative proposal conventions. But in graphic design, this practice works counter to a client's goals and undermines the creative efforts that service them.

LEGITIMATE SUPPORTING MATERIALS

The client may request (or be provided with by the creative group) support materials for a proposal without incurring a fee, such as:

► **prototypes or blank formats** for print to show paper stock and weight, size of piece, and bindery. It provides the client with a "feel" of a printed or fabricated project and can function as a production guide.

► **samples of related work or images**—any visual samples that define parameters are advantageous. But this should not be mistaken for showing designs, as these are very rough preliminary starting points from which concepts can be built.

CLIENT GROUP

The client may request materials to help illuminate the proposal. Whether the creative group charges the client for these depends on the extent of the creative work involved.

EMBELLISHING DESCRIPTION

The client may legitimately ask the creative group to include:

► **preliminary items** needed to form a printing budget, such as paper formats and blank samples (see above).

► **new processes information**—new to the client or the creative group—such as new software applications, effects, unusual illustrative or photographic treatments, new printing techniques, Web applications, etc.

All professional graphic design organizations support the position that creative work should not be done on speculation. Because this is the least tangible of the publishing phases, because it requires the greatest amount of market understanding, because it uses the greatest talents, and because it is the basis for all other fees, the design is the spark that sets the other processes in motion. This service should not be undervalued or given away.

MAJOR DECISIONS:

- set project foundation
- plan rough project skeleton
- assemble category outline
- solicit proposals
- **choose creative group**

SPECULATIVE DESIGN HURTS THE CLIENT

Some clients may need an explanation of why speculative creative work is not in their best interest. Here are the major ways that such practices are detrimental to client goals:

▶ **project parameters** may not be completely formulated when starting, especially if deadline is an issue.

> ▷ *It is difficult to thoroughly brief more than one firm—* solutions after one meeting will be superficial. Effective design results from a relationship that builds as solutions are explored.

> ▷ *Creative concepts need lead time—*to begin a project without a thorough plan is to conduct a time-consuming experiment.

> ▷ *Not comparing concepts developed from equal information—*the client runs a danger of not focusing when receiving more than one spec proposal. Design firms may base concepts on very different approaches and knowledge than their competitors. Concepts may be presented like apples and oranges, and not scaled enough to the unique communication challenge.

> ▷ *Results insufficient—*there are very few documented cases where spec design works out favorably without the in-depth commitment of a collaboration. The firm who wins generally has to begin again with additional parameters.

▶ **best creative efforts can't be applied**. The creative group

> ▷ *can't afford to spend as much time as they would if they were getting paid.* Any paying work they have must take priority.

> ▷ *may be too busy to undertake spec—*better designers must focus on the clients who pay regular fees. The best firm for the project may not be available.

> ▷ *will not have an in-depth understanding* of the assignment, especially after one meeting, unless they have time to do their own research without background; one half a design is like one half a baked recipe.

> ▷ *may be competing* against another creative group, which means that one or more groups will not get paid for their work. The purpose of portfolio reviews, proposals, and interviews is to differentiate one firm from another.

> ▷ *gives initial concepts the most in-depth creative thinking* of the creative group, and, therefore, worth the most. There is a big difference in quality between spec design and commissioned design.

REAL STORY I

The director of publications of a famous museum asked three design firms to submit a speculative brochure cover with their proposal for a new museum and map brochure.

FIRM ONE had great experience and ability. Although their concept was excellent, they didn't have enough time to execute the presentation materials very well.

FIRM TWO had more time to devote to the project, and their presentation was very polished. Yet their concept was not quite right. They needed to do more research.

FIRM THREE was the least experienced. Their ideas lacked conceptual depth and were merely decorative. But they spent a great deal of time on the presentation.

The presentations that the director of publications received were all in different forms and styles. The committee had to choose between apples, oranges, and pears. It is very difficult to look beyond the presentation styles to see the best concept.

The committee ended up choosing Firm Two's work. But because their concept was not right, they had to start over. All the design and presentation work they did on speculation was not used or compensated for. The other two firms were not compensated at all. Perhaps Firm One had the idea with the most potential.

Buying design this way is much like paving a road with no foundation underneath. When a client buys creative services speculatively, they do not allow the creative group to do its best work.

The correct way to hire a designer or firm is to objectively evaluate their finished work in its approach, diversity, sophistication, clarity, and completeness. The design firm's proposal shows their understanding of the project, and rapport can be established.

Creative groups had better enter any speculative opportunity *without* any rose-colored glasses. Often seduced by causes or clients in great need, the creative group should separate speculative work from pro bono work.

MATERIALS WITH NO FEE

Supply formats and related samples to focus a proposal and help a client visualize a project's size and scope. These can:

- ► **show arrangement of components** such as outlines or rough project plans.
- ► **provide samples of work** that are of similar scope.
- ► **show related publication** samples or sites that explore a similar technique or approach.
- ► **demonstrate illustration or photography** samples that are available for use.
- ► **show techniques** such as color or texture treatments, etc.
- ► **demonstrate paper samples** used on print projects,
- ► **be used for presentation** but are the property of the creator.

DISRUPTIVE AND DECEPTIVE

Speculative concepts and designs are not to be provided without a fee payment. All major graphic design organizations band together on this issue.

- ► Speculative design is an **unfair business practice**, providing a valuable service without compensation.
- ► Concepts needed for creating the proposal should be limited in description for purposes of **defining parameters**, not for giving away ideas.
- ► The only creative groups **who will do speculative work**:
  - ▷ *wish to get a foot in the door* with a potential client.
  - ▷ *want to build their portfolios.*
  - ▷ *desperately need business* and are hungry.
  - ▷ *are willing to donate their services.*

RISKING IT ANYWAY

If embarking on this practice, be prepared for:

- ► **not getting paid** for the bulk of the work. It always takes more time than even modest presentation fees can cover.
- ► **having to start over** if rewarded the project. New parameters and considerations always surface between the time a spec project is presented and actually begun.

Pro bono work can be some of the most rewarding creatively, if handled skillfully. Although there is no fee involved, there also can be no competition and limited client retro power. The client, getting professional work for free, doesn't have the usual supervisory role.

REAL STORY II

A large trade association wanted an ad campaign to raise their visibility. One design group had serviced their publication needs for four years and the board of directors asked them for concepts. In the meantime, two board members each went to the design firms that handle their own work to ask for concepts. The first firm laid out a proposal, asking for no money in advance before the first presentation. Though the fees were outlined and approved, after the design presentation, when comparing the work of all three firms, they decided not to go with any, and paid none of them. The board of directors started all over again with a new source, but much time and good will was lost.

- **indecisive decision-makers**—are the only kind of clients who would request speculative work.

- **limiting budget challenges**—because money has already been invested, it may come out of a total budget, taking development resources away from the overall available funds. Meeting any budget expectations is always difficult because there is a negative investment to begin that is rarely recouped.

PRODUCTION GROUP: TECH

The creative group and the tech group work together to fulfill the expectations of the client, as guided through the quotation and proposal processes (see chapter 4, page 38).

BEST WAY TO QUALIFY

Occasionally the tech group will be asked to do sample programming to demonstrate technical considerations (either skill or technique). As with the creative group, qualifications are better screened through portfolio samples and references than through spending time and money on experiments. If used as a way to differentiate firms, sample work for a client should have a fee to cover the tech group's time and investment.

The creative group cannot take the solutions of one tech group and give it to another without appropriate payments (see chapter 5, page 50).

PRODUCTION GROUP: PRINT

Some imaging centers may do sample separations or preparatory work to demonstrate capabilities. Any tests should not be used by the buyer unless paid for. Any printing done on speculation is really pro bono.

PROPOSAL MATERIALS

The creative group generally requests support materials from vendors when preparing the proposal. They receive a quotation (see chapter 4, page 38) on which to base plans and parameters. Prototypes to embellish the proposal become tangible models to based plans and parameters. The printer may provide blank formats for paper, size, or printing technique and are

- **supplied without charge**. The print group rarely requests a fee unless extensive time and costs are involved.

- **not to be used** without permission or mutually agreed payment, where appropriate.

- **the basis for quotations** on specific formats or prototypes. If parameters change, new mock-ups or prototypes should be provided with a new quotation.

"In large-scale advertising, providing 'free' creative solutions in a competitive format is commonplace. In brain surgery, it isn't. Design must be somewhere in between. Clients should determine whether they're purchasing a design, or hiring a designer.

"Designers need to remain open-minded. That's what makes them valuable in the first place. Are there innovative routes which can meet everyone's needs? Perhaps.

"At all costs, the design process has to be valued. Without value, it doesn't matter a hill of beans whether it's done on speculation or paid for."

—Richardson or Richardson
Hopper Paper Company

"Several major studios are now quietly developing new business by working on spec. The end result is a client who expects more for less, or for nothing at all. But more importantly, the small design 'boutiques' (which have always been the heart of industry creativity) usually count on one or two major projects as their lifeblood and cannot afford to work for free."

—Regina Rubino
Designer
Louey/Rubino Design Group
Santa Monica, CA

"Good advertising is conceived in frustration and delivered in agony. It is precious stuff. It shouldn't be wasted and it shouldn't be given away free, no matter how eager an account executive or agency president is to please a client."

—John E. O'Toole
President
Foote, Cone and Belding
New York, NY

"In our society, we express our respect for work by paying for it. When we refuse to do so, we are expressing contempt for the work and the worker. In difficult times, the fundamental rules of human conduct are under attack in and out of business. The only appropriate response is not to allow our own sense of values and self-respect to erode in the face of it."

—Milton Glaser
Designer
Milton Glaser Inc.
New York, NY

"Anyone doing work on speculation is likely to go unpaid. But the fact that one person or firm works on speculation makes it likely that others will be asked to and feel a greater compulsion to answer such requests affirmatively. So the issue of right and wrong cannot be decided on the basis of the individual alone. While work on speculation is likely to damage those who do it, it certainly erodes the creative and financial health of the community. This is why the organizations representing the creative community have drawn codes to regulate fair practices, and why these codes take a strong stand against work on speculation."

—Tad Crawford
Writer and Publisher
Allworth Press
New York, NY

"I don't ask my lawyer, broker, or doctor to do work for me for free while I scope out who might be better at it. To me, that shows disrespect and is a waste of time for all parties involved."

—Amy Strauch
President
What!design
Allston, MA

"I see a difference between a potential client that is truly naïve and needs to be guided along the process and one that is craven, wanting something for nothing. Doing spec work places a mutually agreed upon value of design consultation at exactly $0.

"Requesting design work on spec indicates to me a deep misunderstanding of what a good designer can offer, and by accepting this arrangement designers contribute to this problem.

"Why does it seem that the spec project always ends up going to the firm that had a previous relationship with the client? Is requesting free solutions from different sources a reality check for the client? A chance to get free ideas? An opportunity to goose the incumbent? The one time I made the mistake of doing a spec project, I realized that my charitable efforts should have been going toward helping out an organization that was truly in need. Or toward planting tomatoes. Or doing anything other than reaffirming the misconception that graphic design is cheap, fast, and easy."

—Alexander Isley
Alexander Isley Inc.
Redding, CT

"An understanding of how to reposition [your corporation] in the minds of investors is perhaps the most valuable asset designers bring. We develop positioning strategy based on information gathered through: an audit of industry communications and key competitors' tactics; branding initiatives (current and planned); and discussion to define critical objectives.

"Unpaid competitions are more likely to end in frustration than in good design. The 'winners' are just as likely to wind up frustrated, since they're now somewhat locked into an approach devised before they had a chance to do any real homework . . . in addition to being grumpy about having been put through the hoops. For no money. The spec work approach demeans us all, and perpetuates the myth that design is all about how something looks."

—Lana Rigsby
Rigsby Design
Houston, TX

PRINT PLANNING

Few organizations have marketing plans that do not include print—even if it is only the corporate identity carried into stationery or direct mail promotions. Expanding into cross-media publishing, print is always a viable component. (See chapter 8, Web Site Planning, page 70, for interactive design.)

REDEFINING MEDIA

With the introduction of any new media, the older ones are redefined. Print-based projects (or, more accurately, all non-interactive presentation media) have been refined and better targeted due to the addition of the Internet. This communication technology does not make print obsolete but alters the way it is used. Print will always be valued because it has user-friendly and aesthetic qualities that make it the medium of choice for the backbone of all an organization's graphics. Print will stay viable, as it is repurposed, because

▶ **people don't like to read on the computer**—readers most often print out text from the Internet (which turns it into print).

▶ **portability** of paper is convenient, cheap, and lacks any devices needed to view material.

▶ **hard copy record** allows marking, writing, and a tangibility to documents. Also, print presents higher-quality visual images than can be viewed on a computer monitor.

▶ **signage and displays** will always be needed for commercial spaces; though retail sales blend with online services, people like to see what they are buying.

CATEGORIES OF PRINT PROJECTS

All print-based projects have parallel graphic and development procedures. All require similar project plans, shared parameters, and phases. However, publishing on the Internet has a different process. (See chapter 8, Web Site Planning, page 70.) Print-based projects include:

▶ **corporate identity**—flexible, with cross-media potential and to work collaboratively with an organization's Web site and other marketing or sales efforts.

▶ **collateral**—brochures, booklets, fliers, pamphlets, leaflets, promotional, advertising.

It is said that you can't judge a book by its cover, but people do it all the time. In bookstores, it is the combination of title with design that inspires your hand to reach out and lift a particular cover off the shelf. Similarly, when you receive your mail, it is the more interesting envelope that gets opened first. When you view Web pages, the ones that have the best content and are the easiest to navigate are the ones you return to. How people react to information has changed through time: People read less but see more. They react to patterns, correlate the visual to the verbal, and perceive information in "bite-sized" pieces. Design that compels or motivates is essential for successful marketing.

To attain such a level of design requires a process that encourages the best abilities of each contributor. All design projects follow a general production path. However, in such a custom business, each project is unique in its personality and requirements. Knowing about basic processes helps you to pay attention to those uniquenesses, because it is the way the project is conceived, then produced, that will distinguish your materials and best communicate your message.

- **publications**—newsletters, magazines, journals, annual reports.
- **display**—posters, tabletop, and other point-of-purchase.
- **presentation**—slides, handouts, videos, sales support.
- **signage**—environmental, directional signs, identity markers.

ADDING TO THE STRATEGIC SKELETON

To blend print into an organization's momentum, have a project strategy and overall plan worked out (see chapter 3, page 22). Most print projects have a linear production path with many phases overlapping. The best design is achieved with a step-by-step approach. Ways individual projects differ are in scope, components, and number of meetings (often underestimated during the planning process). Building upon the proposal as a structure, a work plan evolves and should include:

- **Process**—review the major segments and content needed. Divide into categories and determine needs of each.

- **Personnel**—assign specific responsibilities. Good communication, once a project is defined, is the most important component to plan for. Mechanisms to ensure communication can be built in to the overall project procedure.

- **Schedule**—work with team members to develop the best time frame for each component, working backward from the date needed (see chapter 3, page 22).

  In the continuum of the project, a careful schedule feeds activity without the pressure of compromise forced by too little time. It seems all projects escalate to a crescendo of activity at the end. But again, as always, knowledge and planning form the best insurance policy to underwrite success.

TO BEGIN THE PROCESS

Several factors are present in successful projects:

- **the proposal and strategy are approved**—if there are decisions still to be made, it is best to wait until they are made to begin.

- **the project has leadership** through a named project manager and project director (see description of the client champion, page 22, whether part of the client group or the creative group), and an art director.
  - ▷ *The project director*
    - > is usually from the client side and has the authority to approve all steps in the development process.
    - > facilitates decision-makers in the client group.

It is human nature to want to jump right into a project. Many clients feel pressure from their managements to show something as fast as possible. In fact, many publishing projects are really started *after* they are needed. Publishing is a deadline-oriented industry, and this probably won't change. It is one of the major reasons why most printed materials are mediocre. Hopefully, as client skills increase, it will influence quality, effectiveness, and sanity. There is no shortcut for good planning. (If you haven't, please read chapter 3, page 22.)

- ▷ *The art director*
  - \> supervises entire process to ensure quality and that all parameters are met.
  - \> is main liaison with the client and the production group.
- ▷ *The project manager*
  - \> is generally on the creative side and makes sure that all approvals are made before proceeding, always securing feedback from the client (on small projects—the project director may also be the sole client and the project manager may also be the designer).
  - \> ensures all key dates in the project plan are met.
  - \> controls and monitors budget and production process. (See chapter 20, Managing Budgets, page 246.)
- ► **Have initial launch meeting** for the decision-makers and the creative group.
  - ▷ *Let each participant know the agenda* prior to allow them time for research or other preparation.
  - ▷ *Invite a representative of the production group.* Each segment of the team should contribute.
  - ▷ *If a conference call*—have an agenda and a designated moderator to keep discussion moving and to draw consensus.

OVERVIEW OF PRINT PROJECT MANAGEMENT CHECKLIST

- Meet with decision-makers
  - \> check parameters
  - \> check completeness of incoming elements
- Quote project elements
  - \> estimate design and visual components
  - \> estimate rough production fees
  - \> get printing bids
- Receive project
  - \> schedule work flow
  - \> meet with team and make assignment
- Concept development
  - \> monitor project progress
  - \> review designs
  - \> check budget against design

- Presentation
  - \> present to all decision-makers along with creative group
  - \> mention any appropriate budget considerations
- Finished art development
  - \> monitor the project as it progresses
  - \> check all final lasers
  - \> supervise approval process and receive alterations
  - \> manage all approvals
  - \> review budget, request reports from team
- Final output
  - \> check to make sure complete
  - \> review color specifications
  - \> scrutinize each page proof and make final corrections

- \> request budget reports with alterations
- Printing supervision
  - \> review printing proofs
  - \> route proofs to all decision-makers and obtain signature
  - \> communicate costs on alterations
  - \> press-check project
  - \> receive samples
- Analysis and billing
  - \> review actual budget
  - \> check on delivery
  - \> review billing information
  - \> receive results of project and measure against intended goals
  - \> review with team members
  - \> determine how to build on what is learned.

MAJOR
DECISIONS: PRINT

set project foundation
plan rough project skeleton
solicit proposals
choose creative group
finalize project plan

CLIENT GROUP

The linear print process continually becomes smoother in a digital environment. What becomes most apparent are the talents and competencies of the various professionals on the creative and production teams.

CONSTRUCTIVE RELATIONSHIPS

The best design is the result of a strong client/designer relationship where the plans of one inspire ideas in the other. Collaborations vary greatly in character, personality, and rapport, but all these relationships have

▶ **trust**—both sides support the explorations of the other.

▶ **decisiveness**—decision-makers can quickly decipher options.

    ▷ Professionalism involves knowing the difference between personal opinions and professional ones that respond to the economy and environment.

    ▷ It is necessary to be decisive to be fast, and fast decisions are the most economical (unless made too fast by being uninformed). Debating, wavering, or changing all take time that can add to costs.

▶ **inspiration**—both the client and the designer exchange impressions and approaches. Be open to a range of concepts initially, as the most successful marketing strategies might be combinations of several ideas. Take the opportunity to explore all options early so that later decisions can be firmer.

▶ **realistic expectations**—experience is the best guide, so those without should be careful to research, ask questions, and learn from others' examples. One person cannot be an expert in everything. It is important to continually develop a network of resources.

▶ **follow-up**—results become known such as increased sales or membership, and the team evaluates the outcome to plan the next strategy. Always learn from the market feedback and readjust goals or approaches accordingly.

LEAD THE PROJECT LAUNCH MEETING:

It is the client who initiates the beginning of the project, usually with an approved proposal and by organizing for this launch:

▶ **provide agenda in advance** so other participants can prepare.

▶ **bring key items:**

    ▷ *proposal and strategic plan* with the most important issues emphasized (see chapter 4, page 38), materials to give

A fresh and innovative view doesn't mean having to disregard the old ways of company procedure. It means being responsive to change in the market which overshadows doing things the way they have always been done. There can't be a new direction without change. Watch out for decision-makers who verbalize a need for a new direction, but when it comes to taking action or supporting it, they don't really want to change. To be effective, it is necessary to go up against resistance. The only way to achieve something new is through having belief in an idea, planning, diplomacy, and understanding the process.

overview, profile of membership, and market.

▷ *past or related printed pieces* or communications.

▷ *input from other groups* or departments that influence content, development, or specifications.

▷ *corporate identity* or branding usages, such as guidelines or styles, or related communications.

► **assign a key person** to link the client group and the creative group—the project director—who can follow through on approvals and project monitoring.

► **provide information** and context to build a collaborative environment.

CREATIVE GROUP

Beginning projects is the most exciting time for the creative group. Diving into possibilities, exploring variations, and discovering new solutions are their lifeblood. Many can be impatient about planning, but if careful to cover the components, even the wildest creative genius can keep a project on track.

DESIGNER CORE COMPETENCIES

Creative skills encompass a spectrum of expertise (and no one designer can be equally capable in all these areas), but need at least a rough overview in each:

► **concept development**—tied into project strategy, the ability to imagine visual themes and symbols that communicate the client's message to their audience. Sparks for an idea can be a headline, an image, or totally original (see chapter 10, page 100, and chapter 11, page 126). This is where the insight from talent proves one idea more successful than another.

► **presentation**—diplomatic skills are used to show ideas to the client, receive feedback, and negotiate. Client relations are hard to define but critical to the success of the project. Finding design solutions that bridge different viewpoints can make the designer a hero. This requires carefully listening to the client and the market. The best design tends to encompass several levels of meaning, hence appealing to a wide range of viewers. If an idea can mesh viewpoints, it gives initial concepts depth and strength. This is part of the magic in collaborative presentations. Most personality conflicts can be overcome by staying focused on the goals. Most anxieties can be soothed by avoiding them with good rapport.

► **directing**—manage a team of talented contributors who collaborate on a final product. Delegate some tasks to others, direct visual elements, and work with suppliers. Juggling these components all takes good financial, communication, and organizational skills. They require clarity and thoroughness, the correct documentation, and utilizing technological tools for efficiency.

▶ **visual creation**—concepts for generating original photography or illustration. (See chapter 13, page 158.) For some projects this can mean finding stock images (see chapter 14, page 172) or utilizing photomanipulation tools. Many designers like to create composite images from several sources, blending and applying a myriad of visual controls. (See sidebar, chapter 14, Stock Images, page 174.)

▶ **page composition**—with several software packages, platforms, and prepress considerations (see chapter 23, page 266), there are many areas of specialty within this one segment. As the process grows more fluid between production and printing, each print designer needs to decide how far into the prepress area to work that makes project and business sense.

▶ **production management**—on smaller projects, the designer frequently handles the direction of all final production, prepress, and printing processes. On larger projects or in larger creative organizations, directing these processes is a specialty and purchases are made with experience and buying clout.

▶ **database publishing**—find compatibilities between systems and how to convert data into publishing forms. This is in greater demand with Intranets and the Internet (see chapter 8, page 70). Flowing consistent information into either print or online delivery is a way the two media can support each other. Asset management becomes necessary in larger scale integrated environments. (See page 185.)

▶ **presentation graphics**—many of the same design rules apply, but working with a different form of expression with different software requires additional technical skills, becoming complementary to print when cross-purposed to handouts, published papers, or posted online.

▶ **interactive design**—as a separate medium (see chapter 8, page 76), these skills must be complementary print projects.

Having defined the project and its strategy, team members need to be assembled on the basis of required skills. Every project will need a different mix of capabilities, so the best design organizations have a flexible variety of configurations; small groups and freelancers develop relationships with other professionals as needed. Departments interface with other departments in collaboration.

PREPARE FOR THE PROJECT LAUNCH MEETING

Under the client initiation, with the plan building, the art director

► **briefs all members** prior to the meeting to

▷ *research industry trends* or collect impactful background material.

▷ *review project plan* and add details as needed, continuing to evolve the plan through the process.

► **manages proactively** in the meeting:

▷ *supplies appropriate samples* to help define style and tone, either from the creative group's portfolio, stock collections, or other printed pieces.

▷ *reacts to information* and parameters in the discussion.

▷ *asks questions* and gathers the information or research as is needed to begin the concept phase.

MOVING FORWARD

After the project launches, the creative group

► **works with the project manager** to review key decisions and points from the overall project plan, filling in more specifics.

► **follows the project plan** and implements according to the client expectations.

PRODUCTION GROUP: PRINT

The scope of the project is defined as part of the estimate process (see chapter 4, page 38). Care needs to be followed to match these estimates to any new parameters that may be introduced as the project goes through the design and development phases.

PARTNERS IN PROCESS

The printer and other production suppliers can be of further assistance in the planning and preparatory phase:

► **provide samples** of specific materials—later review proposed designs and bring mock-ups where appropriate.

► **give information** on new processes or budget-saving comparable options. This will increase client loyalty and ensure a good use of resources.

► **provide input for schedule** and detail the project phases for monitoring. Then when receiving the project later, the buyer has the correct timing expectations.

"Design isn't just about how something looks anymore. It's about what something is. If you really want to design something, you have to take the time to really understand it and then synthesize it into something new that didn't exist before.

"I find that even some of our really sophisticated clients aren't able to see the patterns and connections to develop effective communications because they're too close to their product. If they bring me in and I'm not thinking about the content and connections, then there's a huge gap where no one is thinking about it. Every designer who touches anything is responsible in some degree for its content."

—Lana Rigsby
Rigsby Design
Houston, TX

"Print is becoming more 'static' in the Web era . . . It's now inflexible as far as latest up-to-date information. Yet there is something necessary about holding a printed piece in hand. I think it is clear, though, that the Internet will take print work away. We can adapt—it is still visual communication—fiber or cyber."

—Mark Misenheimer
Grizzard Advertising
Atlanta, GA

"I'm very pessimistic about the future of graphic design—anyone with a PC and a few fonts can call themselves a graphic designer. The competition has increased in quantity, not quality. And clients now have unrealistic time frames in mind for both creative and prepress processes; everything can be done in a split second, with the click of a mouse, in their minds."

—Jennifer Closner
Closner Design
Minneapolis, MN

WEB SITE PLANNING 8

Interactive projects have unique requirements from older forms of media. The Internet blends aspects of all the old forms such as print, TV, video, etc. and carves out its own expanding communication niche. The Internet also combines professional disciplines, creates new ones, and transforms the way business is conducted.

Web sites for organizations are fundamentally the responsibility of two people: the client and the Webmaster (see chapter 1 for more descriptions of roles). For business sites, these two roles are rarely handled by the same person. But for personal sites, they are almost always the same person. Few commercial sites are written, designed, created, and managed all by one individual. But small Internet-based businesses frequently have one person both deciding about content, functionality, and appearance as well as the construction and maintenance, perhaps serving a less involved client. In those cases, there are still the same collaborations, but oriented to tasks instead of people. The continuum of development is similar.

COMPONENTS FOR BEGINNING A WEB PROJECT

These elements will help anticipate progress, give a guide for decision-makers, and ensure desired results are achieved:

▶ **build a strong team**—Web planning begins with a team comprised of participants with balanced backgrounds. Assemble knowledgeable professionals with skill sets needed for the challenge. For example, the marketer, the database manager, and the programmer may collaborate with the writer, editor, and designer.

▶ **base concept on interactivity**—no site can be successful without taking advantage of the two-way communication, which must be an integral part of the site's subject. First-time visitors rarely respond, join, or purchase and if nothing is learned from them, they are lost. Attract them back to increase the possibility of response. Keeping visitors means maintaining this interactivity:

▷ *update the site often* with fresh content—build a library of materials that can circulate or rotate. Let oth-

ORGANIZATIONS

http://members.tripod.com/
~BestBooksCom/AEP/aep.html
Association of Electronic Publishers

www.iwanet.org
International Webmasters Association

WEB RESOURCES

www.adweek.com
AdWeek Magazine Online

www.ebooknet.com/
E-bookNet.com

www.go.tas.gov.au/infoman/gips/
buildingyourweb4.htm
Government Internet Publishing
    Standards (Governmental site as
    development checklist)

www.hbsp.harvard.edu/ideasatwork/
managersites.html
Harvard Business School Publishing
    for Managers

www.thestandard.com
The Standard Intelligence for
    the Internet Economy

www.wilsonweb.com
Wilson Internet: Web Marketing
    & E-Commerce

The most common mistake for most organizations is to convert their marketing brochure to the Web. The Internet, due to its personalization, interactivity, and one-on-one connection, is a very different medium from print, for which the brochure was created, that often has more of a promotional or sales tone. Users go to the site primarily for information, not a sales pitch. They visit for what interests them, so that should be the major appeal of the content.

ers know through promoting regular updates. Keeping to an update plan and schedule is critical to maintain vitality. Often sites launch with fervor, and even are updated occasionally with the same enthusiasm, only to fizzle due to difficulty in updating. Update what the audience needs to know, respond to why they visit, and provide something extra to keep them coming back. Make sure these promises are within the ongoing resources allotted.

▷ *allow the audience to affect the content* by responding to input and requests. Provide choices and opportunities for individuals to express themselves, comment on the subject, and build a cybercommunity of interest. Keep the site fresh through taking advantage of what interactivity can provide: evolving and involving content. This interaction means that the site can always be improving and fluid.

▷ *develop a forum*—build an online discussion base because people like to participate and be heard. They can develop their names as well as attract viewers for their own sites. Make participation appealing. Ask for moderators from other forums. Publicize forums and build the group. Gather participants from being involved in other forums. Depending on the purpose of the cybercommunity, visibility enhances the core business through developing direct communication with constituents.

▷ *build the community of interest*—exploit the benefits for participants and let this be the basis for a promotional database of those who expressed interest—best served through e-mail announcements or newsletters to encourage visitors. Exchange links with related sites. Build a Listserv from those who express interest.

▷ *have giveaways or competitions*—everyone is attracted to (and has grown to expect) something free. Give rewards for participation both to winners and to the runners-up. Information gathered through offers is the easiest way to gather information on new constituents and be able to determine if the site is reaching the intended audience. Sometimes new audiences are revealed, offering business and communication opportunities. Flexibility is key in building and responding to constituents.

OFFER A TIME-LIMITED SPECIAL

"It might decrease your profits or take more time but can give you great success. This extra offer can be savings, free shipping, or an extra-added product or service to the item they are purchasing. Put a time limit on this special, because it makes the customer act now, and does not give them time to check out your competition."

—Curtis Stevens
1simple.com
Waco, TX

► **form a strong project plan**—when everyone on the team contributes, a good plan will speed up the progress and construction of the site. In many industries, speed into the e-marketing environment is paramount to being a player. The faster and more accurate the process, the better the initial vision and plan need to be. Components that should be included in any Web site plan:

▷ *goals of site defined in writing*—generally profiles target audience, purpose of message, and expected response. (See chapter 4, Estimates and Proposals, page 38.) Be realistic and be able to address specific positive goals such as increased membership, greater geographic reach, reduced publishing costs, delivery of new services, and information.

▷ *outline of site*—with short descriptions of each segment, determine who is contributing what content. Create a site diagram showing major components and resources for fulfilling content.

▷ *development process* and sequence—match tasks to staff through assignments; include follow-up checkpoints. Be sure that the project director has the responsibility to bring together all the segments according to an agreed-upon plan, strategy, and design.

▷ *technical translation time*—for any data that is in different formats. Include testing time for various segments and time for testing prior to launches (often left out of plans). Be sure that all segments and links work and that navigation is both distinctive and easy to follow. Fortunately, many difficulties online can be repaired, as the medium is always changing and allowing for improvement.

▷ *assignment of much of responsibility to list the site* with browsers and portals. This is an ongoing process of checking, updating, and applying to new opportunities. Never assume that once a registration is complete, it can be forgotten about. Knowing status within search engines and major portal industries, being visible to attract reviews and links, and maintaining a cyber-presence requires online public relations skills.

▷ *marketing strategy*—outline for promoting site through various media including print, broadcast, advertising, and online promotion such as the submit-

A CALL TO ACTION FOR BETTER SITE DESIGN, UTILIZATION, AND MANAGEMENT

Everybody should be effectively using the Internet for low-cost/high-reach selling, customer support, marketing and promotion. To get started include:

> YOUR OVERALL MARKETING OBJECTIVES. The Internet should work to help communicate to your customers.

> AN INTERNET SERVICE PROVIDER (ISP) for site hosting— reliability, throughput and uptime requirements.

> A WEB SITE DESIGNER who can understand your message and provide the great look and feel that you want.

> PROMOTE through the search vehicles and partnerships.

> ADD WEB ADDRESS to all company literature.

> EASY NAVIGATION

> CHANGE YOUR SITE as often as possible

> KEEP KEY WEB PAGES GRAPHICALLY LIGHT for quicker viewing time

> KEEP CONTENT USEFUL and interesting

> PROVIDE A METHOD FOR FEEDBACK or comments (e-mail or a guest-book).

—Robert L. FitzPatrick
President
FitzPatrick Management Inc.
Charlotte, NC

—Stephen P. Aranoff
Founder and Principal
Arttex Associates,
Redwood, CA

ting of banner advertisements, links to cooperative sites, press releases, e-mailings, promoting chat group participations, etc.

▷ *updating strategy*—build in flexibility for adding new material or working in new segments. Keep separate records of the time and resources needed for the regular generation of content and production versus new material. Establish a regular schedule, for good organization is imperative in fulfilling visitor expectations. Build and sustain the momentum established. Every site is different in these requirements. What matters is diligence, responsiveness, and the fulfilling of promises.

▷ *interactive system*—manage the responses from visitors in a timely manner. Be sure to have the staff or mechanism set up so that responses can be handled quickly. Visitors expect fast information and attention, so making them wait for information may mean losing their interest.

▷ *maintenance strategy*—to add ongoing data, information, calendars, frequent changes. This can be the most underplanned of the processes. Often viewed with only best-case scenarios, maintenance takes at least twice the time planned in the first quarter. Once the site is developed, staying current is perhaps the Web team's biggest long-term challenge. Frequent changes are imperative to hold an audience and attract new viewers.

▷ *project schedule* of development—the more detailed, the better so that any project deviations are quickly noticed, decided upon, and factors adjusted. Begin with the major phases as the basic skeleton and add details later. The overall direction of the plan is more important than its completeness.

▷ *budget*—separate the start-up or onetime fees from a monthly budget. Schedule regular update reviews and define maintenance procedures.

▶ **think through process**—as in print, the most successful and economically responsible projects share beginnings that have these elements in common:

▷ *project director*—the client assigns one person to facilitate the decision-making and take responsibility for incremental decisions in development.

- ▷ *approve content outline*—whatever the Webmaster needs to do to get all approvals and decisions made before the construction of the site begins, the better.

- ▷ *controling and monitoring*—preparation and development steps. Know key dates.

- ▷ *allocation of detailed budget*—communicate plans on how to maximize resources. Always report quickly on any deviations.

- ▷ *designation of updatable elements*—some elements are very easy to change on the Web when they are planned into the concept. If not planned, other elements can be quite complex as the update might cause a series of other programs to be needed, pages redone or redesigned, and new components created.

▶ **have a team meeting to launch the project.** It is a mistake not to discuss process with those responsible for the different segments—even if they are all done by one person. If there are a lot of participants on the team, it is best to convene in one place initially. If team members are spread apart geographically, a conference call should take the place of the meeting. (See chapter 11, page 26, for tips on Internet presentations.) To make the best use of everyone's time, utilize this discussion to:

- ▷ *solidify expectations*—this assures that the project will get off to a good start. Gather input from all those who will be working with the site as well as any other resources that may interface with its organization.

- ▷ *focus on the most important elements*—participants should come to this session prepared with:

  - > *related organizational materials* such as past printed pieces, profiles, summaries, and samples of format and subject ideas

  - > *description of any content* that needs to be included such as databases or statistics

  - > *outline of the major site categories,* from the proposal. (See chapter 4, page 38.)

  - > *examples of other sites*—demonstrating aspects this site should have

  - > *determination of the critical path*—review scheduled elements and general deadlines. Identify the most time-sensitive components and what can most lead the project forward.

SITE MARKETING PLAN

"No Web strategy is complete without a site marketing plan that defines the intended audience and outlines a strategy to draw them to your site.

> Score a top ten ranking on any of the major search engines, and you're going to get some good traffic. Over 50% of the Web surfing public depends on one or more search engines to locate the content they seek.

> The trick is to BUILD SPIDER-FRIENDLY PAGES that will enhance the spider's ability to review the content of your site and judge its relevance as greater than that of other, competing sites.

> Distill what your audience is looking for into sets of KEY WORDS OR KEY WORD GROUPS. Be certain that these key words appear in several different critical areas within the pages that comprise the site.

> DON'T GET FRUSTRATED if you don't win the first time out. Go on a spy mission, conduct some searches and see what competitors did to earn their top ten spots. Achieving and maintaining a high ranking takes sustained effort and will pay off in increased traffic."

—Nils Menten
Partner
iMarc L.L.C.
Newburyport, MA
www.creativepro.com

▷ *reaching consensus*—gather input and analyze goals, purpose, and the general segments in a content outline.

▷ *definition of specific time frames*—assign resources for each major segment.

▷ *agreeing to the framework and clarify funding*—with the scope of the project in hand, realistic budget numbers can later be attached to each segment. If the budget is preset or tight, break it down into the various components with a clear follow-up procedure. In this way, the whole project will stay on track.

► **integrate with other media**—the site plan should also work with print and presentation materials (see chapter 11, page 134) to

▷ *give cohesion* to all marketing messages—thus build recognizability

▷ *maximize promotional dollars* by sharing elements between projects

▷ *utilize well-designed content* that can be scaled for different media. Print may be used promotionally and the Web might present more detail and background, or the print might give more of a cohesive overview utilizing the Web more promotionally. There are many

CROSSING THE DIGITAL LINE

"Customer service, merchandising, brand building, customization, database marketing, relationship marketing, targeted selling, one-to-one marketing and knowledge management are technologies and marketing disciplines converging to create an emerging cross-media revolution. At the heart of cross-media are money and customers.

"Now the creative director is teamed with the marketing manager and shapes the site's strategy based on ways the site can generate more revenue. Coming up with a cross-media marketing strategy involves a combination of print and online efforts. Both target the same customer.

"When your customer database is linked to your Web site, you're able to track visits, sales, buying habits and customer preferences. All this information is generated by the customers themselves, as they interact with your site.

Armed with this data, you can individually target those customers with the products or services they desire most.

"Customer interaction is now the driving force behind marketing. On the Web, it occurs in three stages.

1. STREAMLINE THE EXPERIENCE. A self-driven site goes beyond the content equivalent of a product brochure. Here, the customer can buy and sell, as well as find quick and accurate answers to the most common tech support and customer service questions.

2. HARVEST INFORMATION about your core customers. Customers want to contribute. In return, they receive advance notification of new products and special offers.

3. TAILOR-MADE MARKETING. E-mail addresses are the lifeblood of Web marketing. The best way to collect them is to ask, offering something in return, such as a weekly newsletter, a regularly updated list of specials, or even tips of the month. These addresses, known as opt-in lists (the recipients have opted to receive messages), form the basis for focused, cost-free campaigns. Target efforts to users who have already expressed interest in products or services. Every outgoing e-mail message can contain personalized content based on the interests and desires provided by each customer.

"People respond well to quick, personalized, highly targeted efforts—as long as you're able to manage your data and respect the needs of your customer base."

—Donnie O'Quinn
Graphic Arts Consultant
Portland, ME
www.creativepro.com

combinations for how they work together. Shared content doesn't necessarily mean same content.

▷ *conceptualize modularly*—the components of a large site, series of publishing efforts, and growing libraries require content management systems. For many organizations, the components of an intranet system comprise compiling databases, image files, product descriptions, service lists, logos, and graphics, all set up in a fluid manner for easy access and repurposing for various applications. Factors to keep in mind:

▷ *include advice from integration professionals* as part of an asset management system (or upgrade). Check the tie into sales and marketing, inventory, and work flow systems. If it is a closed proprietary structure, it won't have the fluid potential to work with a variety of page templates and delivery on a range of computer platforms. The system must distribute content across many applications and output methods. It is formatted to the requirements of each delivery channel.

▷ *share content for consistency and accuracy* throughout applications. It should be easy to update data once and have it flow through the system, eliminating earlier versions.

▷ *integrate the system with all departments*; thus, there is a cost savings in not having redundant content generation, and the information is consistent and available to the entire team, with security lockouts and controls, as needed.

▷ *publish customer service information via the Web*, marketing new products, distribute regulations, or provide subscription/membership services, all using content from one repository.

CLIENT GROUP

The client needs to value Web design and understand what good planning means to a successful project. Too many projects fail at the beginning through the decision-maker's lack of knowledge. The Internet fundamentally changes the way an entire business functions, communicates, and sells services or products. Investigate how the Internet can impact business as globalization and the need for international communication skills continue to grow. This must be considered when defining scope of the design and its potential to keep pace with change.

MAJOR DECISIONS: WEB
- set project foundation
- plan rough project skeleton
- assemble category outline
- solicit proposals
- choose creative group
- **finalize project plan**

Before beginning a Web launch be sure the plan is complete:

▶ **define purpose** of site and have an approved outline and plan. (See page 71.)

▶ **get input from all departments** on what they think the site needs to do, how they use sites, and their expectations for this one.

▶ **assign a Webmaster**—define who is in charge of supervising all development processes. Realize that each site needs its own dedicated Webmaster who is responsible for updating and maintenance (see below).

▶ **define reach and audience profile**—evaluate global potential and make it internationally inviting. (See chapter 11, page 126, and sidebar on page 73.) Even with a subject that seems localized, there can be potential for expanding markets and new business opportunities if the Web is viewed as opening alternative techniques and distribution. Many sub-businesses or specialty companies sprout up as virtual equivalents of larger brick-and-mortar establishments such as publishers, catalogs, or retailers.

BUILD COLLABORATIVE RELATIONSHIPS

Encourage the formation of a cohesive development team. Web development demands facets to work together:

▶ **Marketers** for visibility and exposure—they need to understand both the interests and the technical capabilities of the target market. At the beginning, to launch the project, marketers

▷ *examine examples from other sectors.*

▷ *study effective online marketing techniques.*

▷ *analyze and research audience profile.*

▷ *help define parameters* for site content and style.

▷ *explore new opportunities.*

▷ *present a rough promotional plan.*

▶ **Webmasters** need to be equally skilled in managing both technology and people. Few business sites are too small not to have a dedicated professional if any amount of business or communication is to be transacted. The Webmaster, first and foremost, must be a good communicator—both to a staff to follow directives and in techniques to reach audiences. Because communication is the biggest problem in business, it can easily become exacerbated by new media. Ironically, greater media opportunities open as many, if not more, opportunities for misunderstandings. A good

---

SIX STEPS TO POWERFUL WEB SITE PROMOTION

Address the issues of compelling and consistently updated content, virtual community, and real interactivity so that your site can build a base of regular visitors. Without a solid reason to return, you won't get repeat visits. Spend time building your online traffic.

1. DO DIRECTORY LISTINGS. Get listings in search directories.

2. GET LINKS. . . LOTS OF LINKS. The standard deal is that the other Webmaster will ask you to give a return link in exchange. Approach getting links to your site from other sites:

> Seek out synergy by creating relationships with sites with a similar audience.

> apply to award sites for recognition.

> Develop online strategic relationships with sites that compliment yours.

3. WEBCAST TO YOUR REGULARS. Relationships require communication. Establish an e-mail newsletter, technically known as a Listserv. Subscriptions are generally free and because visitors to your site subscribe voluntarily, mailing to them is not considered spam.

4. ANNOUNCE YOURSELF. Do a press release announcing your site. Be sure to include your industry's trade publications, and distribute to news organizations online.

5. CREATE INCENTIVES. There's little that has the drawing power of a promotional contest or giveaway. People just like free stuff.

6. LEVERAGE YOUR OTHER MARKETING. Use other outreach channels such as print. Buy banner advertising on popular sites that appeal to your target audience or buy key words so that your banner ad is displayed whenever an appropriate key word is selected in a search engine.

— Jon Leland
President and Creative Director
ComBridges
San Anselmo, CA
www.combridges.com

Webmaster knows how to

▷ *understand the audience* and listen to market needs, gathering field information where appropriate.

▷ *mitigate and mediate* between various interest groups.

▷ *plan* through every phase of a project.

▷ *give instructions* to others on the team.

▷ *manage* creative and production processes.

▷ *develop the potential applications*, and operations, of the technology.

▷ *facilitate the collaboration* between team members.

CREATIVE GROUP

Because the area of design has grown too large for one designer to be proficient in every aspect of the discipline, each creative contributor needs to know strengths and acknowledge limitations. Each professional needs to develop resources for additional needs and have techniques for working interactively.

ROLE FLEXIBILITY

▶ **Understand the client's business** enough to participate in content development—this will yield the best integration of form and function.

▷ *Listen to the client* and review information they provide.

▷ *Do your own research* with assistance from the client group, where appropriate

▷ *Site visit the client location* and places where their constituents are active.

▷ *Utilize contacts in parallel industries* and other experience, when possible.

▶ **Have talent and imagination**—bring conceptual skills such as symbolism, juxtaposition, humor, contrast, and mystery. Ideas must be compelling and communicative to the intended viewer.

▶ **Master the visual elements**—composition, emphasis, color, line, and handling images, absorbing software.

▶ **Have background in Web coding** and authoring programs. This develops the ability to design in the vocabulary of the medium.

REASONS FOR WHY CLIENTS HAVE WEB SITES:

> provide information
> establish a presence
> marketing
> customer interaction
> e-commerce
> brand awareness

—Graphic Design USA
New York, NY
www.gdusa.com

WATCH FOR TECHNOBABBLE:

The less talented designers can make up for their lack of talent by emphasizing special-effects skills. Never can technique take the place of content. But it can enhance content to give a message extra power and pizzazz. Special effects *can* be seductive, and pull a viewer in once, but if there is no substance to hold them, or reason to return, the viewer is one click away from moving on and being lost.

THINK IN THE VOCABULARY

Designing for the Web is a fundamentally different way of thinking than designing for print. It has an analytic precision and organization that is quite complex. There are more small files and file management that need a design structure. Print has higher resolution demands and prepress stipulations. So each is mired in technical requirements that are as unique as they are different. In print, the designer who does not understand page construction can't produce a quality product. For the Web, the designer who can't read code is not going to know how to solve many design challenges. In print, the designer needs to know how to present material in a linear progression. For the Web, the designer needs to know how to provide fast information, invite interactivity, and use relationships to affect the design approach. Designers need to *think* within a medium to *create* within it. Each medium is a language with its own rules, exceptions, syntaxes, and poetry. Designers need to

► **understand what is possible technically.** Learning the fundamentals of each medium gives the greatest command of integrating them effectively.

► **work with the client** to define scope, theme, and general direction. Make sure parameters are understood. Web design can be more collaborative than many print projects, as sites with active functionality require professionals in various areas of expertise to add elements to the site's composition and activity.

► **react to information and parameters** in the initial meeting. Try to understand a consensus of various opinions and requirements, bringing participants to articulate ex-

DESIGNING SITES WITH A GLOBAL REACH

"Web designers and content providers often don't take advantage of all that the Web has to offer by making their sites truly World Wide. Globalize your Web site and reach a potential multinational audience of millions.

> PROVIDE THREE LINES rather than two for an address, and add an optional field for District or Region in addition to the State/Province field. Include foreign address terms in your registration or purchase forms: change the Zip form field to Zip/Postal Code and the State field to State/Province.

> INCLUDE THE COUNTRY CODE before every phone number you provide.

> PLACE "USA" AT THE END OF YOUR ADDRESS.

> INCLUDE CURRENCY DESIGNATIONS before all dollar amounts (US$100.00).

> MIND YOUR SALUTATIONS. An incorrect title can be an innocent insult.

> STAY AWAY FROM SLANG. Keep sentences short and your grammar correct.

> DO YOUR RESEARCH: business etiquette . The International Business Resources (http://ciber.bus.msu.edu/busres.htm) has links to sites that address multicultural issues such as business card etiquette.

>TRANSLATE YOUR WEB SITE. Free Web forms creation service, www.freedback.com,

includes selected content in multiple foreign languages.

> PROVIDE ALTERNATE VERSIONS. The most difficult and expensive—though arguably the most effective—way to globalize your Web presence is to create separate sites for visitors around the globe.

> MIRROR YOUR SITE. A mirror is a copy of a site that resides on a Web server in a different geographic location to better serve their local users."

—Greg Holden
Writer
Chicago, IL
www.creativepro.com

pectations. Many comments that can have a major effect on the design may not seem important at the time shared.

▶ **refer to sample sites** for solutions and examples. Make sure that participants use the same vocabulary, such as what is effective navigation, efficiency of segments, various styles, and personalities. What is elegant to one person may be confusing to another.

▶ **help develop and follow the project plan**—create the site according to client expectations (see chapter 11, page 126) by being thorough about putting together factors, considerations, and carefully defined parameters.

PRODUCTION GROUP: TECH

For most product- or membership-based organizations, the amount of information that needs to be included and the condition of that data are critical to how quickly a site can be functional or updatable. If the buyer has a tech professional on staff to compile data, handle translations, and interface with the creative or production groups' technology, the process can be the fastest.

ESTABLISH COMMUNICATION

At the beginning of the project, when attending the launch meeting (see page 68), the programmer should:

▶ **describe existing database construction**—this may encompass a range of marketing or product-related customer information such as membership lists, product instructions, or mailing data.

▶ **set up translation plan for content**—some projects may take some experimentation to verify translations.

▶ **describe development process**—as an overview without going into lengthy detail.

▶ **use examples**—demonstrate various constructions and techniques by work done before or with a comparable site.

The tech group works in collaboration with the creative group, as led by the Webmaster. Together, they develop the site according to client expectations as defined by the approved proposal. (See chapter 4, page 38.) No work should begin without this approval.

WEB SITE DESIGN AGREEMENT

"Limit the amount of work to be done in the creation of the Web site and require additional fees for work exceeding the anticipated scope of the project. Additionally:

> The designer must carefully determine the scope of the work, which may only involve the creation of a prototype or may extend to delivery of a functional site, including testing and debugging, and perhaps ongoing site maintenance.

> Ascertain that the client has registered the domain name or designate who will perform this task.

> Break out the fees for each phase of the Web site. Indicate subsections, sample pages, and any special features.

> Give a description of what maintenance the designer will provide. Decide how compensation for this shall be computed and allow for renegotiation after a specified time period.

> Specify due dates for phases that the designer agrees to undertake.

> Indicate that client delays will increase the designer's time for performance.

> Limit rights to Web usage only.

> If rights exceed the Web, consider elements such as photography and illustrations.

> Ask for advance against expenses.

> Require installment payments and specify when installments will be due.

> Bill for electronic storage media.

>Seek copyright notice in the designer's name and specify authorship credit.

> Indicate any special provisions regarding cancellation.

> Require ready-to-use assets in electronic format or provide for additional compensation to place such assets into electronic format.

> Have the client warrant rights to use all assets that it provides.

> Require the client to indemnify and hold harmless the designer from third party lawsuits."

—Tad Crawford and Eva Doman Bruck
Writers and Publishers
Allworth Press
New York, NY

VIEWPOINTS:
PURPOSE MARRIES CONTENT

"The Internet has muddied the line between information and design. It's become like those personal financial statements you get from your retirement plan, that you cannot understand because an accountant and a mainframe programmer designed it. The Internet has become the territory of the programmer. It's become much more important for the designers to gain control of the front and end design of corporate Web sites."

—Bill Noss
Ari Communications
Windsor, CT

"Clients usually want their name and their 'branded' message on their Internet sites, as on the collateral materials we produce. One difference that we usually propose is that the message needs to be fun to get to. It needs to build quickly and it needs to be memorable."

—Perry Hunter
Barkely & Evergreen
Advertising
Missouri

"Clients not only look for general information on Web sites but I think they also expect extended content/information as well. There are a lot of people hoping to obtain the answers to their questions through Web sites in a much quicker and easier fashion. It seems to be a legitimate source of information for those who want to skip researching elsewhere."

—Kimberly Becker
Carlson Cunico
Saint Paul, MN

"Like traditional direct mail, the Web banner ads that perform best tend to be sweepstakes, fairly shout the word 'Free' prominently or offer the opportunity to participate in a promotional gimmick. Using photographs of famous people is another standard tactic lifted from traditional marketing campaigns. Things like that can double the click-through rate."

—Marlo Silver
USWEB/CKS
San Francisco, CA

"Web sites tend to be product-oriented, and you have to think strategically and bear the responsibility of identifying things that may or may not work from financial and business perspectives. That makes me a business consultant, as well."

— Maria Guidice
Creative Director
HOT Studio
San Francisco, CA
www.creativepro.com

"I would doubt that anyone with imagination in communications hasn't realized that film, television advertising and the new Web media tie together in a new landscape and that a creative practitioner in one section can learn from the others."

—Byron Ferris
Writer
Portland, OR

"All creative content managers share the same challenges: managing staffs of creative producers (and, thanks to the Web, technical experts as well), while keeping up with the ever-changing state of technology. Their goals are the same, too—ensuring a continual flow of visually compelling, clear and functional material, be it printed or on the Web. And it goes without saying that everything has to be produced well enough to satisfy demanding clients, in-house or out."

—Eric J. Adams
Writer
Petaluma, CA
www.creativepro.com

"The breach between webmaker and Web user, along with the wider breach between designers and clients, is driving more and more of us [designers] to the 'safety' of an insider's community. We design for other designers and feel understood at last.

"But this strategy, which is unfortunately supported by awards committees and headhunters, threatens to undermine the subtler creative values of the Web, and the damage could be serious.

"Content sites are the heart of the Web. To ignore them because they're less thrilling than high-end sites sets a dangerous precedent.

"The opportunity to apply quiet design intelligence to a great content site is pitifully undervalued, partly because designers like to 'get off,' but mainly because the creative side of the industry rewards over demonstrations of technique."

—Jeffrey Zeldman
Creative Director
New York, NY
www.zeldman.com
www.alistapart.com

"No matter how many people have been using the Internet for years, just as many are only getting started. The further technology takes us, the more people are left behind when it becomes too evolved for them to comfortably pick it up. Even those who do use the Internet regularly are not necessarily scrutinizing, or cynical, enough to discern the gems from the dreck. Though we should continue to take technology as far as we can, to try new things, and find new ways of transferring information, we should also continue to be aware of our audience, and not create a divide between what we want the Web to do and the people we want to use it."

—Andrea Dudrow
Contributing Editor, Writer
San Francisco, CA
www.creativepro.com

"Brevity is everything on the Web. Short, punchy content is crucial to sustaining visitor interest because your competitor's site is just a yawn and a click away. Captivating feature stories in print may be sleeping pills on a computer screen.

"Of even greater significance are strategy concerns. The new medium requires a radical rethinking of what you want to achieve. One goal might be for your print and online publications to drive traffic to each other. In that case, your content on the Web should complement, not solely duplicate, your content in print. For example, a print publication might direct readers to more in-depth data on a Web site, where, unlike in print, space essentially is unlimited and judicious use of hyperlinks can make large volumes of material manageable. Meanwhile, your Web site might encourage visitors to subscribe to your print publication for material best served by that medium.

"One of the greatest benefits of the Web is its immediacy and the emergence of 'Internet time.' Visitors expect fresh content each visit, or your site becomes history. Only those who learn to move beyond simple porting from print will survive." [Reprinted with permission from *Publish* magazine]

—J.W. Olsen
Technology Writer and Editor
Chicago, IL

"Designing for the Web has different rules, some of which mirror television (the lower resolution) and some of which are unique to the medium (the changing of browser sizes). Some companies perceive the online world to be different from the offline one. Therefore, different design considerations come into play because the competitive focus changes. A Web address is as fundamental as a fax number. But if its business practices are still firmly placed in the offline world, it fails to capitalize on the potential growth offered by a Web presence. The shift has to be organization-wide for it to be truly successful and if it is not a full identity revamp, then it should be something driven by the marketing function and be integrated over time so that the Web component does not look like an add-on."

—Jack Yan
Jack Yan & Associates
Wellington, New Zealand

"The Internet does not enhance our clients' marketing efforts by replacing print but instead by acting as an integrated compound to achieve a goal."

—Simon Lo
Lipton Communications Group
New York, NY

"The importance of the Internet is to provide information. Many clients feel, or are led to believe, that they can make money directly off of their Web site. They try to entertain and get repeat customers. This often leads to complicated pages, slow download times and the necessity to find and download new viewers and players. Most information seekers won't wait and the chance to market and inform is lost. Keep sites simple, logical and easy to navigate."

—Robert Graham Brandon
Webmaster
Portland, OR

"The Web culture that makes for a great Internet development team is simply indigestible for most corporates.

"[Before an initial successful launch] the corporate has only done small things to screw up the award-winning Web team that grew up almost by mistake. But after [the success of the launch], it's obvious that the Internet is vital to the corporate's future. So they send in the adults to run things properly. The adults really don't want anyone doing anything that hasn't been discussed and approved.

"The nature of a large corporate demands conservatism, conformity and sameness—it's needed to preserve what the corporate has built up. The entrepreneurial nature of Internet development does not mix well with the corporate ethos of preserving the status quo.

"So corporates should figure out a way to insulate Internet development teams from the rest of the company. Let them break the rules and get the job done quickly and imaginatively. Try to keep things that way after the site launches. Keep the adults at bay. If you can keep an entrepreneurial team insulated within the corporate space you just may see your company successfully through the wild technological changes that are taking place."

—Bruce Morris
Web Developer's Journal
Internet.com
London, UK

EVALUATE EACH QUARTER:

"After the first quarter of a new site or update, the segments of the plan and calendar should be revised and refined. At this point, the onetime design and construction setup fees are finished. Knowing the update schedule and going through at least one cycle of updating will give enough information about what is realistic, given staff and resources. Each site has a momentum of its own, defined by the first three months of its life. Refining the updatable segments is a necessity if the site is responding to the market. What is learned in the first three months must influence both the design and construction if the site is vital."

—Eleanor Mandler
Web Designer
Mandler & Associates
Chicago, IL

"Personalization should not go overboard, emphasizing customization at the expense of a cohesive look, feel, and identity of the product or service.

"Personalize pages for each of your visitors but maintain a consistent presentation. Use a profile database to store user preferences [and] catalog information; adding personalization rules to templates that select appropriate content for each individual based on the preferences and interests.

"Some Internet users see even the most sincere one-to-one marketing information requests as violations of privacy. To avoid any misunderstandings, state clearly why you are seeking personal information and what you'll use it for. Tell people you won't spam them with e-mail and that you won't sell information about them to third parties. Do not send newsletters and other mailings unless people request them.

"If you're really serious about privacy, you should put a privacy statement on your Web site and have it audited by a privacy policy certification service for continuing compliance.

"A trend where increasingly sophisticated personalization tools, overall strategic knowledge gained from successful one-to-one marketing campaigns, and the growth of Internet-based communication will add up to a future full of opportunity for people looking to use personalized marketing to sell their products and tout their skills."

—Russell Shaw
Journalist
Portland, OR
www.creativepro.com

"Shops that are focused so everyone there knows where they're headed seem to consistently make more money. And these shops also seem to be capable of reinventing themselves where markets change quickly. Strangely, the shops with the long-range plans know what they're changing and why. They're not locked into their plans; they know when to shift and by how much."

—James R. Whittington
Dr. Joseph W. Webb
TrendWatch Partners

"Many clients come to us wanting a Web site with the assumption 'if you build it, they will come.' Many will drop successful print jobs to do a Web page. Being a graphic design firm specializing in corporate identity programs, we try to teach clients about identity programs and how a Web page can fit into their identity and marketing schemes, and not as a be-all, end-all. We encourage them to use both Web and print materials together. As with any design, we teach our clients that having a Web page alone is not very successful. It must be carefully planned and executed, maintained and placed within the client's new or existing identity property."

—Jack DelGado and Daniel Jones
Westhouse Design
Greenville, SC

E-PUBLISHING INNOVATIONS

"Electronic books are inevitable. Someday when we purchase a book online, it will be downloaded to our favorite document reader. It will be enjoyable to look at, following the traditions of page design and typography. It will be colorful with continuous-tone images and sharp figures. It will be enabled with digital features of searchable and selectable text, hypertext links, and animation. Reading will even be a comfortable, natural experience, just as curling up with a book is today. It's a matter of technology, demand, and time.

"As consumers, the first thing we might demand is convenient access to titles, akin to browsing the shelves of a bookstore or the virtual aisles of Amazon.com–one click to purchase and one click to turn a page.

"The second will be when our devices include plenty of storage for a virtual library of texts, a month's (or even a year's) worth of issues of our favorite daily periodicals, reference books, and [a] few dozen novels for those long flights or sunny days on the beach.

"Meanwhile, publishers have a few ideas about what they want in an e-book device: direct access to consumers, reducing the costs of middleman distributors and booksellers; the greater profits from less overhead; trucking heavy paper loads is a significant portion of the price of books today.

"For the distributors and retail booksellers, there are opportunities for market share adjustments. Before long, each retail chain may begin making and selling branded e-book hardware of its own."

[Reprinted with permission from *Publish* magazine]

—Hans Hansen
Developer of Rare Books in
Electronic Formats
Octavo Corp.
Oakland, CA

"As a business tool, Web sites are relatively inexpensive and dynamic—that is, the information on a Web site can be updated, changed, and expanded. Two seconds after your catalog comes off the press, you are locked in. On the Web you can change on a moment's notice and at a low price compared to printing."

—Robert Severn
Severn Associates
Warminster, PA

"The key to reaching today's customers comes from integrating the Web with traditional marketing. In addition to being a standalone marketing vehicle, successful Web marketing supports direct mail, space advertising and in-store promotions. Marketers who only see the Web site as a means to generate sales from new customers may be missing their biggest opportunity. The Web is an excellent customer service and retention tool. When used properly, you can provide a higher level of customized service, reduce the costs of servicing your existing customers and gain vital information about your customers that will help you market to them more effectively."

—Tom Magadieu
Director of Direct Sales
Delorme
Yarmouth, ME
www.creativepro.com

"First impressions are important. We all want Web sites that are visually appealing in order to entice visitors to browse. Just as important is speedy browsing once a visitor has decided to stick around. Everyone has a need for speed. I always tell my clients that impatient visitors to their Web sites are only one click away from heading back to their favorite search engine. Your first line of defense against losing wayward surfers needs to be optimizing your images for speed."

—John Cho
Art Director
DigitalNation Interactive
Alexandria, VA

"To achieve online marketing success requires innovative thinking in a 'three-part' harmony of disciplines: technological implementation, communications, and with interactive design.

"The enormous publicity that the Internet has already received is an artificial beginning, but there is also a real birth process underway that is undeniable and the impact of the Internet on the 21st century should not be underestimated.

"In fact, what may be most revolutionary is the leveling of the competitive 'playing field' for businesses. On the Web, even very small businesses can afford to do a Web site, without the major expenses of distribution and large marketing budgets.

"The result is a whole new medium, not just a new technology.

"The Web is not a panacea, but a complement to conventional marketing like direct mail and telemarketing."

— Jon Leland
President and Creative Director
ComBridges
San Anselmo, CA
www.combridges.com

TRAINING   9

A continual commitment to training is a fact of life in a tech-nologically expanding economy. In essence, training is about keeping pace with change. It is ongoing and inspires the abil-ity to thrive in a shifting marketplace. Although it is human nature to resist change, through training, professionals can learn to embrace it and use it to business advantage. One thing that doesn't change is change itself, which is accelerated by technology. To stay ahead of change requires an attitude of movement, the habit of learning, and the ability to absorb information quickly.

KEEP PACE WITH ADVANCEMENTS

► **Organizations**—although managers understand that the technology is only as good as the person using it, many don't act as though they believe this. With the promise that software gets easier to use, many budget-conscious executives invest less in training employees and rely more on the skills of new hires. Investing in people is investing in a resource that can quit and find another job. Even though the equipment grows obsolete, hardware and software are perceived as more tangible investments. It is true that the benefits of investing in employees may be hard to measure. However, effective and strategic training offers many benefits for the organization who can:

▷ *stay competitive* through its individuals. Having a core competency means the unique blend of specially tal-ented individuals.

▷ *keep current* with the rest of the industry. Production processes demand this, especially as e-business-to-business processes become prevalent.

▷ *gain efficiencies in processes* and save on budgets. New processes are continually faster, larger, and have more capabilities.

▷ *enable employees to choose the most cost-effective meth-ods* to handle various assignments.

▷ *have employees give recommendations* and feedback on how to evolve systems. (However, if not constrained, the tech group often buys hardware and software they'd really like instead of really need. What is done with a system is more important than what kind of system it is. Be careful not to get enamored with capa-bilities over purpose.)

The commitment to a publishing system is not complete with the purchase of hardware and soft-ware. Adequate training is im-perative. Too many companies underestimate the importance and cost of training. But if staff is adequately prepared, then qual-ity and production efficiency will be realized.

"Responding to change doesn't just happen. We need to train our-selves to change in several areas:

1. *technology training* equips us to deal with the incredible constant change in how we do our work.

2. *business training* in our chang-ing industries keeps us current.

3. *environment training* means adapting technology to a par-ticular site or condition.

4. *technique training* means study-ing and implementing 'what works with what, when.'

5. *survival training* is knowing how to deal with your own and oth-ers' change. It means asking questions about what to hold on to, throw away, and add to cope with chaos. Train yourself to do this constant questioning.

"The training that helps us respond to change is sometimes internal, sometimes external. The mecha-nisms vary. Peer-to-peer dialogue, in its various forms, is often the best, and least ex-ploited."

—Norm Wold
Wold Marketing Group
Milwaukee, WI

All the equipment in the world is meaningless without qualified staff to run it. A serious common mistake of decision-makers is to place too much emphasis on software or hardware choices, and then to neglect to fund training plans. Certainly much responsibility lies with the individual employee, as their job security is dependent on their skills. But the employer must reap these benefits and recognize their responsibility in the process. Training is an ongoing commitment. The more software skills workers have, the more valuable they can be to a company, but only if that knowledge is easily applied and expanded. Knowing a little about a lot of software packages is more important than knowing a lot about a few. Online resources require literacy but can be invaluable to developing or expanding skill sets.

PAGE GUIDELINES

Developing and implementing page standards within an organization that produces a quantity of print publications or Web pages has many advantages:

> *ensures that all documents are set up in the same way* and that employees can access the same files, understand their construction, and cover for one another in case of absence or tight deadlines.

> *accelerates the training process* by showing tangible applications for projects, and those implementing don't have to develop the design.

> *becomes a continuing reference* for ongoing work or additions.

> *brings new employees up to speed* quickly because they can train on the guidelines as part of their orientation.

> *reduces output mysteries* and costs by providing solutions for anticipated content and uses.

> *makes updates manageable*, particularly for Web sites, as they can tie into asset management systems. (See chapter 15, page 185.)

▷ *empower employees to help each other*—the software is not as easy to learn as it appears, for knowledge needs to combine with a background in publishing and experience actually using it. Most software packages have become so robust that much can be done easily, but to use it to advantage, greater understanding of the complex features is necessary. Having a mix of skills between employees, allowing each to excel in certain areas, there can be a balance between them and they will better be able to assist others. Plan so skills overlap and staff members can back each other up.

▷ *upgrade and absorb new technologies* easier with an assigned professional responsible.

▷ *encourage the habit of learning*—spread organizationally through example. Employees will develop and thrive in an environment that is conducive to expansion.

▷ *set the environment of change* and help its people stay competitive. Clients create the atmosphere where new ideas can thrive and inspire those they hire to do their best work. Training is about demanding the best from talents.

▶ **Individual professionals**—each person needs to balance how much time is spent on learning new technology. No one can know everything about all related software, especially in publishing. To stay ahead of change:

▷ *learn software and develop skills* to maintain a competitive advantage and greater understanding of processes. Job security, in a technologically-driven economy, is in the skills to take control of that technology.

▷ *take responsibility for self-training*—this is mandatory for professional security. Investing time in self-instruction with tutorials, seminars, courses, conferences, and workshops keeps skills sharp and always is worth the investment.

▷ *apply skills* learned in one software package to help in learning other packages. Proficiencies build upon one another. The more packages learned, the easier it is to add to this knowledge base.

▷ *increase initiative* through knowledge, practice, and staying involved in changing trends. Become part of a community of users, and develop contacts and friendships that encourage expansion. This adds great value to both employers and clients.

FORMS OF TRAINING

There are two kinds of training, whether for organizations or individuals: formal and informal. The range of costs is tremendous, from building a library to getting a degree. Consider the setting for what is the most reasonable and appropriate method to use.

▶ **Formal training** according to a curriculum and structure:

▷ *Classes at a training center* can bring employees up to speed quickly. They allow the student to ask questions and to learn along with others. But classes are only a beginning. There is no shortcut for practice, and the best way to learn is on real projects, not pretend ones. Classes are a good jumping-off point, well-supported by the other methods for training.

▷ *An individual trainer* can be an ideal source for focused needs. After attending a class, a user can benefit by a consultant to work over hurdles and teach efficient techniques that apply specifically to a job or project.

▷ *Books* can augment the other training methods with supplementing both more specific and more general knowledge that can apply directly to the work. Many high-quality "how-to's" are on the market. A regular budget should be established for building a library of resources, accessible to appropriate team members.

▷ *Online tutorials and help functions* are sponsored by every software training company. Registered users can obtain almost every answer for free if they are willing to dig and read a lot of information onscreen. One of the best places for answers is the FAQ sheet, for the most commonly asked questions. There is usually an e-mail function for asking specific questions.

▷ *Seminars* are excellent for overviews. Their range covers business, management, kinds of projects, specific applications, and demonstrations. No matter what other training methods are used, seminars should be encouraged because they are a good way to understand and pinpoint trends. They are a great place to ask questions and get feedback. Most seminars are a day or part of a day and are promoted mostly by associations or training companies.

▷ *Conferences* encompass many seminars together and provide greater in-depth study and interaction. Going online to the sponsoring organizations' Web sites (see Appendix for list) will keep the viewer current with opportunities. Organizations should set aside an annual budget to send staff to key industry conferences and events—especially when the location is close.

"THE INTERNET HAS

> unprecedented ABILITY TO IN-TEGRATE THE MEDIA THAT CAME BEFORE IT. Text and image (print), voice (telephone), audio (radio), and video (film and TV) are available to Internet users anywhere they can find an access point to the telecommunications networks.

> it's a medium that DECENTRAL-IZES THE PRODUCTION CAPABILI-TIES, so that instead of the one-to-many of the mass media we now have many-to-many.

> the nature of the medium itself ENCOURAGES INTERACTION not just with the content presented but also with the producers and with other consumers.

> much of the CONTENT IS AVAIL-ABLE FOR FREE (after the users have made the not-insignificant investments in computers and access fees).

> DOES NOT STOP TO RECOGNIZE NATIONAL BOUNDARIES. Access to content from outside one's borders is not an issue.

"Learning is about curiosity, and by giving us the ability to recognize ourselves as immensely curious beings, the Internet will have a profound impact in our attitudes to what we don't know. For the first time, perhaps, since the printing press allowed the masses access to knowledge and information that had been limited to a few, the Internet will be the driving force that changes educational institutions around the world, and what people expect they will be able to do in order to gain access to learning opportunities wherever they are, at times convenient to them, and—most crucially—regardless of who they are."

—Pedro Hernández-Ramos, Ph.D.
Academic Solutions Manager,
Worldwide Education
Cisco Systems, Inc.
San Jose, CA
www.zonezero.com

> *In-house information center* or help desk—Larger companies may have internal resources to help users learn fundamentals and maintain the system. There may be a full-time training staff or this responsibility may fall to a single staff member.

▶ **Informal training**—which is free-formed with no assigned learning procedure. This ad hoc method is usually driven by project needs and is popular with those who work well alone.

> *Learning on a need-to-know basis* and having resources (such as colleagues, friends, trainers, and suppliers) or a library to meet that need. This is one of the best ways to learn because knowledge can immediately be applied to the specific problem.

> *Background training* (through classes, tutorials, and time spent with manuals) to learn fundamentals and capabilities. (Software manuals are marvelous tools for training when workers have the time to systematically learn. However, manuals generally can't take the place of some professional training, can't answer all employee questions, or address job-specific challenges.)

HOW TO KEEP UP AND STAY CURRENT:

Training is an ongoing individual commitment. Staying current helps uncover areas where more in-depth knowledge is needed. To keep up with trends and information:

> DON'T ALWAYS DO EVERYTHING YOURSELF. For the sake of time, money, and quality, don't do what someone else can do better.

> IDENTIFY OUTSIDE RESOURCES. Develop relations with suppliers as a research and development source. Try to avoid being too much on the bleeding edge, for getting too far ahead of the constituents will lose money due to risky processes. Let others do much of the research and development to benefit from such effort.

> ATTEND CONFERENCES AND SEMINARS to ask questions and interact with peers.

> BLEND SKILLS WITH OTHER EMPLOYEES or experts to complement your abilities.

> SEEK OUT A DEDICATED "GURU," someone who passionately expands their knowledge, who maintains computers, plans system expansions, keeps up with upgrades and licenses, and handles daily problems.

> USE THE COMPUTER AS A TOOL, NOT AS A CRUTCH. It is important to know non-technical methods to find the best solutions. Focus on the idea versus techniques.

> GUARD AGAINST OVER-PERFECTION and becoming enamored with unimportant details

the final viewer will never notice (or care about even if they do).

> SHARPEN COMMUNICATION SKILLS—*how* each person relates to one another, provides instructions, reports progress, handles and controls quality—becomes critical when there is less time to do things.

> MAKE FRIENDS who know things you don't for specific kinds of projects.

> ACCEPT THAT THERE IS ALWAYS MORE TO KNOW. There is no such thing as being caught-up with publishing activities. We will die with full Things to Do lists!

> FIND A SENSE OF COMPLETION from what you *do* with the technology versus what equipment you *have*.

> READ STRATEGICALLY AND QUICKLY. Set aside specific time to skim industry information. Stick to that time limit.

> SET LIMITS ON THE TIME SPENT ON AN ACTIVITY. When you start working on a project, decide when you will stop. This will help prevent over-perfecting and also keep learning-time in balance.

> LEARN TO ASK CALCULATED QUESTIONS for the information needed. Success of information handling has to do with astute questions that prompt efficient sorts and overviews and having a base of well-developed resources.

> DETERMINE HOW MUCH DETAIL IS NEEDED

for any given project and guard against getting more information than you need. Police for the best sources.

> BECOME A GOOD EDITOR of the information you generate for others. Just as you don't want to be overloaded, don't overload them.

> DEPEND ON ELECTRONIC BACKUP-AND-RETRIEVAL system and file naming so that electronic documents are easy to find.

> BE CAREFUL OF CHAT ROOMS. It is easy to spend inefficient hours communicating with others (though fun to do). Keep it on subjects that will help you the most.

>SURF THE WEB STRATEGICALLY. Follow recommendations carefully and get to be an expert at search functions and methods.

> UTILIZE PORTALS ONLINE TO ZERO IN ON AREAS OF INTEREST. They save time from the bigger browsers with too much information that is unsorted and often have strange hierarchy.

> SAMPLE FREE SOFTWARE AND EXTENSIONS ONLINE—help, instructions, and tutorials are easy to follow.

> GET AWAY FROM TECHNOLOGY by making sure you have other activities to balance. The Internet is seductive and no one can comprehensively know anything. Time spent is a better limit than trying to exhaust subject matter.

—Michael Waitsman
Designer
Synthesis Concepts, Inc.
Chicago, IL

▷ *Intuitive exploration*—often achieved through making mistakes or through trying different menus and selections to see what happens. Sometimes new methods are found by surprise. If they are techniques that can be applied right away, this depends a bit on curiosity and luck.

▷ *Asking questions* of other professionals through networking, help desks, phone calls, and vendor services. Chat rooms can be good for considerations shared by other professionals.

▷ *Investigative training* to increase specialization, generally through seminars, research, and trade show attendance (see above).

▷ *Spend time online* doing research, reading reviews, sampling tutorials, and downloading software versions, extensions, or utilities. In the trenches experience produces results, as can an overview of capabilities.

CLIENT GROUP

The client and the project director can create an atmosphere of learning within their organizations and inspire it in those they hire. A company's commitment to training reflects a commitment to excellent work and excellent work derives from utilizing change with fresh new ideas. To promote this atmosphere, employee skills need to be continually reinforced. Many forms of training should be encouraged as part of a corporate culture.

PLAN FOR EFFECTIVENESS

Basic criteria need to be evaluated for a client to understand the requirements, issues, and capabilities of staff (or suppliers):

▶ **demand the basic packages and capabilities**—there's a difference between hands-on experience and an overview knowledge. Some publishing professionals need both—but a manager may only need the latter.

▶ **identify training for specific applications**—many organizations have proprietary systems that need special knowledge.

▶ **determine if the level of hands-on experience needed** matches necessary training to fulfill. For example, newsletter desktop publishing requires different skills than creating four-color brochures or designing Web sites.

▶ **help company executives understand and plan for ongoing training**—the learning curve is always a learning roller coaster, and must be given velocity to have any effect. Efforts need to be continual, both organizationally and individually. Be flexible and supportive.

► **emphasize the initiative for self-training**—yet still remain focused on the organization's priorities to impede wandering experimentation. Keep training focused.

► **encourage specialization among employees**—each group is a mix of expertise, for one person can't know or do everything. Balance the cumulative knowledge of the staff. Hire strategically.

► **plan to upgrade systems** and to help employees adapt to changing technology

  ▷ *attend some formal training classes* with each new upgrade to quickly grasp the changes and new efficiencies that can be gained—grasp an overview of capabilities.

  ▷ *time should be allotted* for informal training methods—particularly access to books, manuals, and online help. This can be very difficult under deadline pressures but should still be pursued because skills developed always speed projects in the end. Applications become apparent.

► **encourage a culture of continual evolution** so that each new technological advancement can be embraced smoothly—don't allow systems to *stop* evolving.

► **outline expectations**—both from the technology and the employee sides—before purchases. Involve staff in the decision and it will help their enthusiasm for learning. Encourage employees to evaluate the system and give feedback to use at the next improvement stages.

► **make the corporate and personal benefits known**—how learning new techniques and methods furthers competitiveness outside and cohesion when balancing with other team members inside.

  ▷ *Most employees want to learn* and welcome the opportunity as a job benefit, though few employees will take initiative on their own. (Only the best and brightest will show such forward-thinking.)

  ▷ *Some employees become afraid or overwhelmed* by technological change. They need to be inspired to tackle these fears.

► **make information easy to access**—such as reference materials, resources, communication with other users (holding memberships to industry organizations makes a statement of commitment), as well as providing for effective system support and maintenance.

Whether a publishing group is within a larger parent organization or a freestanding business, training is always an issue. Watching the seminars and what suppliers are needing for output, keeping pace is best done through those doing the work as led by those who make it their business to understand the work.

CREATIVE GROUP

For the art director and the designer, staying current with technology has become a daily necessity. Like so many other businesses, the Internet affects how design groups assimilate new technology.

METHODS TO ABSORB TRAINING INTO BUSINESS

Incorporating a mix of formal and informal training methods is critical for the creative entrepreneur. Small companies feel the market pressures of applying new technology more intimately than do employees of larger organizations. Those within a larger group need to maintain good learning habits. For both the small and larger organization, several factors will clarify the mechanics that set the foundation for learning vitality:

► **handle software and tutorials online**—knowing how to access help functions and indexes becomes key to advancing. Download upgrades and extensions to greatly save time when advancing software capabilities. Remember that information available online is endless, so find some trusted sources that report regular developments and reviews. Be able to learn from other user experiences as a very exciting aspect of the Internet.

► **find professionals who utilize personal time** for training and form a strong basic foundation in major publishing software. Anyone willing to invest personal time will know the value and appreciate a culture of learning. No employee should become stagnant. Using the packages and keeping an eye on new developments requires juggling and maintaining a momentum of new options.

► **organizations should pay for training on upgrades** or special systems. Thus, if an individual has the foundation, the company and the individual share the expense and investment of the in-depth skills needed to be competitive. Build in regular time for formal and informal training.

► **factor learning costs into overhead**—reflected in schedules and fees. Amortize like equipment costs are amortized and set up a budget for publications, memberships, and seminars. Resources applied to learning are as much a cost of doing business as paying the phone bill and the rent.

FOUR AREAS OF TRAINING

1. COMPUTER LITERACY—Although computer literacy is less a problem now than it was even a few years ago, there really are individuals who have never touched a computer keyboard.

2. PRODUCT-SPECIFIC—The features of each application need to be conveyed to the users. The more complex and powerful the software, the less intuitive it will be.

3. SITE-SPECIFIC—No two corporate computing environments are alike. Productivity questions such as where to print, how to print, formatting and design standards, and file-naming conventions are examples of site-specific information users will need to know.

4. TECHNIQUE—There are two aspects to this:

   Effective technique: knowing which features to apply to accomplish certain tasks.

   Efficient technique: knowing how to sequence features to accomplish tasks quickly.

"Both effectiveness and efficiency are gained through practice, but can be learned through training mechanisms such as user groups, newsletters, demonstrations, or open workshops. Don't underestimate the need for training."

—Beth Chambers
Information Systems Manager
Bellcore
Lisle, IL

"KEEPING UP WITH THE TIMES

Here is a checklist of factors that everyone agrees are critical to keeping up with the technology times:

1. Read at least one high-tech trade magazine each month.

2. Go to the Internet trade shows.

3. Find a tech guru or two and get to know them before you need to know them!

4. When learning new technology, start with some long-time interest, such as music or fine art.

5. Plan the time to explore and research, don't wait for it.

6. Absolutely love what you are doing or change directions now, before it is too late and you feel you have invested too much to do something else.

"If working at the computer isn't 'flowing,' you probably need to learn more. I don't think we need to feel so threatened and inadequate if we don't read every publication that is out there. My desire to use the computer drives me to learn what I need. You will learn what you need to fulfill your creative objectives. The best musicians appear to play effortlessly and their music seems to just flow. This is what I am after in using the computer."

— Maria Piscopo
Creative Services Consultant
Costa Mesa, CA
http://MPiscopo.com

The Seybold Conference reflects the technological advancements of the industry. It is one of the industry events where all the suppliers roll out their new products. See index for Seybold information and comments.

The creative professional must provide an atmosphere conducive to curiosity and experimentation, as part of doing business. If the creative employer can provide a policy of compensation for seminar attendance and individual research, this will further motivate the employee and strengthen creativity. Arrangements can vary between awarding time off to attend conferences or to investigate in-house resources. Every employer needs to have a policy about training, and those wanting to encourage a strong team will utilize it as a benefit.

PRODUCTION GROUP

Training for the production end of the publishing cycle is both easier (because it is more identifiable) and more confusing (because there are so many choices).

CONTINUOUS TARGETED TRAINING

Other than the training suggestions listed above, possible targeted resources for continual training include:

- ► **traveling seminars** on focused topics related to specialties.

- ► **conferences**—local and national—that have training tracks where gaps in knowledge and greater overviews of packages can be obtained. Some have hands-on technical tracks for reasonable tuition.

- ► **junior colleges** for more in-depth continuing education. Good for foundation-building knowledge.

- ► **technical foundations and associations** with courses that can be applied immediately to the work situation and are targeted to increasing skills (see Appendix).

- ► **vendor-sponsored seminars** that are usually free (though biased to certain equipment and software choices) and can demonstrate capabilities.

- ► **consultants** can address specific issues and needs, particularly through activity in associations to examine issues, sponsor conferences, and maintain interactive sites.

- ► **technology management firms** to help with major transitioning and new or major updating system installations. Often hired for onetime enhancements, ongoing third party services are becoming more a part of work flow solutions. (See chapter 18, Workgroups, page 226.)

Production expertise is equally about technical skills and capabilities understanding. Anyone can buy equipment. Only those with a good sense of business, craft, management, and talent can turn equipment into a viable business. Training is a foundational commitment towards competitiveness and viability (not to mention credibility).

"The information systems professional typically views the problems of computerization in terms of the technology rather than in terms of the business. That is, when an implementation falters or fails, it is seen as a missed opportunity to use the technology.

"When the technology invades every area of the business, the technician must be schooled in all areas of the business.

"One of the best ways for an organization to educate its technical staff about the business is through a collaboration between technician and user during systems development. They can train each other in the business of the organization."

—Vicki McConnell
and Karl Koch
Mentor Group
Columbus, OH

"I fear with so much emphasis on digital media, the roots of design will be neglected when training new designers. Although I work with technology and embrace the learning experiences that it offers me, I do have some reservations regarding how fast technology is changing. It is hard for everyone to keep up with the pace it sets. Also, the fact that 'digital' is so accessible means that much of what is created isn't thought through as it should be. I think we need to embrace the advances of digital technology while we look to the past and remember how graphic design has come to be."

—Andrea Nicloy
AN Designs
Pewaukee, WI

"Don't jump on every whiz-bang bandwagon that comes along. It is very important to stay aware of what is out there, but only adopt what you truly need to accomplish your already created goals. It's too easy to get sucked into what the technology manufacturers tell you that you 'must have to survive.' Talk to your peers who have worked with the technology a while before investing your time and money into something that may not help you enough to justify your output."

—Stephen Webster
Photo-Illustrator
Columbus, OH

"Provide employees [with] the opportunity for true participation in management decisions through focus groups. Employees are the true experts; they are now, or will be soon, performing the functions to make the new environment productive. Subscribe to the corollary of the *Peter Principle* that says, 'Decisions rise to the management level where the person making the decision is least qualified to do so.' Not to make use of all available points of view during periods of change will only lengthen the overall process required to make the change work. Change results are inevitable. If we do not plan the results, somebody else will. We might as well get the involved people planning from the beginning. People who are involved in the process tend to be committed to the results."

—Tom Payne
Writer and Consultant
Performance Press
Albuquerque, NM

"The real problem is not if computers think, but if people do."

—B.F. Skinner
Psychologist
Harvard University
Cambridge, MA

"Effective training can't occur until the trainer understands the learning process. Three core factors in learning are the senses, repetition, and exploration. Generally, the more senses you can involve in training, the more learners will retain. Demonstrations, explanations, discussions, supplemental reading, and hands-on experience all combine to produce retained knowledge in the student, and it's important to include these activities in every computer training class."

—Tori Coward
President
Tangent Computer Resources
Dallas, TX

"Decide if your business is client-driven or technology-driven. Those freelancers that are 'client' driven wait until there is an actual need, expressed by job requests, for the new technology. Those 'technology' driven pursue new and developing techniques for the job and competition of the pursuit. The choice of method reflects the freelancer's personality and their client base. There are strong feelings on both sides."

—Maria Piscopo
Creative Services Consultant
Costa Mesa, CA
http://MPiscopo.com

DISTRIBUTED LEARNING

"Learning how to learn is now recognized as a crucial skill for learners of all types, and of all ages. Learners need to master and have readily available a core set of skills and facts, but to be really successful in the future will require mastering new skills and attitudes: How to locate, process, synthesize and communicate information, and an attitude toward 'knowledge' as an evolving personal and social process....

"With the Internet, learners are being asked to reflect on the sources, and to grapple with questions that simply don't arise often with textbooks. Because both the quantity and quality of the information present new practical challenges, our pedagogies are also being challenged. Once the way we teach is open to question, 'what' is being taught is also opened up.

"Our understanding of learning is evolving from a largely personal, isolated experience to one that is social and connected. The task is not so much to recall the fact the teacher wants but to be able to locate it efficiently, understand why a particular fact is important, and how it fits with other related facts. Understanding is enriched by conversations, interactions, and collaborations....

"Learning in a new world will take place beyond traditional education institutions. Distributed learning assumes mediated online sessions and can take a variety of forms. 'Distributed' does not mean isolated, so students in distributed learning situations should have the ability to communicate with faculty and peers, and feel that the experience of working on one's own without (or with limited) face-to-face contact does not limit one's ability to learn....

"There are no significant differences in learning achievement despite the current limitations in the distributed learning experience. Cost savings are often the main motivation in these settings, but learning 'effectiveness' is also high....

"As we get closer to an era where individuals and institutions will be able to afford powerful but capable devices, and have access to relevant content in the context of innovative learning environments, it becomes possible to imagine a future where teachers, learners, administrators, parents, and the community at large can be engaged in the education process in much more meaningful and rewarding ways. The devices that will be used in the future are unlikely to resemble the computers that we use today.... Tools need to adapt to the characteristics of their users, rather than make users fit themselves to inflexible tools."

Pedro Hernández-Ramos, Ph.D.
Academic Solutions Manager, Worldwide Education
Cisco Systems, Inc.
San Jose, CA
www.zonezero.com

DEVELOPMENT

CONCEPTS AND DESIGN

If a project begins with design, it is beginning at the wrong place in the process. Design needs to be strategic and requires preparation. Follow the guidelines in the previous chapters to develop positioning, purpose, and a general foundation from which to begin the development of functional, effective, and recognizable graphics.

Once the parameters are established for a publishing project, the client and the creative group collaborate to determine the creative direction or approach. This begins the design development process. Always focused on the goal of the client, projects that begin with clear communication, an approved proposal (see chapter 4, page 38), and a solid project plan (see chapter 8, page 70) are usually the smoothest and most cost-effective. Starting projects too quickly will minimize the design solutions explored. To maximize on design investment, know the form and genre where the design will function and begin with a solid foundation.

All the best and most effective design approaches aim at the potential recipient's point of view. The collaboration between the client and the creative group is focused on making such a point of view a successful beginning approach.

DESIGN TO COMMUNICATE

The job of the design is to translate a client's message to

► **attract a defined audience.**

► **motivate enquiry.**

► **encourage membership.**

► **sell a product.**

► **convey information.**

Perhaps more than one purpose can be met—the more functions design addresses, the stronger it can be, provided it does not become so complex that the purpose is confused or diluted.

ORGANIZATIONS

www.aiga.org
American Institute of Graphic Arts

www.ac4d.org
American Center for Design

www.dmi.org
Design Management Institute

www.gag.org
Graphic Artists Guild

www.core77.com/obd/welcome.html
Organization of Black Designers

www.segd.org
Society of Environmental Graphic Design

www.spd.org
Society of Publication Designers

www.ucda.com
University & College Designers Association (UCDA)

ORIGINALITY AND INTELLECTUAL PROPERTY

Creative groups are assembled to address specific functions (like publishing) or projects. (See chapter 4, page 43, for kinds of groups.) All creative groups are hired to solve a communication need as creatively as possible, given resources and appropriateness. Originality is part of the creative group's product.

An original concept is necessary to distinguish an organization—to build a visual recognition that helps fortify business momentum. In design and publishing, an original concept is developed for a specific marketing challenge. Originality means a written or visual concept or idea that

▶ **is inherently different** from other concepts or ideas. It can combine previous concepts, provided there is no infringement (see chapter 31), but it must be a new configuration.

▶ **was not seen elsewhere** by the creator (although all ideas are the combination of preexisting elements put together in a new way so there are subliminal influences). What matters most is who had the idea first.

▶ **contains no parts recognizable** from preexisting copyrighted work. (See chapter 31, page 346.) When combining preexisting images, they must be copyright free or used with permission.

▶ **is represented by presentation materials**. It can be a single idea or a theme, a direction, or a campaign. Ultimately it is intended to become finished artwork. If not in tangible form, it cannot be protected. Please note that a verbal idea is not protectable by copyrights. It must have a tangible form to be intellectual property. (See chapter 31, page 346, for copyright qualifications and further description.)

There are cases of a simultaneity of ideas—the same conclusions drawn from similar parameters—although rare. The one to publish first has the advantage. It is possible that two designers may conceive of the same design. For example, two different scientists independently invented the telephone, but Alexander Graham Bell was the first to get it patented and to the market. A famous corporate identity case is the NBC logo: the broadcast company tried to change the peacock image to a more modern-looking monogram of an N form with red and blue geometric shapes. Although they did do a legal search, it wasn't thorough enough and there was an astonishing simultaneity between this N design and a Midwestern educational TV station. NBC had already implemented the new logo on all station identity—both broadcast and print. Losing the case, they then superimposed the peacock behind the N, fading the N out, and bringing the peacock back. Eventually they dropped the N form altogether. It was a very expensive mistake. Perhaps the peacock is a warmer and more accessible form anyway, but that goes counter to their research.

TYPICAL PARAMETERS FOR DESIGN PROJECTS

Most projects encompass more than one medium: At its simplest an organization may use an identity package for a small company (letterhead and logo for word processing templates) or a full branding campaign such as for building visibility or a new product launch. A project can be for a single onetime use or for a comprehensive promotion including cross-media communications. Projects may include brochures, advertising, Web sites, point-of-purchase displays, etc. Design projects need to

▶ **reflect their subject matter**—appropriate design means that the recipient has the desired impression and is motivated in the desired way. The design reflects the personality and character of the content.

▶ **be fast and efficient**—the viewer must grasp the message quickly through visual distinction and clarity. No matter what the medium, all audiences have short attention spans.

▶ **be responsible to a budget** defined by the proposal. (See chapter 4, page 38.) Responsible and efficient design always scales to resources available.

▶ **adapt to multipurpose use**—most design is part of an integrated marketing plan and is inherently cross-media. Even if the context is just a single application, most businesses do use more than one communication medium eventually. Best to have flexible concepts that can adapt versus to have to start all over (unless change is important to the message). Effective advance planning will give visual themes relevance and longevity.

▶ **differentiate the client** from competitors by being visually unique.

▶ **have a strong graphic theme** to give unity, family, or branding to all deliveries of a message. An exception might be design for a single-event targeted for use only once that may need to provide a different kind of message. But if seen on more than one piece, occasion, or medium, consistency becomes important. Develop approaches that are unique, memorable, and express the most about their subject matter while being flexible enough to allow easy variations and additions.

▶ **motivate response** through captivating a need or fulfilling a benefit, grasped quickly by the recipient.

▶ **enhance an organizations' corporate identity** and build recognizability.

THE DESIGNER'S CONTRIBUTION

An organization needs a designer when distinctive marketing is required. For the most visible of graphic materials a company produces, there is no replacement for talent to come up with original visuals. There are many aspects that the designer will continue to do better than the non-designer:

>CREATIVITY—devising solutions that capture attention, making communication unique and memorable.

> EFFICIENCY—proficiency at digital tools speed design processes because designers can integrate all the elements and blend them with correct software utilization.

> QUALITY—the best designers are perfectionists and have the ability to skillfully execute what they envision.

> UNIQUENESS—the ability to create original ideas, communicate, and challenge the process finding new ways to reach an audience.

As the designers develop concepts for presentation, availability of the project director to answer questions or provide information can keep a project moving.

MAJOR DECISIONS

set project foundation
plan rough project skeleton
assemble category outline
solicit proposals
choose creative group
finalize the project plan
select the design theme
approve the design

Organizations committed to excellence do not feel neutral about their corporate identities. They have a sense of pride—even of protection—towards their symbolic image. The best identities are both highly memorable and synonymous with the company name they represent. They are usually simple, elegant, and have a unique personality.

FAILED IDEAS

If an idea the designer presents does not work, it is usually because:

> the client did not communicate the parameters well.

> the designer did not listen or understand needs.

> goals of project were not well enough defined.

> designer was not inspired (not as common as the other three reasons because the designer maintains a standard of work).

CLIENT GROUP

The client must supervise the project to get it off to a good start (see chapter 7, page 62) and to monitor its progress. Ultimately, the client group is responsible for a project's outcome because their name is on it, they commissioned it, and they paid for it. Set up for a strong collaboration:

► **If working with a new creative group**, allow extra time for idea generation as the two groups get to know each other.

► **If working with a creative group on an ongoing basis**, make sure that essential checkpoints are not assumed or even shortcutted.

TO FACILITATE COMMUNICATION IN THE CREATIVE PHASE

It is the client's energy, enthusiasm, and commitment that set a project's tone.

► **Provide the background information** necessary. Site visits with the creative group and research may be necessary. The collaboration between the writer and the designer should begin to give all content development a cohesion and the best use of its potential.

► **Know source of any provided images or content**. Contact owners for permission. When in doubt, don't use.

► **Request documentation** on agreement for any chosen images owned by others to be used. (See chapter 14 for kinds of agreements, page 173.)

► **Understand concept ownerships** (which should be in writing to avoid any potential misunderstanding).

  ▷ *If the client originates a concept* and hires a creative group to execute it, the concept remains the property of the client.

  ▷ *If the creative group originates a concept*, it is the property of the creative group, who gives usage rights to the client. (See chapter 29, page 335.)

  ▷ *If a concept is the result of a collaboration* between the client and the creative group, it is assumed to be owned by the creative group, unless other agreements are made (see chapter 31, page 348).

▶ **Arrange the presentations** with all decision-makers, whether through a meeting or through an online conference call that includes all decision-makers and the creative group (see launch meeting, pages 68 and 73).

▷ *Facilitate all input* to ensure that decision-makers have communication with the creative group during presentations—this allows for questions and viewpoints to be clarified.

▷ *Moderate and negotiate decisions*—bring a consensus. Coordinate all decision-makers as one of the most important functions of the client manager. Good communication results by including everyone in the presentations. It ensures that all parameters will be included and all questions answered.

▷ *The designer gains feedback through responses.* Each decision-maker will convey a slightly different view, so it is the designer's job to interpret all those views into the project. It is best to have someone from the creative group present at all presentations to decision-makers, even those that the client group makes to its upper management. The creative group may learn a great deal from reactions and comments to concepts that can be very useful in the rest of the work. They are also present to answer any possible questions. Timing in these presentations is best planned as a definable agenda item.

▶ **Approve all concept and design directions** before work continues. Understand the various phases and key checkpoints from the project plan (see chapter 7, page 62).

LEGAL SEARCHES

The client is responsible for all appropriate legal searches (such as copyright, trademark, or corporate identity) for any concepts they select to use. Unless this process is delegated to the creative group as part of the project scope, the default of responsibility is automatically with the client, not the creative group. Whether the client or the creative group commissions the search, obtaining legal clearance is essential before any publication. Legal searches establish:

▶ **originality and legal use**—protecting the design as exclusive. This often relies on proving date of origination.

"Design is a long-term investment. It is generally paid for at one time, but has positive effects into the future. If it is not well thought-out in the beginning, it can be very expensive in the long-run. If marketing materials are not adequately geared to the business need, they will need to be redone. This hurts in three ways: the same services are paid for twice, time is lost, and the market becomes confused. More simply, effective design is an investment, while ineffective design is an expense.

"Good design, stemming from clear corporate goals, creates the atmosphere of receptivity to services or products. Then it is up to the client company to follow leads with results."

—Michael Waitsman
Principal
Synthesis Concepts, Inc.
Chicago, IL

Appropriateness of use is the best guide for judging ideas. Design always has specific problems to solve; creativity involves finding the best solutions. Good ideas often say more than they appear to: they convey harmony, distinction, and dynamism—qualities that attract the viewer. Above all, the best graphic design expresses the message and style of the organization it represents.

► **intellectual property ownership.** Utilizes intellectual property attorneys to handle various levels of searches from national to international. International is necessary for use on the Internet. Different levels may require different schedules and fees. Be sure to what extent the search is conducted.

► **simultaneity**—where a concept is too similar to another that belongs to someone else—is rare and unpredictable. What is not rare are similarities that may cause some design adjustment. If there is concept simultaneity:

▷ *the designer is not paid* for creative fees, and must create another concept for replacement that does clear the copyright.

▷ *the designer is reimbursed* for out-of-pocket expenses related to development of the first concept and pays the agreed-upon project fees for the second concept.

As the client works with the creative group, decisiveness and skill at keeping a project moving are very valued. The biggest way to lose money on publications is to change directions. No one can be completely certain any design will be successful until it is tried. But there can be no good design without a good client because the client is the one who makes the choices upon which designs are realized.

CREATIVE GROUP

Designers are translators—giving impact and expression to a client's marketing message. The best design instantly communicates. The more the design team understands the client and the marketing plan, the greater the effectiveness of the resulting work.

The best way to launch a project is through a meeting of the entire team—particularly the decision-makers with the creative group. (See chapter 7, page 68, and chapter 8, page 73.) After reviewing the project plan approved by the client, the creative group, led by the art director, proceeds with design processes.

DESIGN DEVELOPMENT

Every designer approaches concept development differently: Some work in spurts, some through doing something else while thinking, and others in a flurry of creative activity, generating a lot of ideas and honing down to a few. But all creative groups, led by the art director, need to have certain ingredients to set the stage for good idea creation.

▶ **Contact with the client**—if information provided is incomplete and questions develop during conceptualization. Even a nuance of meaning can make a difference in a visual balance.

▶ **Create original approaches** to the assignment—everyone works differently. From brainstorming as a group to individual searches for the best solutions, talent, imagination, and insight are the basis of the designer's contribution. (See sidebar, page 102.)

▶ **Present concepts** to the client in the form of comprehensive presentations ("comps"—which may be sketches, mock-ups, or prototypes—see Glossary for further list).

▷ *The form and size of comps* should be specified in the proposal submitted by the creative group. (See chapter 4, page 38.) Expectations and ability to visualize are key concerns.

▷ *Ownership of all comprehensive presentation materials* is retained by the creative group, even if in possession of the client. Many presentations of alternate ideas may be kept in the client files, even though those ideas are really owned by the creative group. (See chapter 29, page 330, for issues of ownership.)

▷ *Review parameters to start presentation discussion.* Reinforce the goals of the project and the directions that everyone had agreed to when launching.

▷ *Most groups show more than one idea initially.* However, the first presentation may require the combining of ideas or the introduction of new parameters—needing further development. The "first presentation" may be a phase of several, more refining, presentations.

▷ *Any concept, comp, or design that the client does not choose remains the property of the creative group.* Developing any outtakes would need to be negotiated, adding to project scope. Sometimes more than one idea has potential—and may even work in series. But this adjustment *will* require budget reallocations. Money can be saved if two ideas are developed at once.

▷ *All decision-makers from the client group need to be present* during key presentations. If not, unpredictable factors can later derail progress.

▶ **Use the Internet and conference call** for presentations (or videoconferencing) if not all participants can meet in one place (also, see more tips on page 64). Be careful

SIMPLE DESIGN

From the design point of view, the simplest design is the hardest to create. The best pianists make their performances look effortless to perform. The most quotable phrases in literature poignantly capture the greatest essence of meaning using the fewest words. The best athletes make their game almost musical, poetic, graceful, and, again, easy looking. Simple design is elegant, inseparable from its content, and seems almost obvious!

But simple is not easy because it is a condensation, a crystallization. It epitomizes meaning. This is achieved through endless trials and errors, revisions, editing, experimenting, combining, and sometimes starting all over again. When the image is done, it is reduced to essentials. During creation, the elegant idea is usually born from a decisive moment of recognition. The creator *knows* it works. Some experts call this the moment of the "a-ha!" Those who study creativity continually try to quantify how to get to that moment, but any artist will say that it takes a lot of work, concentration, and transcending the challenge—it takes forgetting about it to achieve it.

▷ *that there is not a lot of time between posting images and reviewing with participants*—this means each will bring their initial reactions to the discussion rather than a reaction from reviewing it privately with colleagues first. This is more streamlined and will save time.

▷ *to describe the differences in color viewing space* (RGB versus CMYK) when monitor viewing is approximating print. Adjust color in videoconference facility and explain variances. Give examples, where possible.

▷ *to have all decision-makers in the conference call*—if not, new parameters later on may cause expensive delays.

▷ *trying to reach consensus for direction* by the end of the call. Letting decisions go unmade, even if the decision is to do more research or develop more ideas, means new opinions can influence the discussion without being in it.

▶ **To include a framework for the production plan**—build upon the approved design (see chapter 7, page 62 and chapter 8, page 70).

▷ *Concepts should fit within the budget range* agreed to in advance. If a design is presented that is more expensive than the budget, the client should be told at the presentation.

▷ *Any additional costs* for a given design are to be ap-

WORKING COLLABORATIVELY
(See page 137, Working Alone)

Mining for new ideas requires participants who are open-minded and willing to not only listen to new approaches but try them. As a participant on a creative team these tips should help the process:

> LEAVE EGOS AT THE DOOR. Ideas are independent from those who think them up. An idea is either valuable or not, a spark in a new direction or not.

> DON'T BE AFRAID TO THINK UP CRAZY IDEAS. Loosen up. Use techniques to jog thought in new directions such as random input, role-playing, or turning the problem into a story.

> BE A DIPLOMAT. Listen to others before jumping to conclusions. Initial reactions may miss a hidden element that might be the spark to a new approach.

> WORK SOMEWHERE ELSE—change locations by brainstorming away from the office or usual space. Being away

from familiar surroundings and with a different pace can loosen thinking.

> LOOK BEYOND PERSONALITIES to the shared goal. Dealing with people can be harder than dealing with ideas, so keeping a prioritized focus helps meeting the overall needs of everyone, and then the personalities become secondary.

> KNOW WHAT INFORMATION TO SHARE and what to develop first before getting feedback. Some data is proprietary either to an individual or to an organization. Some ideas are better saved for different circumstances. Discretion is important when dealing with ideas that become or are already intellectual property.

> DON'T STOP AT THE FIRST IDEAS or the first ones the creative team agrees upon. Areas of disagreement can be fruitful.

> BE CAREFUL OF "RED-LIGHT" THINKERS who just shoot down ideas. It is very easy to criticize and much harder to originate. Some critics believe they appear smart if they find what is wrong with every new approach. Be wary and look beyond this limitation because new results will not be achieved without trying something new. Address negativity as it occurs and move past it.

> DON'T COMPETE WITH COWORKERS. The real competition is outside of the team. Sharing skills, knowledge, and information within a team, based on trust, is necessary for standing up to the external forces of the economy and marketplace.

> INCLUDE DECISION-MAKERS in the final choices and through their feedback; expect to further develop concepts.

proved by the client *before* the work is performed (unless the creative group wishes to pay for it).

▷ *If resources are necessary for the selected design*, such as photography or illustration, research may be needed to find the right artist or image. Quotations should be firmed up along with the design approval.

► **Solicit suggestions on the most efficient production methods** for the approved design from others in the production cycle, which will vary from project to project but usually include:

▷ writer

▷ photographer

▷ illustrator

▷ page composer

▷ production coordinator

▷ prepress specialist

▷ printer

ROLE OF THE ART DIRECTOR

To manage projects throughout the development process and to lead the creative group, the art director has several roles and responsibilities. This professional needs both to have an overview of the project, as well as perceive its details and be:

► **an idea generator and editor**—working closely with the client and the market gives insight to experience.

► **facilitator**—handles team meetings or brainstorming sessions, ensuring that all those with creative input are given the opportunity to experiment and help further develop the concepts. Including all creative minds and blending the results is a rare and special talent of the art director—someone who can set ego aside and pull together concepts that are the most relevant to the creative challenge.

► **the client liaison**—good at listening, explaining, and reporting. People skills are necessary for understanding the project requirements and strong leadership to convey that vision to the creative team.

► **a critic**—developing the best ideas can be a process of sifting through possibilities and choosing options. Keeping creative momentum going can keep new possibilities emerging.

► **coordinator** of all opinions and factors into the project—is able to maintain the vision and enthusiasm to keep the project moving, relevant, and profitable.

Designers need an ever-expanding skill set to best serve their clients. There will be a savings for clients if the images developed can also be used for several media including print, online, presentation, broadcast, and display purposes.

The challenge for the graphic designer is to be conversant in the parameters of other media. In the past, design education was implemented during college and through apprenticeship. Then, the designer perfected skills through experience. Today, designers needs to know almost everything they needed to know in the past, but their education never ends. There is a never-ending adventure with developing technology, keeping up to date, utilizing it, and being a resource for clients.

► **a team leader/ coach**—when working with other designers or a team of creatives. Responsible for beginning a creative atmosphere that is picked up by the team, and encourages a momentum that fuels the enthusiasm until completion.

► **a referee or diplomat**—where differences over various concepts occur. Use courage to support new ideas. But at the same time, make sure those new approaches are sufficiently supported by sound reasoning and marketing reasons. Filter through all the agendas and keep everyone focused on the same goals to help mediate derailing delays and conflicts.

DIVIDING UP ROLES

The designer or design team either works with the production group or handles all design implementation themselves, such as for small projects. Large projects may have a formal and defined process for the two groups to interface. Progress will be slower and more defined. Channels of communication and reporting are part of the project plan. Whatever scale the team is working on, the more that each member understands about the roles, processes, and expectations of the others, the smoother the process. With specialization also comes the responsibility to understand how the parts fit into the whole. If the designer is also the art director, make sure that the concepts are tested against a target market before locking into new directions.

The complexion of the creative team is most influenced by the talents of the art director. (See chapter 4, page 43.) When skills are well matched to the challenges and communication is easily facilitated, the best work can be achieved.

Similarly, the creative and production groups can work for different firms or be departments in the same organization. It is rare to have a project where *some* aspects aren't outsourced. Each project is different and many creative groups are consortiums of various talented individuals who bring a portion to the whole. Always managed by the art director, who forms a strong and inspiring collaboration between design and production, the creative group can find some of the most creative and innovative solutions.

PRODUCTION GROUP

Ideally, when a project is ready to be designed, the production group has had involvement in the planning. Often, though, a design is brought to production after it is already developed. Then, the production group must evaluate the project requirements against the approved design structure and find the most expedient procedure. If the production group has been involved in the planning, their input can influence the way the design is prepared prior to final processes. They can

▶ **suggest time-saving methods** if involved before final art is created.

▶ **recommend modifications** that may make the design more efficient to produce, saving costs and speeding the process, especially when a project has a short time frame.

▶ **give estimates** as part of the proposal process. (See chapter 4, page 38.)

RX FOR LOW-ENERGY SYNDROME

"How does a brainstorming facilitator revitalize the energy level? Try using these energy boosters:

>CHANGE THE GROUPS. If your brainstormers are working in large groups, break into small groups, or vice versa.

> REFER TO THE MATERIALS YOU'VE BROUGHT FOR INSPIRATION.

> LOOK AT THE ANTITHESIS OF WHAT YOU'RE TRYING TO ACCOMPLISH. Ask 'What would be the one thing we don't want to create or accomplish? What would we NOT want to name this product?'

> CHANGE THE SCALE OF THE PRODUCT. If it's a large package, image it as something you can hold in your hand. How does this alter your thinking?

> CHANGE THE RHYTHM OF THE SESSION. Get up and do something physical or introduce a new stimuli, like music or video clips."

— Cailin Boyle
Writer
San Francisco, CA

"BOTTOM-LINE BRAINSTORMING

Producing a great brainstorming meeting is like producing a Broadway show, complete with a set (decorated meeting room), props (notepads and inspirational think toys), costumes, lighting, sound plot (theme of the session) and director (the facilitator). To get the most out of your idea generation?

1. GET A GOOD MIX. The ideal group is diverse, including experts, non-experts, people from different domains within the organization and even people from outside the organization, such as clients. A diverse group of people allows you to look with fresh eyes.

2. FIND A FEEL-GOOD PLACE. Hold your meetings in a risk-free zone where people can speak their minds without fear of criticism or ridicule.

3. BRING A FACILITATOR. Without one, old habits might pull the group toward judgmental thinking and away from productive, creative thinking.

4. INSPIRE THE EYES. Make the environment visually stimulating.

5. LET THE MUSIC PLAY. Music can help set the tone and heighten a session.

6. HAVE A LITTLE FUN. An environment of playfulness and humor is highly conducive to creativity. Playing relaxes the tension in a group.

7. TOY AROUND WITH THE PROBLEM. Having a box of toys will change the mood and invite people to be more open and playful.

8. ELABORATE SOME MORE. Encourage participants to elaborate on existing ideas. Apply the SCAMPER checklist of nine creative-thinking principles:

9. DON'T ASSUME OR JUDGE. Require everyone to suspend all critical comments until after the idea-generation stage.

10. GO FOR THE NUMBERS. Quantity breeds quality. Every participant should generate as many ideas as possible.

11. CLARIFY THINKING. Any speaker or listener can call for a full hearing and understand the idea.

12. RECORD AND CLUSTER IDEAS. Record ideas visually with cartoons, diagrams, mind maps, graphics, or printed phrases illustrating the group's thinking.

"After you've created the maximum number of ideas possible, change your strategy to practicality thinking, where ideas are assessed to find those of most value."

— Michael Michalko
Creativity Expert and Author
Churchville, NY

"Everything new is an addition or modification of something that already exists. When you're brainstorming an idea, product, service or process, start with a subject and engage in some 'mental manipulation.' Suggested by Alex Osborn, a pioneer teacher, SCAMPER is a mnemonic checklist of nine ways to re-envision a subject.

SUBSTITUTE. Develop alternative ideas for things that already exist. Can we use other materials or a different process?

COMBINE. Can we combine the purpose with something else? What can multiply the possible uses?

ADAPT. Does the past offer a parallel? What different or unusual contexts can we put our client in? What can we adapt from sports, television, books, politics, movies or religion?

MAGNIFY. Take a subject and add something to it. Can we maximize existing strengths? What can be magnified, made larger, or extended? What can add extra value?

MODIFY. Can we give it a new twist? What can we do differently? Can we change the perspective?

PUT TO OTHER USES. Change the context. Are there new ways to use it? Can we make it do more things? Can we find other benefits?

ELIMINATE. Subtract something. What should we omit? What ISN'T the problem? Can we streamline it?

REARRANGE. What if we interchanged components, changed the pattern, or change the pace?

REVERSE. Look at opposites and you'll see things you normally miss. Can we turn these negatives into positives?"

—Michael Michalko
Creativity Expert and Author
Churchville, NY

"An outside facilitator should bring several skills to the group:

> OBJECTIVITY. A facilitator should be objective but not too far removed from the creative process.

> THE ABILITY TO LISTEN AND COMMUNICATE. A facilitator has to assess the energy level of the group, monitor it constantly, and adjust it when necessary.

> A NURTURING CHARACTER. A good facilitator draws people out of their shells and ignites their creativity.

> A SENSE OF EXPLORATION. A good facilitator has to be willing to wade through the muck and mush of the early stages of brainstorming, when good ideas are sometimes still buried. Eventually, the gems will surface, but they're bound to need a solid buffing before they truly shine.

An unfacilitated, internal brainstorming session typically generates 10–15 ideas, while a brainstorming session run by a facilitator might generate as many as 200 ideas."

Cailin Boyle
Writer
San Francisco, CA

FOLLOW PLAN

No matter at what point in a project, the production group assembles according to instructions provided by the creative group who:

▶ **should have a team meeting to begin** the production planning process, no matter what phase the project is in.

▶ **gives context to the project**, provides parameters and design philosophy, and shares comprehensive presentation materials.

▶ **explains variables** that are most likely to change and requests new estimates if parameters alter during development.

MIND THE STORE

When the creative and production teams convene after designs are approved, a rough plan should already be in place. The production group, before starting the project, needs to

▶ **revise their estimates** as needed from initial planning. As design develops, the specifications may shift. Prepare final quotations according to the final design direction. Quotations that were prepared *before* concepts were created need to be verified. The financial agreements almost always need to be reviewed after designs are approved.

▶ **work out any differences** in versions or technology before feeling deadline constraints. Compatibility of variables may influence the design and project construction, but do testing and conversions before moving too far into final processes.

Design should not progress to production without the buyer (usually the art director) assuming the approval of the client.

"I like designers who beat me up. I don't respect people whose response, when I say, 'Let's do it this way,' is 'That sounds good.' That's not the point of the whole process. If I knew what I were doing, I wouldn't have to ask designers to come in and help, and be willing to pay good money for it. But when you're beating me up, I want to know why. I want to know why my idea is really not good. I want to know what your idea is going to contribute. What's really important is how much designers bring to the party—the totality of their contribution. The best projects that I've done involved designers who have really contributed a lot beyond the look of the thing. And the more the designer brings to the party, the more likely it is that we're going to party again."

—Robert Moulthrop
Marketing Director
Deloitte Haskins & Sells
New York, NY

"There are a couple of reasons that top management's concern and participation ought to be mandatory. Communication design (like design in general) is crucial to a corporation's success, yet not likely to be understood throughout the corporation. Until it is understood, it needs support from the top. And because it depends on a combination of broad concepts and extremely fine details in execution, top management must be mindful of the vigilance required to keep a good program from eroding. Top involvement is the irreducible minimum requirement for long-range communication effectiveness. We can think of no exception to the rule that where there is a sustained communication program of distinction, top management is directly involved."

—Ralph Caplan
Consultant and
Writer
New York, NY

"If a committee has a strong leader, the members of the committee can inform the leader and it can work well, particularly if the leader is in an area they are not familiar with and people on the board are. That is where a lot of designers are lacking in skills of managing a presentation and organizing material so that you can manipulate their attention. You are on a stage and just as an actor will manipulate his audience to evoke responses, you are doing the same thing. You can't simply let your work speak for itself because it won't. You have to set the stage and that is a whole art. A designer has to be good at both coming up with the ideas and also presenting them—and most people are good at only one or the other. There is a danger, though, in someone else botching a presentation who didn't actually do the work. You have to get access to the one who is making the decisions. If not, good luck."

—Kevin Shanley
Designer
SWA Group
Houston, TX

"To communicate effectively and build a distinctive identity, a corporation must develop its own voice. A voice that is instantly recognizable. A voice that conjures up strong images and associations. A voice that establishes the tone of the company's communications: classic or innovative, glamorous or functional, aristocratic or irreverent, elegant or just simple and down-to-earth. A voice that fuses language, design, and content to convey the company's distinctive personality and culture."

— Ronald R. Manzke
Designer
Siegel & Galee/Cross
Los Angeles, CA

"Most designers will be disappointed when I reject a particular direction, but the real professionals realize that I don't do it arbitrarily. There are solid reasons why something won't work. I don't just say, 'We have to change this.' Instead, I'll say, 'It would be better if we emphasized this aspect because, for these reasons, it's more important than what you've shown.' Ego is always involved in design. But when you can separate your ego from your design—when you don't take it personally, and you realize there's a reason for making changes—then you become a much better designer, certainly a better business person."

—Kathleen E. Zann
Marketing Communication
James River Corporation
Richmond, VA

"I provide designers with all the technical background they need. If they're not out in the cold, if the purpose of the project and its audience are thoroughly understood, they should be able to solve the problem. I don't want to go through a lot of iterations and be presented with tons of sketches or comps. If there's enough verbal interaction and agreement before we actually execute something, there's no need for them. All I need to see is one design concept or maybe two."

—John J. Dietsch, Director
Corporate Communications
Booz-Allen & Hamilton Inc.
Bethesda, MD

"We want to see eight to ten logo ideas. It would be unacceptable if someone came in and announced, 'Here it is; here's your logo.' We need to see options, but they don't necessarily have to be eight to ten completely different solutions. A few variations enable us to say, 'Yeah, we like this one,' or 'Maybe you need to go back and refine these two.' Sometimes if we like several designs, we might ask the designer to take a particular aspect from one and combine it with another."

—John and Bill Schwartz
Owners
Schwartz Brothers Restaurants
Seattle, WA

"Shaping perceptions is the heart and soul of design. Color, image, type, paper texture, shape, and size work together to convey one or more messages, which can reinforce, supplement or even replace text. Designers are becoming increasingly sophisticated in using perceptual tools—by choice and by necessity. They recognize the growing importance of perception in communicating with today's global audiences, with their shared and divergent values and changing ways of processing information."

—Frankfurt Gips Balkind
New York, Los Angeles,
Washington DC
Simpson Paper Company
Fairfield, CA

"Graphic design can help the world: when we're providing information, we are hopefully enlightening others; when we're providing directions, we are helping people find their way through the world; and even when we're selling widgets, we're boosting the economy and providing jobs. Furthermore, when created with the right heart, graphic design can have a powerful social impact."

—David Sterling
Designer
Container Corporation
Chicago, IL

"In an era of ever-increasing media stimulation and decreasing attention spans, creating and conveying both image and message has become more important than ever. Today's designers have to be very cognizant of their audience. It is not enough to create 'nice' work; a designer has to do 'smart' work. With only a few moments to capture the target audience's imagination, a designer has just seconds to make that connection. Their design has to be focused, appealing, and right for that consumer. Someone has spent a lot of time and money defining that consumer, and now they need to see that their advertising dollars are getting results."

—Barbara Krouse
Amber Design Associates
Hackettstown, NJ

"Designers will have an impact on the direction of society by exploiting and extrapolating from this incredible information explosion and assembling pieces of information into cohesive patterns that can be understood. It's a more challenging job today than ever before."

—John Massey
Director of Design
Container Corporation
Chicago, IL

"The key to delivering the best solution is heavily dependent on a well thought through strategic platform and execution plan. If all you're emphasizing is the 'creative,' then expect your clients to look elsewhere for help in solving their business problems. No grid can solve a bad starting point. Because these are not just 'design' problems. Design is an approach.

"If practiced with strategic and creative understanding, along with heightened proficiency in different media, it naturally rotates to an integrated approach to business problem solving: print, film, digital, events, collateral, whatever."

— Steven Bagby
Designer
Bagby and Company
Chicago, IL

"Because designers have the pulse of popular culture, we are very valuable to clients. As designers our 'feelers' are always out there, whether we realize exactly what we are picking up or not. We are GETTING IT. Most clients are not as 'tuned' into popular culture; they look to designers to lead them into the future. BUT, and I mean a big BUT, because they don't understand the popular culture, they have a hard time letting go of traditional marketing values. Instinct is a difficult concept to articulate and almost impossible to sell!"

— MaeLin and Amy Jo Levine
Visual Asylum
San Diego, CA

"Graphic designers are communicators and brand strategy experts. Thinking design today is able to understand or anticipate the needs and aspirations of a very savvy public, a public prone to boredom, looking for content stimuli, excitement and experience. Brands have transcended their original meaning of sending messages around product characteristics to become, instead, the total expression of a company, product, service or personality. In marketing terms, the word 'branding' has extended its traditional definition to include a sensorial component that is relevant to the personal needs and lifestyle of an international audience. Craft the brand's message in a way that courts the consumer, to enhance the emotional bond, the pact between the consumer, the brand, and its products. Consumers want products that answer their needs but also fulfill their desires. To achieve this goal, graphic designers and clients need to become 'Brand Captains'; they need to become masters of entertainment, innovation and service, they need to have a vision that will embrace the changes in society and answer consumers' expectations and fantasies."

— Marc Gobé
Director
d/g* worldwide
New York, NY
www.allworth.com

"We are challenged to somehow seamlessly continue the excellent creative work we do, learn a new set of skills to integrate the computer into new areas of design media, and master the expertise in production tasks once reserved for prepress professionals. Though technology may offer tools to do much of the work, it is forever linked to the human controlling it. Should one person be responsible for the skills needed to control such a complex sequence?

"With the computer, the designer is more likely to carry the creative steps into the mechanical/production phase directly. This creates an atmosphere where no one person is relegated with the responsibility of doing only mechanicals, and no individual can strut one's ego claiming to be superior to such lowly tasks. Managing jobs and egos takes on new meaning in this age of technology."

—Alyce Kaprow
Computer Graphics Consultant
Newton, MA

"Skills a designer needs beyond design: communication skills; organizational abilities; business savvy; team work; psychology behind the image; technical knowledge; partnering with suppliers."

—Lynn Finch
President
CADTEK Services, Inc.
Maitland, FL

"The technology crunch has stunted the creative process. Real concepts are becoming endangered species, threatened by an increasing number of designers who mistake tools for creativity and clients who happily trade quality for Photoshop. And I hear of too many clients that put their agencies in the 'design while you wait' mode, not appreciating that great ideas, like great wines, need some time to breathe before they are ready for consumption. Just because you can do something doesn't mean you should.

"So that leaves us with the obligation to provide our clients with powerful communication based on solid conceptual and strategic thinking and to avoid meaningless decoration, no matter how cool or 'edgy' technology-driven design might seem to be. If we as a business can do that, there's no limit to what can be accomplished."

—David Ellis
Vigon/Ellis

"It's our skill as designers, not as computer users, that will set us apart. We will go further concentrating on the craft of design and less on the tools. Quality is ultimately tied to design. The computer only allows you to express the talents you already have, and gives you speed and control in executing projects."

—Robert Vann
Designer
Lake Mary, FL

"Do we complain about technology? No. We revel in it. We find that we have what we've always been missing: control. We control a project from beginning to end. We can produce design in record time. What is the price for this control? Losing design time due to learning software, debugging software, and updating software. More time is spent learning to typeset, scan photos, color correct, composite, and trap than on design. On the up side, the more we learn to do well, the more we can do in-house and bill for it. On the down side, the more we do in a shorter time, the more is expected."

— Eric White
Designer
Binary Arts
Winter Springs, FL

"If you get a good designer, it's easy to teach them the tools. You can't take a techno-head and teach them design."

—Thom Dupper
Bennett & Dupper
Orlando, FL

"Today, the success of a graphic designer depends upon creative ability and expertise in the hands-on basics of production and design, combined with the skillful use of computer technology to explore new artistic avenues. There is a real need for a graphic designer to be constantly aware of any changes in technology, not only in the design field, but also in the printing/production processes.
"There is a great amount of pressure on a designer to implement the best technology to enhance production speed without sacrificing design integrity."

— Nancy Miller
Designer
Studio One Graphics
Irving, TX

"Technology is dangerous because it allows emerging designers to imagine that things are more exciting than they are. Because it looks professional, as if it has been thought about. But, in fact, it hasn't been."

— Ivan Chermayeff
Chermayeff and Geismar
New York, NY

"The wonderful thing about traditional design is that it is slow—it gives you time to think. The bad thing about digital design is that you can do bad design so quickly.

"Amateur designers don't trust themselves to keep it simple. They believe they can prevent a piece from looking boring by 'dressing it up.' So you have a combination of utterly standardized two- or three-column structures filled with an unpredictable variety of oddments tossed together like fruit salad, trying desperately to be original and innovative and—heaven help us—*creative*. It would be far better if they'd forget about being different and concentrate on making the message understandable. Go for clarity, not beauty. We live in an era when all the concentration is on the medium. But the medium is not the message. The message is the message."

—Jan White
Designer, Author, Consultant
Westport, CT

"I have always believed that design is about the idea and about the problem and about solving it. We have the best technology and everything state-of-the-art in our studio, but it has never, ever been about the tools for me. It is about the idea. People don't solve problems by picking up a pencil and exploring ideas anymore. I will say, okay, show me 35 concepts. What I get too often is one concept with 35 variations. Those are not concepts. All the emphasis on technology is changing our processes and I think what happens right now is we have an entire generation of young designers who don't know the process or the history."

—Mark Sackett
Sackett Design Associates
San Francisco, CA

"Computer and/or digital technology has, by its nature, allowed a migration of underdeveloped artists and designers into the field. Our industry is [suffering] and will suffer from overpopulation due to this effect. I see creativity dropping, as I believe it has in the last 15 years. What I see now is 'me-too-ism' at an all-time high. Scanners, copiers, do-it-all software have taken the skill level or proficiency level of a professional to the basement. A high school senior can produce presentation grade illustration and art, with nothing more than keyboard experience and a rudimentary understanding of the software. What we are faced with is an unbelievable number of non-professionals flooding the market with low level, low prices work. Client awareness of this is climbing."

—Mike Hruby
Art by Hruby
Elgin, IL

"We should all become a little humble with the tools of computer technology, but take solace in the realization that design principles stay steady and that the artistic foundation is secure. Creativity is still the domain of our personal human choice. Our minds and hands still rule the future of design; it's the wild stallion of technology whose reins we have to keep learning how to hold onto."

— Gregory F. Golem
Designer
United Services Life
Insurance
Arlington, VA

"The three lessons I learned about design at Apple were: (1) design is not about creating a thing; it is a process; (2) the Macintosh is more than a tool; it is a medium; and (3) there are no new ideas; technical innovations are nothing more than a series of ideas linked together by association—much like the design process. Technology has always changed the way we experience the world. And, therefore, the way we design and create. And, ultimately, the things we design and create. And the way we work with one another. If the computer is the great equalizer, where do we as designers add value? First, we cannot be specialists anymore. We need to understand the macroview of the client's needs. Perhaps the best combination is to be a generalist as to types of projects but be industry specific so that clients do not have to pay for a steep learning curve. Second, be ready to take on one-of-a-kind projects or systems projects. Third, work more collaboratively internally and externally, with projects being more process driven and less 'free form,' and get used to working with a broader range of people (programmers, animators, interface designers) than in the past."

—Clement Mok
Designer
Sapient Corporation
San Francisco, CA

"Clients demand writers and designers who understand how to use different kinds of media effectively to communicate a strategy to different kinds of audiences. Our clients desire a design partner who understands their business and can do great design that works. There is not room for an 'art for art's sake' attitude. This has not changed."

—Donna Eby and
Lisa Sanger
Partners
Sanger & Eby Design
Cincinnati, OH

"If you leave the research up to someone else, you probably aren't going to like what you get. You really should be there leading and creating the methodology to get what you need from it. If you don't, you aren't in a good position to defend your strategic and design solutions. If you haven't done your homework up front, what should be an objective business process turns into a subjective aesthetic one."

—Martha Bowman
Executive Vice President
of Marketing
DeLor Group
Louisville, KY

"Ultimately a designer's success or failure is based on their credibility and people skills. It's one thing to design an effective publication, it's quite another to 'sell' it to management who might not understand why you made the design decisions and tradeoffs that you did. Ultimately, successful designers are those who educate their bosses and coworkers. Designers must be educators. That's why I think even corporate designers should teach a course or two at a local college or present seminars and workshops in order to hone their education skills, so they become comfortable and confident defending their decisions to management and clients."

—Roger Parker
Designer, Consultant
Dover, NH
www.newentrepreneur.com

"Good typography depends on the visual contrast between one font and another, and the contrast between text blocks and the surrounding empty space. Nothing attracts the eye and brain of the viewer like strong contrast and distinctive patterns, and you only get those attributes by carefully designing them into your pages."

—Patrick J. Lynch
Yale University School
of Medicine
New Haven, CT
http://patricklynch.net

"Creativity isn't dictated by environment, but rather by one's attitude, approach, and efforts to stay inspired.

"Exposure to the arts, continuing education, and consulting with colleagues for ideas further promote creative flow.

"From museum visits, movies and games at work, to 'trend trips,' guest speakers, and celebrations, the possibilities are endless for keeping creatives motivated.

"Creativity exists regardless of setting. A designer's approach and effort determine the reward and outcome. [Where you work] isn't the issue; inspiring and challenging oneself is."

—Adriane Lee Schwartz
President
The Creative Source
New York, NY

"New Media—with its nonlinear pacing, multiplicity of viewer choices, and customized content—has placed a new demand on print design. It is no longer enough for a brochure to be engaging, compelling and informative. In order for it to work, it must be interactive.

"What constitutes interactivity on a two-dimensional page with finite pacing and no active buttons? Interactivity is a kind of experience. It is not a simulated experience guided by fake Internet icons. It's a genuine, emotional experience. The meaning or message is not a predetermined answer that we're uncovering for our viewers. But instead, it is a journey down a two-way street—created with an awareness that meaning is mutually constructed and exists somewhat between the reader and the printed image. Maybe its story is told by simulating a road trip, maybe the story begins from both the front as well as the back. Maybe it asks questions that don't have specific answers, but reframes the viewer's perceptions. Successful print must be interactive."

—Wendy Blattner
Belk Mignogna Associates
New York, NY

"Influences are not only inevitable but desirable; art proceeds from artist to artist, from idea to idea, in a golden chain of contingency that endows culture with its character and coherency. The history of art is rife with subtle appropriations, gentle filches and sly echoes."

—Julia Keller
Tribune Cultural Critic
Chicago, IL

"Originality is becoming a passe concept. Because of the Internet and the easy acquisition of images, there's a thought of, `It's all fair game.' There's a feeling of, `If it's out there, it's in the public domain.' And it's not."

—Thomas Shirley
Digital Imaging Instructor
Chicago, IL

"Digital technology is great. It's like opening up a huge box of goodies, especially if you're like me and not an illustrator. Having these things available digitally is wonderful. With Photoshop you can take a photo, change the color scheme or combine two images. It's like bringing home a whole new paint box. Your creativity can just go wild. My only regret is that a job I used to do with an Exacto knife, a drawing board, and design markers now takes a $25,000 computer."

—Ron Lovell
Writer and Professor
Oregon State University
Bend, OR

"E.B. White said people end up in advertising because they like to write and draw. Today, people who like to write and draw end up doing more than that. We end up learning Photoshop and Illustrator and Flash. We end up learning HTML and JavaScript and Style Sheets. And the rudiments of Web serving. And how to write an invoice. And how to write a follow-up invoice. And when to contact an attorney.

"All in the service of that original impulse to write and draw.

"Our pages drip with technique but are barren of content, because we no longer have time to think about anything but technique.

"Exploration is good. The bad comes when we allow technique to obliterate the concept it was intended to deliver; the very, very bad, when we forget that we were supposed to have a concept in the first place.

"And likewise for the Web site that loses audience share because nobody knows where to click, let alone what the site is supposed to be about.

"We forget this cardinal rule all the time, and as a result we frequently spend hours on work destined for the trash bin.

"There is a hardcore audience that designs sites, reads *A List Apart*, and flocks to edgy, often-cryptic Web tours de force, where we may be enthralled by execution for its own sake, if the work is original and well-rendered. For us, the question 'how the hell did they do that?' is its own reward, and we frequently return to sites that would reduce average users to tears. There is a place for such sites, even a need for them; but we must remember that we are a minority audience, and that what works for us as viewers is not necessarily the best procedure for all sites and all audiences. (And if we don't remember, we have clients and CEOs to remind us.)

"Before we put pen to pixel, we must ask ourselves what we are trying to communicate, and to whom... if the goal is to reach out and touch someone besides ourselves."

—Jeffrey Zeldman
Creative Director
zeldman.com
alistapart.com
New York, NY

"Design for your reader, not yourself and not your client. Do whatever it takes to balance 'attention' with readability. Use typographic and color contrast to attract attention to your publications and make titles and headlines stand out. But strive to make your body copy as 'transparent' and easy to read as possible."

—Roger Parker
Designer, Consultant
Dover, NH
www.newentrepreneur.com

"Creativity isn't dictated by environment, but rather by one's attitude, approach, and efforts to stay inspired."

— Adriane Lee Schwartz
President
The Creative Resource
Creative Recruitment Firm
New York, NY

"There's a huge misconception about the difference between 'creativity' and 'talent.' In any endeavor, there are tremendously talented people who aren't creative—people who can sing someone else's songs really well or do a totally beautiful painting, but who don't necessarily bring in new ideas. Creativity is bringing new ideas.
"While I believe talent is genetic, I think everyone has creativity."

— Tom Monahan
Creative Director
Before & After
Tiverton, RI

"Exposure to the arts, continuing education and consulting with colleagues for ideas further promote creative flow. From museum visits, movies, and games at work to 'trend trips,' guest speakers and celebrations, the possibilities are less for keeping creatives motivated."

— Adriane Lee Schwartz
President
The Creative Resource
Creative Recruitment Firm
New York, NY

"Sometimes we don't have two solutions because we either didn't have enough time or because one is just so right that we don't think we can top it. When this happens, we tell the client exactly why we're only bringing one solution and hope he believes in our idea as much as we do. Often, there's another way. By looking for more than one solution when we're stumped, we come up with new and better ideas than even we expected."

— Bruce Turkel
Executive Creative Director
Turkel Schwartz & Partners
Coconut Grove, FL

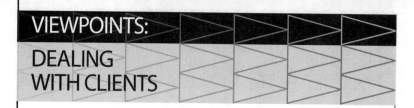

VIEWPOINTS:
DEALING WITH CLIENTS

"Excitement is the key. It's much easier to do a great job when you have what you need to do the job well and can count on your client to back you up. When your client is enthusiastic about a job, he's more willing to commit time, money, and authority to nurture and protect the project's outcome. For this, you need to remember that you're creating the job for your client—not yourself—and that your client has certain needs and requirements that are quite different from your own."

—Bruce Turkel
Executive Creative Director
Turkel Schwartz & Partners
Coconut Grove, FL

"Tell me what you want and I'll give you what you need. Even if a client comes to you with a silly request, you must be able to find something positive in it. I never say 'no' to the client in a meeting, even if I think he has a bad idea, because it's the designer's job to experiment first. My job is to define and refine the client's concept to make it more expansive."

—Supon Phornirunlit
Supon Design Group
Washington, DC

"When we first built our firm, we believed that we knew what our clients needed, so we generally presented only one solution to a problem. We soon learned that what we thought was a show of confidence was really forcing our clients into a decision they may not have been prepared to make.

"By presenting only one idea, we were basically telling our clients to 'Take it or leave it,' then wondering why they chose the latter. Now we generally present our clients with a few different ways to solve their problem, then gently try to convince them that our favored idea is right for them."

—Bruce Turkel
Executive Creative Director
Turkel Schwartz & Partners
Coconut Grove, FL

"Nurturing creative chemistry with clients takes practice, patience, persistence, and planning."

—Susan E. Davis
HOW New York Editor
New York, NY

"Not bothering to show clients how and why a piece satisfies their needs is almost the same as the piece not satisfying their objectives. In this business, perception IS reality. If you fail to create the perception of your solution's utility in your client's mind, you'll probably also fail to create the piece you envisioned."

—Bruce Turkel
Executive Creative Director
Turkel Schwartz & Partners
Coconut Grove, FL

"Most anyone can figure out a conventional response to any given problem. Even the spouses of clients, as it turns out. In fact, one of the hardest things to do in client relationships is getting them to embrace something that is, in fact, truly original. Clients are always more comfortable with a solution if it walks and talks like something they've seen before. That's because, by and large, people think reproductively. When confronted with a problem, we regurgitate what we've been taught. Taught by teachers, by peers, by the media. What's worked in the past will work now."

—Paul Matthaeus
Chief Creative Officer/CEO
Digital Kitchen
Seattle, WA
www.d-kitchen.com

WHEN THE CLIENT CHOOSES THE WRONG DESIGN

"E-mail the client or send him a memo to reiterate what he said he wanted and what you think needs to be done. Say something like, 'Let me remind you that we feel this project would work better if steps 1, 2, and 3 were done, but understand that you want to go forward with the project as you've described it.' Then keep a copy in the project file. If the client comes back to you and says that the piece didn't work like he expected, you can pull out the memo and ask if he might like to try one of the other suggestions instead. And in spite of any strong urge to say 'I told you so,' don't."

—Jim Reynolds
Senior Media Director
Cracker Barrel Old Country Store
Lebanon, TN

WEB SITE DESIGN 11

The Internet is unique from all other communications because of two-way interaction. Success is measured by viewer involvement. Communicators have to conceptualize as differently in this medium as they do in print and television. It is a *medium*, whereas presentation graphics, though it can involve the viewer, does not communicate back to the originators. (Presentations and point-of-purchase are extensions of the print medium due to the similarities of passive audience and process of project development. (See chapter 19, page 236.) The Internet provides factors that make the design vocabulary unique from other media:

► **Choices**—the Web is so vast that any subject can be researched. There are *too* many choices, hence the need for good editors and portal services.

► **Speed**—getting information is instantaneous, allowing 24-hour access. Visiting a site is as easy as one click.

► **Fluidity**—content is always changing. It is not a library, but more like a periodical. What is lasting needs a structure as a foundation for what always changes.

► **Communities**—new relationships are formed through interaction with others of similar interests. There is an international potential to build communities based on subjects or causes. This requires intercultural considerations.

► **Entertainment/education**—content to keep viewers interested. A viewer must feel a benefit to return to a site.

► **Sampling**—everyone likes free giveaways and the Web allows for great availability. It has developed a culture where viewers *expect* to get something.

► **Profits**—making money on the Web is not its priority — instead, its priority is information more than sales or subscriptions—with the exception of high consumer products (many with very narrow product niches such as www.justballs.com). The cybermarketplace interfaces with the physical, and when used together, both can become stronger.

► **Visibility**—having a site is much like having corporate stationery, a catalog, or a sales presentation—all available instantly anywhere at any time to anyone interested in

ORGANIZATIONS
(see also chapter 10)

www.aaim.org
Association for Applied Interactive
   Multimedia

www.amcomm.org
Association for Multimedia
   Communciations

MAGAZINES

www.3dartist.com
3D Artist

www.dsphere.net
Design Sphere

www.newmedia.com
NewMedia Magazine

WEB RESOURCES

http://webdesign.about.com/compute/
webdesign
About.com

www.creativepro.com
Creative Pro

www.creativesight.com
Creative Sight

www.graphic-design.com/DTG/
index.html
*Design, Typography & Graphics
   Online* Magazine

www.digitalwebcast.com/Htm/
Features/copyright_webcasting1.htm
Digital Webcast

www.awdsgn.com
The Interactive Design Forum

www.graphic-design.com
The Internet Design and
   Publishing Center

whatever depth may captivate them. It can present to individuals easily over time and distance.

► **Libraries and publishers** have fundamentally changed the way research is done and data is shared. The Internet governs the transference of information from sales presentations to sending documents to developing new communication opportunities.

► **Printing is evolving**—led by the software vendors promoting a smooth, trackable, and interactive work flow— connecting designer to printer and printer to suppliers with new services and business models. (See sidebar, chapter 25, page 290.)

The Web alters the ways that businesses communicate with audiences (revolutionizing business-to-business relationships) and the way resources are allotted and used.

This technology alters training, operations, financial transactions, and many other regular business functions. It has become an integral part of every aspect of publishing, and business life in general.

INTERNET DESIGN ISSUES

The basic rules of good communication still apply online, but new concerns challenge creative teams to extend ideas into flexible branding or identity systems. A Web site:

► **needs to convey information**. The purpose, subject, and navigation must be very clear from the first moment the viewer is confronted. If choices and navigation aren't clear, leaving the site is done with a mere click of the "Back" button. Holding the viewer is the job of the content developers who collaborate with the designers to conceptualize a site that will be compelling and take advantage of the medium to get to know constituents. If it is not interactive, it does not utilize the medium.

► **is never done**—but always ongoing, so the design must accommodate expansion. It must be flexible enough to establish a visual language that can be built upon consistently. Unlike print, a Web site is never static and always requires design attention to be most dynamic and effective.

► **is developed by both a continuous and an evolving team**. The Webmaster brings a vision of consistency to the site but encourages collaboration and new approaches, responding to the changing needs and perceptions of the marketplace.

- **needs to have a Webmaster who keeps ahead of the site**—controls its evolution and always is looking forward to the next incarnation. With a fluid media, there are also daily issues with site management and promotion that require a structure for decision-makers.

- **measure speed in weeks versus years**—there's an in-bred fast pace. Developing projects quickly and updating sites in a timely manner presents intricate scheduling challenges—often the greatest challenge of the development team.

- **uses movement and audio** as well as visual elements—demanding new skills and areas of expertise. The creative group profile will determine the extent of blended media.

DORSEY'S RULES OF NEW MEDIA

"Ten ground rules for working through interactive projects:

1. SEARCH FOR ONENESS. We gather all the content, and put it into one interface system. The user can stay in one place and see everything there is to see. Create a structure that'll give everyone in the organization a place to put every kind of information they might need to present. Make that structure available, and the content will develop naturally, over time. Think of the interface as a suitcase: you don't know how many hats, shoes, pants, and socks you'll need, but you can make a suitcase that accommodates all of them.

2. REUSE TECHNOLOGY. You don't have to start from scratch on every project. Though the message may change, the tools that present the message can be the same. This begins to give the structures you design value as products in themselves.

3. FONTS ARE YOUR FRIENDS. Pick a couple of favorites, and get to know them intimately—each typeface has its own set of rules. Use the font that becomes a unit with everything around it, fits the shape and space available, and lets the message come through.

4. DESIGN IT FOR A KID. The fact is we're all big kids, so keep it simple.

User testing is baloney. You can't account for all the contexts and ways of reading and ways of thinking that all users will bring. If you try to be that comprehensive, you'll just get confused. Design it the way you would love it, the way you would use it, as opposed to wondering what they want. When you replace 'they' with 'I,' your chances of giving users something they think is cool are ten times better.

5. RATION THE INFORMATION. Use small, concise chunks. If you can't make your point in a paragraph, you probably don't have a point to make. Set those mini-paragraphs in large type so that people can scan them easily. Sometimes good interface design isn't as much about design as it is about knowing what to put in and what to leave out.

6. USE THE VISUAL LANGUAGE THAT WORKS. Find a set of conventions, devices, windows, and so on that best serves everything in the site, and don't deviate from it. That way, no matter what section you're in, you're using the same exact interface.

7. SAY IT WITH WORDS. Always lay out an interface in words first by creating a typographic system. Sometimes you don't need icons

at all. Simple words can communicate better.

8. STAY AWAY FROM REALITY. Design around the content you're serving. Don't recreate something that people have seen before, because they'll compare it to the real world experience, and it won't measure up. Instead, create something that no one's ever seen before, and make it easy to use. Content reveals visual structure. Break the information down to its most basic data types.

9. SKETCH IT OUT. Do all the initial conceptual work on paper, because the thinking should be quick. By making quick drawings, you can run through a lot of iterations of your idea, and test your thinking. If you present your first conceptual ideas to the client as sketches, they can imagine what the design is going to be like instead of being distracted by issues of color and typeface.

10. TRUST YOURSELF. Do a lot of sketches and use the best of your knowledge. Then put the whole process into a concept book. Show the client the one idea that works. When the client asks 'Did you try...' you can say 'Here are the studies we went through to figure this solution out'."

—Bryan Dorsey, Pasadena, CA
Mateo Neri, Agoura Hills, CA
Original Context
www.quickface.com

► **requires careful procedure and file management** systems, with uniquenesses to each workgroup and each project. Setting guidelines initially for image, file, and document conventions will ensure an efficient process. Every worker can access current or archived files with implemented guidelines.

► **uses flexible design and templates**—to adapt to various platforms and browsers, tested on different systems that mirror the intended audience.

► **relies on interactive navigation** to inspire participation and to build the community—utilizing forums, surveys, competitions, chat rooms, and other online participation techniques. The purpose of any Web site is to not only convey information, but to get information back. The more exchange, the more possibility to attract membership, sales, or education.

MAJOR DECISIONS: WEB

set project foundation
plan rough project skeleton
assemble category outline
solicit proposals
choose creative group
finalize the project plan
select the theme and plan
approve the design scheme

CLIENT GROUP

Clients maintain a similar supervisory role as in the print process, but lessons can also be learned from the experience of video, animation firms, or TV production facilities. In keeping current, be sure to:

► **keep a list of sites that effectively communicate**—these examples will be useful when working with colleagues and the creative group to help define what might be appropriate or possible for a new approach. The collection should be continually edited and added to as a resource.

► **not simply translate print graphics onto the Web**—it makes no use of the medium and will have no ability to entice viewers to return if they are merely presented with information. Choices need to be compelling and the reader must be enticed to participate.

► **emphasize the interactive nature of the medium**—what would the visitor most like to know? What information would be beneficial to collect from them? View it as an exchange—more like the telephone than the television.

► **encourage change and propel momentum**—everyone on the team needs to contribute new ideas and ways to accomplish site goals. Don't ever consider any site "finished."

► **keep communication flowing** throughout the development process to keep everyone focused on the same goals, even though how to reach those goals constantly will evolve.

- ► **set up clear lines of responsibility, monitoring, and communication**—these should be outlined as part of the project plan. (See chapter 8, page 70.)

- ► **establish design guidelines** if content is contributed from many sources. (See also chapter 9, page 89.)

- ► **support the Webmaster** (whether inside the client group or the creative group)—handle day-to-day issues with development, updating, maintenance, and general site management. The Webmaster also will

  - ▷ *coordinate and promote the site through browsers*, search engines, and industry portals. Portals grow in importance to organize the huge amount of information available. Making sure that the site is included in all the industry's major search and membership engines is an ongoing responsibility. Schedule regular updates to check on browser and portal status, being sure that registrations are kept current and accurate.

  - ▷ *coordinate team activities* and keep budgets on track, making sure the site continues to evolve according to plan.

  - ▷ *report progress to decision-makers*—making sure the work meets all parameters and expectations.

  - ▷ *mediate any conflicts* or misunderstandings between participants, and promote experiments that can be cost-effectively tried on the Web.

  - ▷ *encourage and collect new ideas*—always with an eye for ways to improve the site—both through reach and functionality.

CREATIVE GROUP

Giving the graphic designer more tools is like taking a child from the candy store to Disney World. The possibilities are only limited by imagination, ability, and time. For the strategic designer, who is able to synthesize a lot of information and who embraces the technical, the Internet is one of the most exciting inventions since Gutenberg's printing press over 500 years ago. It is a force to bridge people together in new ways. Designers can be in the driver's seat of these changes if they understand the strategic potential and use of the medium. If they are willing to work with clients on redefining ways to communicate with constituents, blend media solutions, and be a part of defining a new techno-communication landscape, the field of design offers a huge growth area.

"Although American TV shows, movies, and advertising have carried this country's design sensibilities around the world, you still must understand the trends and culture of the countries in which you intend to work. Designers who can introduce new concepts while incorporating the social, political, and cultural habits of local populations are well-positioned for success.

"People living outside North America don't think like Americans or Canadians. Color is one example: colors have a variety of meanings. In Brazil, for instance, purple represents death, while in Hong Kong, white is for funerals. In Mexico, yellow flowers signify death; in France they suggest infidelity.

"The world is an open market for talented designers. Understanding the cultures, markets, logistics, and legalities of the nations in which you expand your services are the keys that unlock the door to your global success."

—Michael H. Sedge
President
The Strawberry Media Agency
Naples, Italy

DESIGN CONCERNS

Here are some earned-in-the-trenches pieces of advice when designing for the Internet:

- ▶ **content is more important than anything else**—viewers will forgive much in the way of visual finesse if they are anxious for information. However, it is effective design that will make them anxious to learn more, revisit, and become a respondant to the site. If content is not interesting or useful, it should not be cluttering up the bandwidths and diluting the message.

- ▶ **don't make visuals too large to load quickly**—the Internet first has to be scaled for information (unless it is an artistic site). The nature of the site dictates the design, and possibly the extent visitors will wait for images to complete. Fast graphics promote happy users. Do provide textual alternatives for graphics that do take time to lead so viewers have something to look at while they wait.

- ▶ **be an advocate for the user** during the development process. Ease of use is the most important design parameter. As the designer focuses on the end-user, decisions will follow the right priorities.

- ▶ **think of structure**—part of the designer's job is to simplify content into segments so that the site's organization is clearly comprehensible.

TEN QUICK TIPS FOR BETTER SITE DESIGN

"1. KEEP YOUR GOAL IN MIND. It may be to distribute information, to sell or advertise a product.

2. DESIGN SITES, NOT PAGES. Worry about the site's overall design. Designing on the site level also means that your work becomes far easier for others to extend. Most of a site's life is maintenance, rather than design. Create a site with a well defined and codified page structure.

3. NEVER DESIGN IN HTML. By drawing your prototype, or using a mature design program, you ensure that the only obstacle to designing a great page is imagination.

4. WRITE DOWN YOUR DESIGN, including typeface attributes, graphical effects, and anything else you might forget.

5. GO FOR CONSISTENCY. Create a site where a user, if they have ever been on any other part of the site, instantly knows where they are on the site when any other section loads. The consistency of design of a site is one of the factors that differentiates amateur sites from professional ones.

6. TEMPLATES. Use templates for your pages with the basic tags, including common graphic elements such as navigation bars and site logos, already in place.

7. SEE IT LIKE A USER. If your site isn't easy to view, you've failed to achieve your goal, regardless of how cool your design is. Users appreciate a site that loads fast and gets them where they want to go in as few clicks as possible.

8. ASK YOUR USERS WHAT THEY SEE. Proactively research how viewers can be thrown by things such as confusing hierarchies, OR unclear icons.

9. CONTENT IS KING. Only with *both* good design *and* good content can you have a truly great site. Don't ever sacrifice content on the altar of good design.

10. DON'T LINK FOR THE HECK OF IT. Hypermedia can be extremely confusing for users. Putting in many many links doesn't help users find what they need. It's important to link in a consistent, well thought out manner that users can learn to navigate."

—Tom Karlo
www.pixelfoundry.com

▷ *Break down* into discernible and logical categories with a hierarchy of content importance. Prioritize content.

▷ *Understand emphasis and pacing* of content and use redundant elements strategically.

▷ *Lead the viewer through* with visual techniques and a consistent placement of elements and navigation scheme.

▷ *Use rollovers*—lead the viewer to what you emphasize and then reinforce action of the viewer. Most viewers are anxious to know instantly what a site is about and how to find the information they seek. To orient, having the site respond to their actions helps them to become comfortable and willing to explore further.

▷ *Be complete*—include contact information that is easy to use and works without glitches. It is surprising how many Web sites lack essential contact and location information.

▷ *Streamline*—a site should be agile and flexible with nothing extra. All element should work easily and be directed by logic with a consistent design that makes it hold together as one entity.

▶ **utilize the Web site for:**

▷ *experimenting with new ideas*—learn and possibly apply results to future designs.

▷ *portfolio showings*—frequently updated with most recent samples.

▷ *concept presentations to committees* or decision-makers that are widespread geographically. (See pages 27 and 64 for tips.)

▷ *sales presentation* to prospective clients—geared toward specific sectors and as interactive as possible.

▷ *representing design capabilities*—through this example of structure and design.

▷ *communicating with clients and suppliers*, facilitating construction process, and using in work flow processes. (See chapter 18, page 226.)

DESIGN PRIORITIES

Software is making design for the Web easier. The less code designers need to know the happier they are. Still, here are some things that a designer should consider specifically:

Some sites don't understand the strength of visual recognition. With updates, the whole look and operation of the site changes. Extensive changes should only be appropriate every few years, as an identity takes a while to build. Hard to measure, an identity can be abandoned prematurely and thwart marketing potential. If a site changes too often, it indicates faulty market responsiveness.

MEET YOUR CLIENTS IN CYBERSPACE

"Experts in design, education, architecture, and business are beginning to use the Web for virtual storyboarding, text-based chat conferences, electronic markerboards and slide presentations.

"Collaboration on the Internet falls into two categories:

> store-and-forward principal that's similar to e-mail, allowing people to leave information on a site for others to read.

> live conference, which can take the form of a simple chat session or involve 3D spaces and different 'avatars' that represent each speaker."

— Mary Jo Fahey
Writer
New York, NY

NONLINEAR ANIMATION

"To keep your viewer's attention, you need to learn the secret of 'being engaging' rather than just 'being natural.' Use *nonlinear* motion patterns. To avoid monotony in motion:

> accelerate or decelerate.

> use curvilinear motion paths. When moving an object along a linear path, rotating it during the shift will help enliven it.

> add a third dimension.

> color and texture effects.

> any transformation such as shape morphing or gradual unfolding, growth, or inflection.

> make linear motions short; avoid frustrating the viewer by pushing an object too far in a too monotonous manner. For short bits of motion, it is much more difficult for the eye to notice their linearity.

> use a number of short linear movements of different objects instead of a single object's complex motion path."

— Dmitry Kirsanov
www.kirsanov.com
Dartmouth, Nova Scotia

► **Learn to read HTML code** even though there are more and more ways to avoid it. By learning code, any design that doesn't function properly can be remedied easily, even if the solution is a cut-and-paste. Understanding the mechanics is essential to design a sleek, fast-moving, compelling site that fits all parameters.

► **Emphasize navigation design**—this key component ties a site together and a visitor needs to understand and move around freely, yet still absorb essential messages.

▷ *Include major navigation elements on every page* so viewers can get from one page to any other. The more comfortable a viewer is with the site's subject and navigation, the more likely they are to revisit.

▷ *Use three-second/three-click rule*—make sure information is obvious to the visitor in three seconds and don't allow any page to be more than three clicks away from any other.

► **Design for choices**—make sure that key messages are prevalent throughout the content and utilize repetition advantageously. All important elements should be easy to find and follow a hierarchy of relevance.

THE ART OF ANIMATION

"Evolution has trained our eye to constantly watch for moving objects that can signify a danger or a change in the environment. This makes animation highly efficient in guiding the viewer through the key points of your composition. A moving object makes us feel that the entire composition is about to go from one state to another. This instability adds dynamic flavor to a design.

> REALISTIC DYNAMISM—use a *naturally dynamic* object with an expressive motion theme, such as a car, a running man, or a flying bird.

> ABSTRACT DYNAMISM—all forms are inherently asymmetric, unstable, and therefore dynamic:

• *Visual simplicity* is an essential prerequisite for any object's dynamic function.

• *Diagonality*, rotation, non-horizontality and non-verticality all have a strong motion implication.

• *Slant*, or skew, is a rotation with some parts of the object's outline still horizontal or vertical, and others, rotated at some angle. The slanted parts add force to it and make it more expressive.

• *Curvature range*—expand the range to add dynamism to the shape.

• *Blur themes*, notably by the 'wind' and 'motion blur' texturizing effects.

• *Asymmetry*—the shape *could* be symmetric but isn't.

• *Proportion*—differ dimensions considerably.

> DYNAMIZING A LINE—a straight line has dynamic power. It drives our eyes from one part of the composition to another.

> CONTRASTING OBJECTS generate streams of perception between them.

> EYE FLOWS—whenever we establish a contrast link between two objects, we encourage the viewer to pay a special attention to this pair of elements —the viewer's eyes will jump from one object to another, inducing a *flow* of vision—

eye flows. The denser and more prominent an object, the stronger its 'gravity force' of attracting viewers' eyes. The habits of reading text (from left to right) and watching objects fall (from top to bottom) make these two directions more natural for the visual perception.

"If an element belongs to more than one contrast link (which is most often the case), eye inertia will turn the path that we travel from a sequence of linear fragments into a smooth curve.

"A 'perceptual map' of the page, with eye flows visualized by lines, shows the popular pathways that viewers are most *likely* to use wandering across the page.

"Scan the page in various directions, try to identify areas where eye flows feel uncomfortable due to running into an obstacle or having to turn at too sharp edges.

"The smooth flow of perception that's achieved by a thorough dynamic analysis is well worth the additional trouble."

—Dmitry Kirsanov
www.kirsanov.com
Dartmouth, Nova Scotia

- ► **Learn about animation and sound**—continually expand capabilities with the software upgrades. Care should be taken not to venture too far ahead of the common viewer's base technology. Be sure to understand the level of systems that are most used by constituents.

- ► **Stay involved with update production**—as part of the overall project plan (see chapter 8, page 70). There needs to be an ongoing relationship with those who maintain the site whether inside or outside the client or creative organizations. Collaboration ensures that the look and feel of the site will maintain its integrity as it flexibly changes with evolving content. The strength of the design will be tested through new situations, opportunities, and content changes. Consistency in navigation and branding message needs to be maintained as they build.

- ► **Design for multiple browsers**—be sure that elements are flexible enough to adapt to various user conditions.

  - ▷ *Design for users who have the lowest common denominator systems*—know the technology of the audience.

  - ▷ *Provide two versions* when using plug-ins (such as Shockwave or Flash) and give the user a choice.

  - ▷ *Test the site* in possible browsers and comply with standards through validation programs (often a part of Web authoring software or available via the Web).

Any graphic designer who has no Web experience will become more and more limited. Granted, there will always be the craft of the traditional media and a demand for print, but branding and integrated marketing is cross-media and will demand those skills of creators.

Interactive design, expanding with the use of animation and sound, has to follow the same design rules as any other form of communication: clarity, impact, emphasis, contrast, and seduction. Design's purpose is to attract the attention of an audience to solve marketing challenges. That means cross-media concerns affect every design idea and decision. (See chapter 8, page 74; chapter 11, page 132; and chapter 19, page 236.) When the Internet is available on everyone's television, the graphic reach for the designer is awesome.

EXPANSIVE CONCEPTUAL THINKING

The designer's education needs to cover a smattering of traditional print media, as well as presentation, interactive, and online ingredients. For any organization, these media need to interlink in a marketing strategy. This is where the designer

COMPLEX DESIGN ISSUES:

> personalization
> customization
> globalization
> accessibility
> data integration

WHEN BAD DESIGN ELEMENTS BECOME THE STANDARD

"No Web site is seen in isolation: users come expecting things to work the same way they are already used to.

"Web design is difficult because the main issues concern information architecture and task flow, neither of which can be standardized. Some aspects of information architecture are starting to reach the level of design convention:

> most corporate sites have a PRODUCTS CATEGORY and an about the company category.

> NAVIGATION TABS are used for rapid switching between alternative views of the same information object.

> LEFT-JUSTIFIED NAVIGATION RAIL—a colored stripe down the left side of the page to contain the main navigation links.

> BREADCRUMB TRAIL across the top of the page situates the current page relative to its parent nodes and to allow users to jump up several levels in a single click."

— Jakob Nielsen
www.useit.com

can be most effective. Internet design requires these approaches:

▶ **personalization**—the viewer is one individual, so the voice is singular and the appeal feels inclusive. Many template designs can house personalized data so that each viewer sees pages configured just for them based on their profiles and responses.

▶ **modular construction** and a tree-formation outline of possible choice scenarios—enables various paths to be anticipated.

▶ **navigation clarity**—entice the visitor from anywhere to anywhere within the site. Hold their interest so they don't instantly leave the site when following a suggested or related link. Care with links is important when building viewership relations. Don't make it too easy to want to leave the site.

▶ **brevity**—remember that viewers don't like to read online. Allow text-heavy information to be easily downloaded for hard copy printouts. Keep all other text segments that are read online very short and easy to read—condense information wherever possible. Don't confuse the viewers by bombarding them with too much information or too many choices. This dilutes the purpose and impact of the site.

CLIENTS NEED TO BE TAUGHT

"Clients need to know what you are building for them, how it works, how you are paid, what work you will be doing, etc. By keeping clients 'in the know' you will be keeping them happy. Speak about concepts they understand and use analogies that they can relate.

"My favorite analogy to use is: Building a House is just like Building a Web site

> INITIAL CONTACT/MEETING. The best thing to do in a first impression is begin educating the client. Ask, 'When you ask someone to build you a house, do you call the person up on the phone and say 'how much will it cost to build my house?' The builder will need more detailed and relevant information.

> ASK QUESTIONS THAT WILL HELP YOU PUT TOGETHER AN ACCURATE ESTIMATE. Also run through the process of building a Web site. Along the way, you have to teach them what you are going to build, why they need it, how it will work and how much it will cost.

"Asking questions will also get the client to really think about the Web site and what they will be doing with it. If you are counting on the client to supply things that you can't build the site without, he/she needs to be aware that they are an important player in the building of the site. If a builder was not given the building permit or if the roofer that was hired did not show up, the home could not be properly built. The estimate should also be signed so that you can begin your work. Having the client sign something insures that he/she reads it and is aware of what you are building.

> PLANNING THE SITE. You can't build a house without a plan. Planning of a Web site should be explained and classified as a separate type of work. Most sites that I have built are 50% planning, and I bill the time as a line item so the client sees the amount of planning necessary. Planning includes: how a viewer will navigate through the site, how many pages will there be, how will the information be broken down and organized, where will the information be found, how the viewer can contact the business, etc. This plan should be signed off on by the client so he/she is certain that you are building what they want.

> BUILDING THE SITE. A foundation to the site is built with the structure. It is the way each page links and the basic layout. Once the foundation of the site is laid out the details can be added

> GOING LIVE AND MAINTENANCE. As people browse the site and use it, it will become obvious that there may be flaws or areas that should be updated. Maintaining a site is critical to its success."

—Adam Strong
Strong Visuals, Springfield, IL
www.StrongVisuals.com and www.lunareclipse.net

▶ **visualize structure of the whole site** as it will be reflected in the clarity, consistency, and navigation. It gains its own momentum and identity through the balance of change and consistency.

BEGIN THE DESIGN PROCESS

With the team assembled and planning phase completed:

▶ **participate in initial launch meeting** (see chapter 8, page 73), having conducted any appropriate research in preparation.

  ▷ *Understand categories*, goals, and client intent.

  ▷ *Examine intended audience* and point of view—get into their shoes, as much as possible.

  ▷ *Have clear parameters* and detailed scope of site.

  ▷ *Make messages cross-media* so they must be flexible in translation, serving several platforms. Themes with a clear strong graphic message do best.

▶ **work with the tech group** by involving them early in the process. They should already be included in the planning phases. (See chapter 8.)

  ▷ *Discuss initial design scheme* and navigation concepts before showing to client to ensure its technical viability and that the team all agrees on the solution.

  ▷ *Collaborate on presentation* to client:

  > designer presents visual scheme and outline.

  > tech professional presents structure of data and systems and other areas of specialty.

  ▷ *Be clear on file formats and technical specifications*—include who is expected to do what.

  ▷ *Set up agreed conventions* on file names and folder construction—be sure client doesn't pay the designer to set up files and then repay the tech to redo them.

  ▷ *Keep communication open*—especially through updates and maintenance. Relationships are ongoing as a site is never really completed. A Web site is always unfinished and is most vital when it is changed through input and involvement from its visitors.

Design can't be static but must be continually reevaluated and questioned. It is a great challenge to maintain a consistent and recognizable identity amidst a demanding visual environment that thrives on variety and novelty. Building a brand and a

WORKING ALONE: (see Working Collaboratively, chapter 10, page 107)

Not dealing with a team means all decisions reside on one person's evaluation. The greater that person's market understanding and experience, the more informed and on target the content, provided the creator is skilled at finding fresh approaches. Being able to both get an overview and change and refine is important when working alone:

> DON'T STOP WITH THE FIRST IDEA. There are always other options, especially evident when looking at the situation from another point of view. To develop a concept into an engaging series that will sell a product or give membership information, content must give the viewer something or provide them with enough curiosity to return. Free offers, educational information, series that change or build, and entertaining animations are all devices that will inspire a viewer to return and, provided timing and marketing are targeted, respond. By not stopping with the first idea for site concepts, new dimensions, variations, and ideas can result. Staying open and moving on allows for the integration of new approaches.

> DEVELOP A GOOD OUTLINE that allows for flexibility in adding new material. Plan in new features or announcements in series to keep the site changing and fresh. Be sure that the home page shows a frequent change so that the viewer knows there is something new upon return.

> PUT MATERIAL AWAY FOR A FEW DAYS AT A TIME. Come back to it with a fresh point of view. This should be scheduled into the development plan. While not looking at working concepts, investigate new opportunities or resources. Then, when returning to develop content, different aspects may be added as inspired by new influences.

> SOLIDIFY TECHNICAL RESOURCES, whether through classes, purchasing books, online training, or getting help from associates, everyone needs to have places to get answers.

> RESEARCH OPPORTUNITIES ONLINE to keep current with market and preferences.

> PLAN ONGOING SITE IMPROVEMENT that responds to viewers' feedback. Make sure the project is developed interactively by seeking response and opinions from viewers.

> KNOW COMPETITORS' SITES AND DEFINE DIFFERENCES. Content creation is slowed down greatly when found to be like another site. It's hard to break into the competition with a 'me-too' idea unless there is a clear price, time, or quality advantage. Know this early in site development and it will save a lot of time later.

> TRY SMALL TEST MARKETS FOR REACTIONS TO IDEAS. Ask for reviews of the site before launching to get a variety of feedback. One of the trickiest aspects of Web design is the navigational system. It must be easy for visitors to find their way around the site. This level of design requires simple, yet memorable, navigational systems that the visitor can learn instantly.

structure that will last, utilize budgets to invest in wise directions, and utilize the various talents of team members to make and keep sites dynamic.

Additionally, the designer must always be prodding the client to try new ideas and techniques online to keep the site vital and fresh. The Internet graveyard is full of sites that were fantastic when introduced, but not sustained. A design that is great to see once but loads slowly, or is impractical to update, will be abandoned as surely as last year's fashions. Design makes the site easy and inviting. But if the subject is not of interest to the audience, the site can be beautiful and work well, but die as surely as if it were ugly and not cared for. It is a combination of relevance and design that will sustain the site through the evolution and updating needed throughout its lifespan.

Work closely with the designer and help develop a collaboration to increase the depth and effectiveness of the client's investment. The tech group carries out the plan, applies content, and produces the functional and visual aspects of the site. The tech group is responsible for developing the technical parts of the design specifications.

APPROPRIATENESS OF SKILLS

Led by the Webmaster, the technical group may be one programmer or a whole team of interactive and database specialists. The nature of the site will determine how much tech-support is needed and the profile of the team.

To have the best integration of content with visual dynamism and ease of use, the tech group needs to work very closely with the creative group. If able to give input during the formulation of the design structure, the visual and the functional aspects are allowed to intertwine. Good communication between the groups is essential to develop a site that will perform to its maximum.

TECH GROUP RESPONSIBILITIES

Setting up a delineation between what aspects are the purview of the creative group and the tech group will interlink the talents and skills of the two groups. It will avoid any redundancies and misuse of resources. Overall, the tech group

► **ensures functionality** in construction and user ease— understands the user needs and capabilities so that solutions to communications challenges can be offered from a technological base.

▷ Makes sure the *best technical solutions* are applied to the information challenges.

▷ *Writes code or scripts* for all segments according to interactivity and user expectations.

▷ *Integrates all data* from all sources and implements them into overall structure. Gives input to creative group on the complexion and requirements of various functional aspects and services.

▷ *Interfaces all segments* together and ensures links— checks for correct formats, accuracy, and structure. The Tech Group also handles refinements.

▷ *Tests to have aspects work smoothly*—checks to be sure there are no broken links or dead ends. Other sites

**ELEMENTS OF
EFFECTIVE WEB DESIGN**

> "SIMPLICITY. Use scrolling text and gif animations conservatively or not at all. Too much repetitive movement is distracting and sometimes annoying.

> CONSISTENCY. Use *one* design throughout the site. It is confusing if pages within a site are not consistent in design.

> IDENTITY. Reflect the purpose of the site in the design. A Web site should reflect the *philosophy and identity* of the business or entity for which it was designed.

> CONTENT. Make sure that as much information as possible is made available.

> USER-FRIENDLY NAVIGATION is one of the most important elements of design—take time to lay out a logical and effective method.

> VISUAL APPEAL. Images must be properly anti-aliased to the background of a site.

> COMPATIBILITY. Make your Web site platform, resolution, and browser compatible. If the site is not compatible to all viewers, consider creating alternate pages for those without the capability to view the enhanced site."

—Jean Kaiser
http://webdesign.about.com

that link may change and need to be checked periodically. Adding material may make some links obsolete or broken.

  ▷ *Implements interactive structure* for responses. Keep responses in top shape and monitor daily. Make sure any responses needed are handled promptly. It is best if most e-mails, orders, and requests are handled within two days.

▶ **organizes and manages site elements**

  ▷ *keeps track of data coming in* from various sources and keeps up with changing data. Most business-to-business sites and work flow management tools depend on timely information accessibility. Be careful to only undertake commitments that will work within the record-keeping habits of the group.

  ▷ *designs site conventions* such as file naming and modes of construction to allow team access to both current and archived information.

▶ **implements design plan** by working with creative group, led by the art director, and works most closely with the designer to

  ▷ *understand identity guidelines* or branding issues that are established to give consistency and a visual language to all pieces.

  ▷ *advise on construction of elements* and file management during development. Being part of the planning teams allows the tech group to give input and to understand the purpose and direction of the site before construction.

  ▷ *follow templates* for incoming combinations of information; design is flexible enough to handle ranges of smallest to largest, text versus visual, access to libraries, etc. Take advantage of cross-media publication technological advancements and strategically use the Web to repurpose content. Scale messages to fit the profile of the viewer asking. The closer that connection, the more successful and targeted the Web effort.

  ▷ *consult when there are format changes* or additional segments that may require the creation of some new elements, and conveys appropriate questions back to the designer. The initial rapport, if set up carefully, comes into the most important camaraderie if under deadline pressure.

- ► **handles maintenance and updates** of evolving content
  - ▷ *keep update methods simple and easy* by working with the creative group on elements that change. If the site is not easy to update it runs the risk of becoming obsolete and unvisited quickly.
  - ▷ *adds or changes posted information* (such as listings, dates, or calendar entries) regularly as part of plan.
  - ▷ coordinates with *new versions* of software and stays current with industry. Most tech groups actually lead the industry, but be careful if spending too much time at the "bleeding edge:" the vendors love the beta testers, but it does take time and money away from more lucrative client work.)
  - ▷ keeps up-to-date on *industry developments* that affect functioning such as new cost- or speed-saving methods or capabilities-expanding new equipment. This requires time on the Web to preview demos, examine reviews, and investigate options. It requires time for keeping up with industry publications, releases, and system upgrades. It may require training (see chapter 9, page 88).
  - ▷ *handles site registrations* and portal listings following marketing strategy, and works on key words and the hierarchy of browsers. Monitors status and uses techniques to raise visibility to match the queries of potential constituents. Building the audience of those who will potentially respond economically is a process that needs diligence, consistency, and a lot of attention to detail. This area is best if within the purview of a key employee to be responsible for upkeep.
  - ▷ *develops archival systems* for past online articles, publications, images, or other data that some viewers may wish to access.
  - ▷ *monitors incoming communications* and site-related e-mails; each site will have a different interactive relationship with its audience that must be managed for fulfillment and response.
- ► **alters project plan** (see chapter 8, page 70) as needed with technology and market changes. The plan should not be a static document, but flexible to alter course yet meet the same goal. If the goal changes, then the entire site should be reevaluated.

Planning maintenance and site updates as part of the design phase will yield the most effective solutions because evolution, rather than revolution, becomes possible.

"The link is the heart and soul of Web interactivity.

"Two items—interactivity and linearity—are concepts we need to work with to achieve our linking goals.

"Begin with a linear or hierarchical structure.

"Forms, interactive games, multimedia events such as virtual reality or live cameras, community bulletin boards, and real-time chat rooms all enable the end user to interact with a Web site, or with other individuals on the Web.

"Not only can Web designers offer these opportunities on their sites, but they can also extend that interaction to other people via community forums, guestbooks, mail programs and chat rooms—with real people, all over the world.

"When a person interacts with information in this way, he or she can, and often does, depart from this linear structure into one that allows for a more free-flowing, nonlinear event. The nonlinear experience allows and encourages user choice. Every person coming to this site can conceivably surf it differently."

— Molly E. Holzschlag
Executive Editor
Tucson, AZ
www.webreview.com
www.molly.com

"As a big, bad Web designer, you face a number of challenges when creating Internet banner advertisements. You have a very small space—usually about 468 pixels wide by 60 pixels high—in which to convey your marketing spiel.

"Guidelines in Action: functional components, browser-safe colors, easy-to-read text, a layout that draws the reader's eyes into the ad, and boundaries formed by graphic elements instead of a static box."

—Meredith Little
Contributing Columnist
Element K
www.creativepro.com

"One of the most important elements of good site design is good, consistent navigation.

"The more freedom you give your viewers, the less anxious they will feel, and the more likely they will be to stick around and give you more pageviews. Make it easy for people to visit you, and they will. Make it difficult, confound them with too many incongruous options, let them bury themselves deep in your hierarchy without hope of rising back to the surface, and you'll probably find them on MarthaStewart.com, learning how to knit baby blankets out of pine needles."

— Andrea Dudrow
Writer
San Francisco, CA
www.creativepro.com

"We are witnessing the birth of a new form of media—perhaps the ultimate form of media. The new medium will get some of its genes from television and some from the Web. But the revolution is not about technology. Rather, it is about connecting with people. The human quality we call 'talent' will be the most precious commodity in this new order—because talent is the source for making that human connection.

"Visual communication is becoming the critical factor in the success of most enterprises. Graphic designers are uniquely qualified to play a vital role in this revolution because of our ability to enhance and facilitate communication.

"The designers who have the most success will be completely at ease with rapid change and complex technology. They won't try to substitute technology for ideas. They won't abandon traditional image making, but, rather, [will] add new techniques to their repertoire. They will be team players—collaborating with others who have special skills and share their spirit of innovation, adventure, and thirst for new frontiers."

—Billy Pittard
Pittard Sullivan
Culver City, CA

"Graphic content should be firmly rooted in communications and marketing strategy. For this reason, I always team up with at least a writer/strategist. We work together to effectively communicate the resulting message to our intended audience. We also work closely with the client to achieve the company's independent goals. Maintenance is either carried out by me (the designer) or by the company since many clients prefer to bring maintenance in-house. When this occurs, I continue to work with my client on a consulting basis. This has the secondary benefit of keeping a consistent look for all the company's marketing communications vehicles—both print and electronic."

—Marty Olver
Diseno International
Portland, OR

"In this day and age, if you can't do it all or at least put together a team than can do it all you will not be able to attract any significant, long-term business. This applies to every area of advertising and marketing."

—Marsha Fay
CommuniGraphics
Vista, CA

"Part of the conflict on the Web is between the way typography is used in print and the typographic needs of links. The typographic hierarchy of print media is pretty well established. Designers who play well with type in print media may feel an instinctive need to downplay links, to subordinate them to text, just as typography per se generally subordinates itself to the text. Writers, too, are usually concerned with the flow of words, of moving readers through ideas to a conclusion, rather than pushing them off in mid-sentence to an entirely different discussion. But the Web wants to link, and links want the reader to know they're there. When faced with this type of conflict over goals, it's best for all parties to sit down at the table and prioritize. Designers and writers coming to the Web from other media need to recognize the Web itself as a partner in the design process. Unless you take its needs into consideration, there will be no peace."

—Eileen Gunn
Writer, Editor, and
Web Producer
Columnist
www.itcfonts.com

"When entering an experimental site I am pummeled with messages about updates that I need to download, platforms my browser won't support, and images that won't load. These projects are self-indulgent. They are luxuries for the entitled class; the complicated layouts and pop-up boxes mean that many home users won't be able to experience them. That's why experimentation in any art form will never disappear. As subcultural agents, even the most underground Web designer is taking part in the discourse. Of course, we value the winners, but let's not forget about the avant garde: those hardy fools sent forth to be mowed down in battle so that others may tread the same ground with impunity. The fringe does not exist to achieve ascendancy, but to leave a mark of its own. Let us not be daunted by that which we don't understand; being confused and disoriented is part of the fun!"

—Joseph P.M. Foster
Production Assistant
Portland, OR
www.creativepro.com

"The introduction of the animated GIF marked the onslaught of the gratuitous moving picture on the Web. When is animation appropriate on the Web, and when is it just, well, cheesy?
Just because all the bells and whistles are out there, and tools for animating your site are readily available, doesn't mean using them will make your site better. Conversely, just because you eschew animation in favor of a simpler, cleaner design, doesn't mean your site will be better either. The important thing is to use the medium most appropriate to your message. For a site about cartoons, animation is great. For a site about the government, maybe you should stick to something else. Something like words. After all, the Web was built on them."

—Andrea Dudrow
Writer
San Francisco, CA
www.creativepro.com

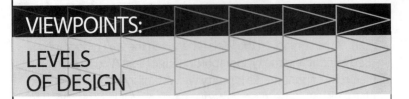
"As software becomes more versatile, designers are able to create visually appealing art more simply. However, we've gotten to the point of needing our technology to be bigger, faster, more complicated. How many Photoshop filters, special fonts, and must-have design programs do we need? Not to mention the fact that anyone with a computer and a word processing program with built-in clip art thinks they're automatically a graphic designer. Basically, technology has brought out the very best—and very worst—in design."

—Elise Korn
Ray V. Anthony & Company
Knoxville, TN

"As a Web designer, you end up becoming a jack-of-all-trades in a lot of ways—illustrator, designer, typographer, video person, editor, writer in some cases, and having people with that wide skill set who can at least think in all those different things if not execute—is really important."

—Auriea Harvey
Designer
Entropy8
Gent, Belgium

"It is important that designers not be hired just to make buttons for a Web site, but be hired to create a site. The education of graphic design focuses heavily on the processes of design, its development and documentation. These models of thinking about visual design are the same models used in information design. Hierarchy must be established to facilitate the communication of a message. Data needs to be categorized and significant relationships discovered among the information. A framework should be created upon which all of this may be draped, to create a whole.

"Design should be our tool to not just create the face of the Web, but the body as well. Technology may seem daunting, but as designers come to understand their larger role within Web development, they will come to understand the function of the technology, and subsequently how to build purpose beyond the surface."

—Don McCants
Media One Web Developer
Integration Technology
Richmond, VA

"As designers, we're trained to design things rather than what I'd call places or experiences. We tend to design a single thing as opposed to a range of experiences that can happen with a number of people.

"The pieces that create experiences, that give some control over to the user, that allow the content to be manipulated or changed or interacted with in some way, that go beyond being superficial, are the most successful ones."

— Chris Edwards
Vice President of Design
Art Technology Group
Boston, MA

"Quite a few of us were really excited about the content abilities of the Internet and what we could do with bringing compelling content and content experiences down the pipe. But, what I've found is most people don't want to spend the time on the Internet to explore content that way; they want to get a job done and go out and play.

"So, what we can do to make their experience really fast and easy and simple—that's kind of my perspective. But then there's the other direction, the broadband technologies or hybrids, something that works with the Internet and CD-ROM that will make exploring content and more compelling experiences more viable down the road. But, for now, it really looks like it's e-commerce, shopping, communication, those types of things."

— Bill Flora
Creative Director
Learning Business Unit
Microsoft
Redmond, WA

"I think designers in general are control freaks in a lot of ways. Having to work on these multidisciplinary teams, which can be quite large—on a big project you can have up to 30 people—it really changes the way we have to approach design. The thing that's interesting is seeing different skill sets emerge: people who are specifically good at motion graphics and visual design vs. people who are really good at user interface design vs. people who are really good at what I call conceptual design or being able to invent new forms for this medium."

— Chris Edwards
Graduate Design Instructor
Yale University
IIT Institute of Design

"The good [Web site] projects have a pleasing visual style or use sound, motion, and video to excite the user. Whereas the great projects really made it more of an emotional issue where the users were really being able to live the content."

— Stephen Simula
Associate Vice President
Fitch, Inc.
Columbus, OH

VIEWPOINTS:

DESIGN AS DIFFERENTIATING

"A Web site is a reflection not only of the professional expertise of the individual or individuals actually developing the site but also of the professional knowledge of the people making up the institution. On the more immediate level of the designer this entails an eye for visual coherence; it entails the ability to arrange the various pieces of a page or site, be they text or image, in a way that successfully expresses a purpose; it entails the ability of the designer to translate this purpose successfully, given the tools at his or her disposal."

—Leo Robert Klein
Library Web Coordinator
Newman Library
Baruch College
City University of New York

"Companies have to understand that Web sites are not just public relations stunts or electronic billboards. The number-one mistake being made on the Web is not taking the medium seriously. The second mistake is trying to design sites that are all things to all people. The best Web sites are an inch wide and a mile deep—if you don't focus, you're just part of the circus."

—David Siegel
Chairman
Studio Verso
San Francisco, CA
www.creativepro.com

"Now we are all designers in a sense. That is what the Web's success has made of us. We're likely to have an opinion either as developers or users on good design, bad design, no design. It's in our blood. We are all connoisseurs.

"People love simplicity. Simplicity sells. When we go to a Web guru we don't want to hear that the world is complicated and full of contradictions, we want to hear what it takes to make our Web site a success, and if it takes ten steps instead of ten thousand, so much the better. Thus we naturally gravitate to those with a simpler message, if for no better reason than we are beginners and looking for a place to start. We also naturally gravitate to those whose message appeals most to our way of thinking, to our own tastes and predilections. The Web has a wonderful way of being all things to all people."

—Leo Robert Klein
Library Web Coordinator
Newman Library
Baruch College
City University of New York

"When you plan your Web project allow time and budget for the design of an overall information architecture, the creation of appealing graphic design, engaging writing, and, of course, the expertise to handle the technical issues."

—Jon Leland
President and Creative
Director
ComBridges
San Anselmo, CA
www.combridges.com

"You have this opportunity as a designer to make a thing that is connecting potentially millions and millions of people together into some sort of a shared or personalized experience. That's a very different way of thinking than most designers are used to: CD-ROMs, and any other sort of disk-based media, don't have any way of being part of the network, of being able to create social experiences. And that's something that's very, very interesting about the Internet: it's the extension of the telephone, right? It's what we like to do, we like to talk to people, we like to understand where other people are coming from."

—Chris Edwards
Graduate Design Instructor
Yale University
IIT Institute of Design

"Clients want something that will get your attention and if it's measured by the amount of hits made by visitors, then that would be the measure of success."

—Bryan Smith
Mill Stone Graphics
Woodstock, MD

"The lines have blurred as we all enter cyberspace. Projects no longer begin and end. They are all part of a continuum called the Web. Web pages continually grow, designs expand, and everything that was once tactile is now just another image in an expanding universe of images. The Web is nothing more than a new medium. It does not reach everyone. At the end of the day, it's just a printer who neglects to use paper. In the long run, substance and intelligence will outrun the fluff and dazzle we are seeing now."

—Paul J. O'Malley
Schmid Advertising
Bloomsburg, PA

"The designer's main objective is to communicate to the browser at a glance all relevant and/or interesting information available on the site in the concise and functional manner. With the advent of more and more powerful Web authoring tools, the design team is playing an ever increasing role in the overall interactivity of the site as well as the content."

—Michael Koehn
Envision Design Group
Modesto, CA

"There are many more integrated projects. Designers need to think in terms of all media, even when working on a project for one. They need to think about how it will translate to other media, and they need to think about this at the development stage."

—Debbie Gersahberg
Brinkman Instruments
Westbury, NY

WRITING 12

Conceptual direction and writing may either begin the project or follow the design concept, enabling an integration of verbal and visual messages. Make sure that the writing and design mesh, which is best accomplished as a collaborative process between the client and the creative group. The writer may either be part of the client group or the creative group, but is usually in the latter.

STAGES OF WRITING

Most projects follow a fairly linear content development. The designer collaborates with the writer at any point within this process. Sometimes the visual and literal aspects of a project are begun at the same time and have tandem development steps, working back and forth so the two aspects of the project integrate. Every professional writer works differently. Writers approach their subjects, break down research, handle the writing and editing, embellish content, and mesh with the other team members. Most writers

► **originate concept development**—done in collaboration with the designer, under supervision of the art director

  ▷ *explore themes*—an idea can start with a phrase or a visual depiction. Ideas can come from anywhere, and the more talented are able to mine them and fresh approaches most readily.

  ▷ *pace and story*—having a narrative often gives a concept an instant momentum. A story can be told in one image or in a whole series.

  ▷ *visual content*—any design concept needs to be made visual as soon as possible. A sketch is a protected idea because it is tangible and can then be developed.

  ▷ *cross-media possibilities*—pursue ideas that can expand to encompass several media or be scaled to a specific need. This flexibility should be part of the initial conceptual process. Parameters scale the concept, so explore how an idea can be recognized and test it in several genres to make the most of graphic investment.

► **research source material**—this phase can sometimes take longer than originating an idea, depending on the scope of the assignment. Research can include getting facts from several kinds of sources:

  ▷ *read previous publications* and appropriate industry

ORGANIZATIONS

www.copydesk.org
American Copy Editors Society

www.asja.org/index9.php
The American Society of Journalists and Authors (ASJA)

http://awpwriter.org
The Associated Writing Programs (AWP)

www.publishersweekly.com/aar
Association of Authors' Representatives, Inc. (AAR)

www.authorsguild.org
The Authors Guild

www.webcom.com/registry
The Authors Registry

www.the-efa.org
Editorial Freelancers Association

www.hwg.org
HTML Writers Guild

www.diac.com/~capune/html/welcome_page.html
National Writers Association

www.nwu.org
The National Writers Union

www.pen.org
PEN American Center

http://stc.org
Society for Technical Communication

www.spawn.org
The Small Publishers, Artists and Writers Network

www.sfep.demon.co.uk
Society of Freelance Editors and Proofreaders

www.wga.org
Writers Guild of America

overviews, including material from competition. Be careful not to be too swayed by what others are doing. A fresh approach can only come from a willingness to be daring or to experiment.

▷ *investigate sources* for information such as libraries or online Web sites—understand ownerships when using quotations or excerpts.

▷ *visit locations* that pertain to the audience and their interests or constituents. Develop profile.

▷ *interview* appropriate participants or experts to examine both sides of issues, or to resolve contradictions.

► **develop an outline** that contains several key components:

▷ *suggested segments and categories* of material represented.

▷ *pace and flow of content* and delivery mechanism.

▷ *visual approaches*—even if the idea is as simple as one bold word or a short description.

▷ *how concept relates* to other visual material—sometimes as part of a series or a progression.

▷ *addressing each parameter* or specification as defined by the client and the project purpose and direction.

► **write a first draft of the manuscript** for the client to edit.

▷ *Highlight any missing information* or places that need more research or additional information.

▷ *Encourage the greatest amount of changes* needed at this point, for it is the best time to explore unusual approaches. It is the time to be expansive and not to always follow the most conservative or safest solution.

▷ *Relate to samples of visuals or sketches* of proposed visuals from the designer. (See design development processes to parallel, chapter 10, page 100.)

► **revise the manuscript**—incorporate client's or editor's edits, depending on scope. Receive client feedback and incorporate it to determine when the project can move to the next phase.

Finalizing copy can be enormously difficult for many clients. Each time the words change form, they look different. Changes need to be expected on the first drafts of pages (whether for print or Web) because seeing copy in a new form is a chance for those already familiar with it to see it fresh.

► **prepare the final manuscript**—the writing process generally ends with text in a final approved form. The final manuscript incorporates all appropriate proofreading. The more final this is before beginning production, the more efficient and cost-effective the production will be. (See chapter 16, page 196.) Even with the fluidity of digital processes, having strong control of the verbal content and being able to edit and adjust as copy develops will help efficiency from this point on.

Because the client initiates communication projects, the client often initiates the writing and always serves as a resource for its development. The creative group may also commission writing as part of visually-oriented projects, or projects that must fit within the corporate identity or branding.

Writing always follows project parameters already established in the strategic development. (See chapter 3, page 22.)

SET FOUNDATION FOR INSPIRING PROCESS

It is the job of the client and the art director to match the writer's creative skills with the assignment. The collaboration:

- ► **commissions the writing** to start the creative phase. There are several ways that the writer is given an assignment:

  - ▷ *the client contracts directly* with the writer or else the writer is on the client's staff. Most organizations of any size have a marketer and writer are key to developing publications and promotions.

  - ▷ *the creative group contracts or provides* the writing in-house sources, covered by the project plan.

    - \> Most small design firms may not have a staff copy-writer, but have relationships with freelancers.

    - \> Larger creative firms may have several writers on staff or a whole department of writers.

  - ▷ *the client handles the writing*—often when content is very technical. Then copy is provided to the creative group which is part of the client's project preparation.

- ► **initiates the writing process** with preparation and planning:

  - ▷ *assembles resources* such as statistics, crucial facts, photographs, studies, and any other relevant information. This background material can be in verbal, handwritten, typewritten, digital, or printed form.

  - ▷ *assembles supporting background* which may include:
    - \> company information
    - \> marketing plan
    - \> mission statement
    - \> articles on the industry
    - \> competitor's literature
    - \> market research data
    - \> slides and videos
    - \> sales presentation materials
      - • samples of products or services
      - • tech specification sheets
      - • biographies
      - • testimonials from customers or member.

The decisions list in the sidebar:

- set project foundation
- plan rough project skeleton
- assemble category outline
- solicit proposals
- choose creative group
- finalize project plan
- select design theme
- approve design
- **review writing outline**
- **edit first draft content**
- **review revised content**
- **review final content**

One of the greatest potentials for extra costs that clients incur happens when they prematurely turn a manuscript over to the creative group before needed approvals. This is tempting to do in the interest of getting a "head start." But time is generally lost later through alterations. It is worth taking time at the beginning to have all decision-makers "sign-off" on the content and the manuscript.

If they understand that changes will add significantly to the project costs, decision-makers will be careful to approve the writing at an early point. This also ensures that the decision-makers are part of the process and will better anticipate the next phases.

It's easy to think that typographical alterations are easy and cheap with digital publishing, but it's the *ramifications* of changes that can affect the project in major ways—from placement of design elements to the number of pages needed.

► **assign the project director** to

　▷ *facilitate development* by first providing research materials and getting the project elements together.

　▷ *set up any needed interviews* with organization employees, members, and customers, where appropriate.

　▷ *manage phases,* address all deadlines, and facilitate approval processes.

DEVELOP COLLABORATION

No matter who generates the writing, the writer and the editor need to work closely with the client, art director, and designer from the beginning to ensure the integration of writing with the design. Often visuals are being developed or found at the same time the writing progresses. Collaboration is fed by good communication and the sharing of ideas. The client establishes the character of project development:

► **sets up environment** that inspires the creative team and facilitates

　▷ *openness to new suggestions*—be willing to experiment—the beginning of the project is the time to try various ideas, exploring the level of difference, chance, and radicalness.

　▷ *decisiveness about audience needs* and to gauge concept decisions based upon this interaction.

　▷ *good communication* between the art director, writer, and designer for the best integrated concepts. Camaraderie always shows in the finish and on the final invoice.

► **directs the process**—expresses enthusiasm and any initial ideas or approaches to the creative team. It makes the process much richer if the client is also

　▷ *available for questions* as the creative group begins to conceptualize, especially as first draft manuscripts are being developed. A thorough writer is bound to have questions.

　▷ *able to review the drafts,* sometimes progressive versions, and evaluate against expectations and parameters, providing additional insight and suggestions as needed. The clients who coach and encourage, engage in the process, and contribute their own ideas are usually the happiest with the results.

　▷ *requesting more work,* such as research, exploration, or rewriting, if not quite on target. This is the most important and most cost-effective time to make changes, add material, or steer the project in a new direction.

- **understands responsibilities**—completes the writing phase in the most efficient and cost-effective manner:
  - ▷ *obtains all approvals* by decision-makers before the design proceeds. (See chapter 10, page 100.)
  - ▷ *proofreads the final manuscript* (checks the text for errors). Errors are ultimately the client's responsibility. (See chapter 26, page 311.) This works best with several people to check for errors, typos, and clarity.

CREATIVE GROUP

The writer is considered part of the creative group, due to the need for close collaboration, whether

- **in-house with the client**—resources are readily available.
- **inside the creative firm**, such as ad agency, marketing firm, or graphic design organization. Sometimes more than one of these agencies work together.
- **an independent contractor** such as a freelancer or ad hoc assembly of professionals.

THE CREATIVE PROCESS: DEVELOP THE COPY

No matter who commissions the writing, it is best to review with the art director and the designer each of these major stages in the process:

- **Research phase**—the writer begins by gathering appropriate data and information, often provided by the client. Further, the writer
  - ▷ *reads background materials* such as history, overviews, or data about the products and services.
  - ▷ *searches for information* online or through industry sources—finding facts and statistics to embellish the message.
  - ▷ *collects data* and verifies factual content, contacting original sources for confirmation and permissions.
  - ▷ *works with the project director* to set up interviews, coordinate meetings, and keeps the project on schedule.
  - ▷ *interviews appropriate people* such as client management staff, customers, suppliers, and constituents.
  - ▷ *visits location sites* for visual components and for experiencing the atmosphere the writing shall reflect.
- **Outline of content preparation**—proceeds when research is complete. All segments are identified and input is received from the client on hierarchy of information and what content needs specific treatments:

EFFECTIVE COPY PROOFING

Six pairs of eyes should look at each project before it prints or is finished:

> WRITER—who generates the information initially.

> EDITOR—who reviews the content, logic, style, and grammar.

> NOVICE—someone who knows little about the subject who can provide a fresh viewpoint and understanding, who can ask penetrating questions.

> CLERICAL—format, punctuation, and cleanup.

> DESKTOP PUBLISHER—who places the copy into pages.

> PROOFER—who reads for correctness.

> COPYEDITOR—who checks over content and correctness.

Each project has its own order through these various individuals. Skipping any is risky. Most projects go through the last three professionals a varying number of times.

▷ *submit outline to the art director* to verify how it complements the design theme.

▷ *review with the client*, confirm content, and agree on scope.

► **First draft manuscript**—once the client approves the outline, the writer

    ▷ *expands* and fills in all the major segments and sub-sections—adding meat to the bones of the plan.

    ▷ *creates the copy* and works through several drafts and their own corrections, perfecting and polishing the conceptual approach.

    ▷ *reviews rough copy against the design*—the art director may request any needed revisions.

    ▷ *presents to the client for discussion*—fills in missing facts, and secures approvals. The writer may need to do more research or explore additional directions after client evaluation. This is the most important, efficient, and cost-effective time to make any changes.

► **Revised manuscript**—that incorporates all changes from the client and creative group. To complete this phase:

    ▷ *incorporate all appropriate feedback* and any requested additional information.

    ▷ *work with the editor* on polishing and tightening:

        > develop any additional segments.

        > massage content flow and pacing.

        > refine logic of structure and ability to explain message.

        > control length of segments and balance between subjects.

        > correct all typographical errors.

        > verify grammatical correctness and copyediting.

        > check over for stylistic consistency.

    ▷ *allow for several levels of revision*, depending on the exactness and the complexity of the project.

► **Final manuscript**—important to complete before production begins. Make sure that:

    ▷ *all feedback is incorporated.*

    ▷ *general text is approved by the client and the creative group.*

    ▷ *all proofing is accomplished and errors corrected.*

    ▷ *any additional drafts or research may be requested* by the client and require extra costs if beyond the scope of the defined project plan. (See chapter 7, page 62, and chapter 8, page 70.) The client must be apprised of additional costs before work begins on expanded scope.

► **Completion of the writing phase**—design phase begins when the client approves the final manuscript

  ▷ *even if the visual is developing at the same time* due to deadline demands or to ensure the integration of the verbal and visual messages.

  ▷ *if the design progresses too far* before the final text is completed; extra costs may be unwelcome and incurred for unavoidable alterations.

  ▷ *a final proofread version is supplied digitally* to the creative group with matching hard copy (even if as e-mail with accompanying fax).

  ▷ allows the creative group to *transform the manuscript into the page design* and review it for conversion errors. The creative group is not responsible for typographical errors. (See chapter 26, page 310, for who is responsible.)

  ▷ *any information about ownership rights*, which are similar to the design rights. (See chapter 29, page 330.)

The challenge to writing for digital media is both in generating the content and the *form* you put the content into. Certain kinds of information like biographies, price listings, and service descriptions need to be modular and expand or condense easily. How it's written depends on context. How writers work has to change when writing content for online and flexible presentations. With content management software that can direct tagged content into appropriate templates, how that content is written (style, context, formality, tense) is critical. How it is written affects how concepts are formed and how others respond.

PRODUCTION GROUP: TECH

While the outline of categories for a Web site is being developed, the writer

► **is likely to have a tech background**—this can help scale the outline to what is best programmable.

► **works closely with the Webmaster** on verbalizing the strategy. (See chapter 1, page 2, for description of role.)

► **acts as a representative from the technical side** when any of the managers on the team do not have tech experience. It is wise to review capabilities during strategic development.

PRODUCTION GROUP: PRINT

The print group is rarely involved with the content development of publications or sites, unless integral to certain materials or printing techniques. To utilize a writer and a creative group for the creation of promotional materials for the print sector, the printer then becomes the client. Most printers do utilize print and Web sites for business-to-business development and marketing relations.

"The PC has made a huge difference in how enjoyable it is to write. The ability to manipulate text the way you would clay in sculpturing is vastly encouraging to a writer. It's one of my great puzzlements that the quality of writing in the world hasn't improved in line with the opportunity computers have given for people to improve."

—Penn & Teller
Comedians
Framington, MA

"We all have an innate sense of what we like and don't like. Often, though, we can't really say what it is we like—what works versus what doesn't work, or what it takes to make what we don't like become what we *do* like. Without visual skills, we don't know how to fix it. So we start over and scrap what may be good ideas. No time savings. If you need something typed, ask a typist. If you need something written, ask a writer. If you need something designed, ask a designer. If you need a newsletter, ask all three."

—Betsy Shepherd
Writer
Chicago, IL

"In many cases, desktop publishing has put design in the hands of writers whether they want it or not, particularly for presentations and in-house documentation. At the same time, the threshold of expectation has been raised.

"Writers need to understand that their words are only one dimension of the message. Since they create the message, they should be highly motivated to learn how to use type, paper, illustrations, photographs, color, etc. to enhance rather than impede their message.

"Although a lot of designers disparage the 'untrained' who are using digital publishing, in the long run the technology will make the visually illiterate more aware. When the 35mm camera was made available to the masses, there was a great outcry by those who believed that flagrant use of the camera would destroy photography as an art."

—C.J. Metschke
Writer and Trainer
Monterey Press
Vero Beach, FL

"Writers should be concerned with visual literacy so they can communicate more efficiently and effectively: not only among themselves, but with designers and others as well. They will also begin to understand the value of visual literacy and the effort it takes to achieve it.

"Digital publishing can't help but increase visual awareness. Whether it produces literacy is debatable. But it does make writers and others aware of what we do as designers.

"Most designers I know are not concerned about writers learning to do layouts and design on the computer. It really isn't any more of a threat than our having word processors makes us writers or poets."

—Hayward Blake
Designer
Hayward Blake & Company
Evanston, IL

"Keeping computers in perspective as a design tool is a challenge for our industry. Computers open infinite possibilities for the designer, but allow bad design to happen more frequently. If our function is to communicate, we must base that communication on sound concepts and apply good design principles to make that communication effective. I see a lot of work that is aesthetically very nice but says very little."

—Randy Messer
Art Directors Association of Iowa
Des Moines, IA

"We are challenged to somehow seamlessly continue the excellent creative work we do, learn a new set of skills to integrate the computer into new areas of design media, and master the expertise in production tasks once reserved only for prepress professionals."

—John Stewart
Orion4Media
Fairfield, CT

"The biggest challenge facing designers now is the optimum use, not the *maximum* use, of computers. Faster production, numerous drafts, and copious changes are not design solutions. We must be careful not to dedicate too much time to the mechanics of a project and short-change the thought process."

—Gail James
Advertising Production
Association of Puget Sound
Seattle, WA

— Arthur C. Clarke
Author and Scientist
Sri Lanka, Ceylon

"Computers do not give more leisure time to authors. By removing the sheer drudgery of writing, they encourage them to work twice as hard and produce five times as much."

"Computer-aided design and production have enabled us to streamline the production process, but this means that increased project management skills are in demand. And keeping up with the technology is costly. We also see an information gap in designers' skills. Experienced designers are not fluent in the computer, and the learning curve is painfully long. On the other hand, computer designers first entering the profession are not skilled in client needs, business savvy, or managing their time and budgets.

"There are opportunities emerging for designers to become strategic partners with their clients, providing sophisticated visual communication skills to guide the mass of in-house digital publishing efforts. 'Good is the worst enemy of great,' and responsible graphic design doesn't stop at good."

— Miranda Moss
Designer
Yamamoto/Moss
Minneapolis, MN

"The best way to find interesting content on the Web is to go through somebody else.

"A Weblog is a page of links, updated frequently. Often the blogger will add commentary, providing some context for his/her hypertextual pointers. It's this context that makes Weblogs so much fun. Most bloggers share personal insights or interesting anecdotes on their Weblogs; even those that don't write commentary in their blogs share a bit of themselves via the links they point to.

"perfect.com.uk is the most minimal of Weblogs. Perfect guides us to a wealth of brilliant Web design each and every day. Anytime I'm short on inspiration, or feeling like the Web is starting to look the same, I go to Perfect, my faith in humanity is restored."

— Dan Blaker
Quality Assurance Engineer
Extensis
Portland, OR
www.creativepro.com

"Desktop publishing does not stop visual illiterates from doing terrible things on paper—it can make it easier for them. What is wrong with computers is *not* the computers, it is the people using them."

— John Rosberg
Rockwell International Corp.
Chicago, IL

COMMISSIONED IMAGES

Original artwork is one of the greatest visual marketing tools for establishing distinction. It can become synonymous with the organization, product, or service that it represents.

Purchasing creative services is best as a collaboration between art director, designer, and the illustrator or photographer. Usually the illustrator or photographer enters the project cycle after the proposal and design phases (see chapters 4 and 10, respectively)—and sometimes is part of the strategic work. Most typically, images are commissioned most often in the development processes.

DEFINE DECISION-MAKING AUTHORITY

The authorization to create visual images, commissioned specifically for a defined use, is usually given to the art director by the client group, sometimes by the client's project director. The art director

► **chooses the illustrator or photographer**

  ▷ *according to the design concept*—often already developed and presented (see chapter 10, page 100). Sometimes a sample of illustration or photographic work is used to demonstrate style and approach. Rarely are visuals created prior to an approved design concept.

  ▷ *by artistic style*—sought through portfolio reviews or through knowing the artist's work.

  ▷ *through reputation*—style and how the artist is to work with.

  ▷ *by budget*—if the artist can meet the financial demands.

  ▷ *by availability*—schedules can be tricky because artists are almost always individual creators, limited in workload by timing and human hours.

► **supervises the development of the work** according to the approved content and design.

  ▷ *A commission awarded* is usually project-specific with unique working relationships.

  ▷ *Gives good instructions*—source materials, and back-

ORGANIZATIONS

www.apanational.com
Advertising Photographers of America

www.asmp.org/index.html
American Society of Media
  Photographers (ASMP)

www.icp.org/home/index.html
International Center of Photography
  (ICP)

www.aipress.com
International Freelance
  Photographers Association

http://thenafp.org
National Association of Freelance
  Photographers

www.pmai.org
Photo Marketing Association
  International

www.psa-photo.org
Photographic Society of America
  (PSA)

www.ppa.com
Professional Photographers of
  America (PP of A)

www.arcat.com/arcatcos/cos37/
arc37761.cfm
Professional Women Photographers

www.rps.org
Royal Photographic Society

www.societyillustrators.org
Society of Illustrators

www.WomenInPhotography.org
Women in Photography International

www.wcpp.org
World Council of Professional
  Photographers (WCPP)

MAGAZINES

www.pdn-pix.com
Photo District News

WEB RESOURCES

www.theispot.org
The ispot Showcase

ground to the artist to increase the depth and appropriateness of the creative solutions and scale to output.

▷ *Judges appropriateness*—the better the artist relates to the subject through experience, the more easily this experience can be translated into a visual expression.

CLIENT GROUP

Usually working with the creative group for resources, clients delegate the activity of working with talent for project components to the managers of the creative team: the project director approving the activities of the art director.

FINDING TALENT

Research the best photographer or illustrator for a project by drawing on several resources:

▶ **Review portfolios**—most professional photographers or illustrators are handled by a representative who manages the business, pricing, and marketing of the artists' work. Most have samples available online.

▶ **Receive illustration and photographic catalogs** from stock agencies that may also represent the talent for assignments. Many of these firms have good online representation and selection of styles, approaches, and price ranges.

▶ **Get recommendations** from colleagues or contacts by asking of experience and results. Research what has worked or not worked for competitors.

▶ **Respond to visuals and images** from daily activities—collect any graphics or treatments that stand out through content, concept, and style. Then contact the organization that produced the piece and ask for the creator's name(s).

▶ **Utilize directories**—through organizations or online, previewing samples and investigating choices

If an artist is unavailable for an assignment, it is illegal to hire another artist to mimic the ideas or style or content of the original artist's work. Rather, find another artist who can use an approach to meet the parameters in a different way.

THE ART OF DEVELOPING ART

As part of the design process, the art director presents the concepts for visual images to the client usually as part of the design presentation. (See chapter 10, Concepts and Design, page 100.) The client should:

▶ **respond to the design and concept presentation, whether it consists of sketches or fully developed prototypes.** Understand the stylistic approaches to be explored and the intent of the designer.

- ▶ **Select a specific design** or concept for the visual direction, usually by reviewing sketches or preliminary prototypes. (See chapter 10, page 106, for various presentation methods and the decision-making process.)

- ▶ **Approve the visual style** and the selection of illustrator or photographer based on portfolio samples. Do not expect to see an exact example of a specific project, as every project is different.

- ▶ **Review the budget** for change from the estimate in scope or vision—using the plan to show early deviations. It is very important to check the design and the image supplier's fees against the budget. The proposal or initial estimate can only approximate costs, for the estimate is made before the design concept exists. (See chapter 4, Estimates and Proposals, page 38.) Most proposals should contain a not-to-exceed figure, but occasionally the best concept may require additional resources, such as more involved photography, illustration, manipulation, etc. The creative group advises the client of this *during* the concept presentation. It is up to the client to decide if they wish to spend the extra money.

- ▶ **Review any usage rights** with content owners and provide ownership information to the art director. (See chapter 29, page 330.)

  - ▷ *Purchase the use of images* only for the purpose specified in the proposal. If other uses will be needed later, negotiate them up front for the best arrangement.

  - ▷ *Purchase unlimited use* for wide-audience distribution. (See chapter 29, page 330.)

  - ▷ *Define agreements* before allowing work to proceed.

- ▶ **Be available for approvals** during the development process, especially for photographic sessions if location or sets are involved. For projects under heavy deadlines, the decision-maker availability is crucial:

  - ▷ *check proofs*—Polaroids® or developed proof sheets are helpful for checking composition and details while changes are still easy to make.

  - ▷ *define clear technical specifications*—look at the photographic set for details, whether location or studio.

  - ▷ *have staff available* as needed, especially if shooting on the client's premises. Assign staff assistance.

For large publications that are commissioning a series of illustration or photography, often more pencils are requested than later needed while content is under development. The budgets should include some money for cancelling of unused images.

ANSEL WOULD APPROVE

"Only with a heightened awareness, brought on by the notions of what digital photography can accomplish, are we beginning to discover what photography was all along: the very act of deception.

"I recently asked Sarah Adams if she thought that her grandfather would have taken to digital photography, to which she responded:

'Yes, we believe Ansel would have been immersed in digital technology.' Several possible reasons:

a. MORE ENVIRONMENTALLY SOUND

b. ARCHIVABILITY / restoration of older negatives

c. GREATER ACCESS to color realm

d. THE NEWNESS OF NEW TOOLS; recently learned that at the 1915 Pan-American exposition in SF at the age of 13 he taught himself quickly the art of typing and taught others at a booth!!?!

e. Access to photographic manipulating tools: dodging and burning, etc."

—Brian Masck
Technology Coordinator
Muskegon Chronicle
Muskegon, MI
www.zonezero.com

> ▷ *review pencils* for illustrations and all visual components before production phases begin.

▶ **Pay cancellation fees** if work is stopped during progress. (See pages 154 and 155.)

CREATIVE GROUP

The photographer, the illustrator, or both join the creative group under the direction of the art director. They may enter the process:

▶ **in the conceptual development phase** at the beginning, especially if a sample of the artist's work is used in the presentation to demonstrate direction, approach, or style.

▶ **after initial strategic designs have been presented** to the client and a direction has been selected. The artist's works are defined by specifications and parameters.

▶ **after production has been completed** by working to tighten layouts and plans—specifications exact to concept.

▶ **by providing stock images** utilizing library archives or stock photo agencies. (See chapter 14, page 172.) Often projects are a combination of commissioned and stock work. Occasionally a picture researcher may be employed for very large projects with specific subject needs.

▶ **as subcontractors or on staff**—either with the design or client organizations. Many photographers and illustrators are independent freelance talent, bringing their focus to

---

PROTECT YOUR IMAGES

"Digital watermarking makes you easy to find and your images difficult to steal

"Avoid watermarking woes. Follow these pointers to make sure your images are watermarking-ready:

1) CHECK THE LICENSING AGREEMENTS for all the parts of your image that are taken from stock sources. You cannot copyright an image if someone else already holds a copyright to it. This can get tricky, so be careful.

2) USE A VARIETY OF COLORS to create the image; the watermark will not embed correctly if the image is made up of only one color.

3) ANNOTATE YOUR IMAGE with all file information, including credits, contact information and captions.

4) Make sure your image has the MINIMUM NUMBER OF PIXELS REQUIRED to be watermarked.

5) LEAN TOWARD IMAGE QUALITY over file size when compressing your images.

"Digital watermarking is the most widespread way to guard against image theft on the Web.

"Unlike encryption, which scrambles text messages or credit card numbers so the recipient Web site cannot decipher them without your permission, digital watermarks could be called securities of conscience. Simply put, digital watermarking is a means by which creators and owners embed digital data into an image to facilitate copyright protection.

"Some people think that because an image is digitally watermarked, no one can steal it. Yet a digital watermark is more of a trust-based solution, a means of identifying the owner of a particular image, and then getting in touch with him or her to license its use."

—Russell Shaw
Journalist
Portland OR
www.creativepro.com

the project for its duration. Illustrators and photographers specialize in style and approaches and should never be asked to mimic or copy the style or approach of another. (See chapter 10, page 100, for issues of originality.)

ROLES OF PARTICIPANTS

The development of visual images can be a fairly linear process once the creative resource is chosen:

► **The art director** commissions the creation of visual resources, and

▷ *shows designs and gives specifications* to the illustrator or photographer prior to beginning work. If there needs to be communication with the client, this is facilitated.

▷ *specifies all financial agreements* and arrangements in advance, including payment terms. Secures understanding for any work-in-progress or cancellation fees.

▷ *receives and reviews all preliminary work*, such as pencil sketches or Polaroids®.

▷ *shows work-in-progress to the client* for approval (variations specific to project needs). If disagreements occur between the art director and the illustrator or photographer, it may be best to explore both the directions from both opinions. But if time and budget do not permit, the art director has the responsibility to represent the client in the dispute and has more knowledge of the marketing considerations. So the art director's judgment must prevail.

▷ *secures client approval* on concepts and budgets before and during the process. Mediates all input and gives one set of instructions back to the illustrator or photographer if changes are needed on the preliminary.

▷ *supervises the direction of finished artwork* or photographs (which often includes location supervision of photography sessions).

▷ *provides approvals if the client is not available*—especially under deadline conditions. The client may agree to delegate approvals to the art director at the start. (If so, the client accepts the financial implications of the art director's decisions as long as it fits within approved parameters of the project.)

EVALUATING PORTFOLIOS

"Beyond technical skills and creative vision, here's a checklist of added factors that clients have in evaluating freelancers they will hire:

> range of media experience

> number of active clients

> turnaround time on typical projects

> other related services available

> project production management skills

> flexibility and the ability to take direction

> meeting deadlines and the ability to change direction

> communication skills with management and vendors

> knowing how to ask the right questions on a project

> reliability and follow-through are very important

> balance of creativity and business sense."

—Maria Piscopo
Creative Services Consultant
Costa Mesa, CA
http://MPiscopo.com

▷ *manages the process* and receives final materials. It is easy to lose track of original images (such as photographs, transparencies, original artwork, historical materials, product samples) during the production processes, especially if under heavy deadlines. Therefore, any transfer of images from one place to another should be documented in writing by the creative group. It is the art director's responsibility to keep track of images unless turned over to the client.

► **The illustrator** works according to the predefined specifications provided by the art director and

▷ *provides rough pencil sketches* of the concept to the art director for approval prior to creating finished artwork. Be clear about intended style and colors.

▷ *creates finished artwork* under direction of the art director, according to the project parameters and expectations.

▷ *guarantees originality of work* and legalities of any sources used—illustrators are responsible to know copyrights and not utilize or "borrow" any other images or proprietary techniques or styles without securing permission.

SELLING YOURSELF AND YOUR PHOTOGRAPHY

"Unless you are one of the very select few, you won't just spend your time taking photographs. There will be lots of other tasks that come your way. You'll be a technician, a secretary, a bookkeeper, messenger, furniture mover, bill collector, and, most importantly, a salesperson.

"So, how do you sell?

"Step One is to relax. Step Two is to find your own style. We each sell well because we do it our own way, consistent with our own style and personality. Don't get the idea that there are no rules, or that anything goes when it comes to selling. There are, in fact, eight rules that will help you streamline your sales success:

RULE 1: *People Buy from People.* Don't give your customers any reason to want to buy from someone other than you.

RULE 2: *Suspects Aren't Prospects.* Not everyone who calls you up or visits your shop is going to buy. The best way to sort out suspects is to ask questions.

RULE 3: *Prospects Must Be Qualified and Listened To.* Determine that the prospects are ready to buy, aware of what you offer, and able to meet your price.

RULE 4: *Price is in the Salesperson's Mind.* If you think the price is a good, fair price, it is. No matter how high, no matter how outrageous.

RULE 5: *Never Offer Anything 'Free.'* Nothing is ever 'free.' Everything has a value.

RULE 6: *Sell the Benefits, Not the Product.* 'Sell the sizzle, not the steak.' If you appreciate what that means and if you can apply it to your photography and your services, then you'll never starve.

RULE 7: *Don't Forget to Ask for the Sale—the Close.* There's very little you can't change once you understand what the customer really wants.

RULE 8: *Know When to Stop Talking.* Don't get so hung up in reciting the benefits of your product that you forget to stop and see if it's time to close. Listen to your prospect's needs, fears, and desires."

—Chuck DeLaney
Dean
New York Institute of Photography
New York, NY

- ▷ *protects work*—utilize technological processes to help guard against work being copied:
    - > encryption devices—check the latest software developments as several are quite reasonable.
    - > scrambling devices—good when transmitting secure information.
    - > watermarks—many stock agencies use them to mark low res images for comping purposes only.
    - > encapsulation—when exporting out of the original program, the formats needed for page layout or Web authoring systems prevent the original from being altered. However, derivative copies can still be made if the digital encapsulated image is obtained.
- ▷ *has a cancellation fee arranged in advance*—usually, if an illustration is cancelled after work is in progress (for reasons other than the illustrator not working within the parameters), the client pays a 50 percent cancellation fee plus expenses. If the illustrator does not meet specified parameters, there is no payment.
- ► **The photographer** works to the predefined specifications provided by the art director, and
- ▷ *often provides Polaroids®* (from the set or site visit) to the art director for approval prior to finishing the photo shoot. Exceptions to this would be live assignments or at locations where spontaneous images are part of the concept. A thorough briefing of priorities for images, needs of design, and directions from the creative team all work into the approach of the images.

▷ *creates finished photographs* to specifications provided by the art director, who usually is on set or location, providing direction. It is advisable that the client be present during the initial part of photographic sessions, in case content decisions need to be made. If the photograph does not meet specified parameters, there is no payment. Work will be redone at no cost to the client until it does fit the project parameters.

▷ *guarantees originality of work* and gains permission for any other images or styles that may be used in helping to develop a set or situation to shoot.

▷ *obtains model releases* for any people used in images.

▷ *handles all equipment, processing, and technical issues* surrounding film, prints, digitizing, and photographic manipulation unless permission is given to another creative professional on the team to alter. No image should be altered without permission of the photographer (see chapter 31, page 348).

▷ *has a cancellation fee arrangement in advance*—generally

> if a photographic session is cancelled by the client more than twenty-four hours before the scheduled time, the client will not be charged, except for large guaranteed bookings involving travel, equipment, or model fees.

> if the client cancels the photographic session less than twenty-four hours before the scheduled time, the client will be charged a 50 percent cancellation fee and incurred expenses up to that point. This is because the photographer may be unable to book any other projects for that time slot. (This is according to general photographer's trade practices. See the individual photographer's estimate form, as these practices may vary.)

PRODUCTION GROUP: PRINT

The print or tech group can give advice on the best production process for a given approach or effect that the design dictates—*if* their input is included at the conceptual stage. If brought in too late in a project, some work may have to be adjusted for technical tolerances. This can cost time and money if not correctly anticipated. The art director facilitates this by showing designs to the print or tech group in advance of creating finished artwork or photographs.

"Twenty years ago, no one except fashion magazines took art directors seriously. Now, the role of design in publications has become crucial. Most magazines put art directors at the top of the masthead. Many publications still make the mistake of assuming that 'content' refers strictly to what's written. Visual content is critical now, and print media's future depends on a merging of the written and visual. We haven't seen many major magazines edited by former art directors, but I think that will change soon.

"Baby-boomers grew up with television, and that influenced publications. But even TV has trouble grabbing our attention now. It's paradoxical, but the future of print actually depends on creating more visual content."

—Roger Black
Designer
Roger Black, Inc.
New York, NY

"The illustration industry is in terrible shape. Commissions are fewer; deadlines are shorter; fees are lower. Creativity is valued less; competition is up from stock, royalty free or photographic images. Large media giants continue to be swallowed by even larger ones while illustrators are told it's illegal to compare their pricing. Intellectual property, the coin of the new millennium, is becoming so valuable that more and more clients insist on work for hire or all rights contracts.

"The prevailing fear is that little can be done to confront these threats. Many artists are finding more easy and fun ways to support their families.

"Perhaps it is the solitary nature of the business or an obstinate resistance to structure, but the 'industry' has not responded effectively."

—Graphic Artists Guild Online
www.gag.org

"The ethical side of photo manipulation can be all but forgotten when it comes to art photographs. Here the ability to alter images by use of new techniques in image processing has given art photographers a flexibility that in the past belonged only to painters and sculptors. Photographers used to be constrained to produce only faithful reproductions of the reality that lay in front of their cameras. This limited the ability of artists to interact with the medium. While a painter could return to the canvas and revise, remove, restore, an art photographer was usually limited to slight modifications in the darkroom."

—Ron Lovell
Writer and Professor
Oregon State University
Bend, OR

"You have to do your work as a photographer and separate yourself from the pack. I can't emphasize strongly enough how much hard work is the answer. And this hard work continues even after you're making a living as a photographer. Because you have to go out and do the visual pushups every day. You have to make time to shoot for yourself."

— Jay Maisel
  Photographer
  New York, NY

"Illustrators are more than picture makers or stylists. They come up with the ideas that bolster my art direction. My design is, therefore, a frame for their illustration. I know that this is not fashionable to say at a time when graphic designers have asserted more creative independence—often combining design and illustration into a single typographic manifestation—but the best illustrators offer more than a design framework. They are storytellers.

"An illustration—like a painting or sculpture and even a short story or novel—can trigger a variety of emotional and intellectual responses that a great piece of typeplay can never achieve. Even an illustration that is keyed to a very specific subject or event can, if the artist pushes all the right metaphoric and symbolic buttons, transcend its timeliness and become an icon of the particular event and a metaphor for other issues.

"What makes illustration a constantly renewable resource for designers is the huge variety of formal and stylistic perspectives prevalent today. Some designers select an illustrator on the basis of how a style fits an overall concept. But I believe that the most effective use of an illustrator occurs when the designer/art director's aesthetic and conceptual preferences are in sync with the artist's. Then all the designer needs to do is give the illustrator the freedom to do what he or she does best. The illustrators whom I use reflect my concerns in some way.

"An illustrator works best when he or she has the freedom to breathe, to interpret. Over-art-directing an illustrator's image or images is often counterproductive. The joy of illustration comes in seeing the artist's rendition for the first time—like opening a present on Christmas morn-

— Steven Heller
  Art Director, Author, Cochair
  School of the Visual Arts
  MFA/Design Program
  New York, NY

"The newcomers see opportunities, while traditionalists see threats. Rather than embracing the change and using it to our benefit, we find fault with the new technology and defend the status quo. We leave giant opportunity for those without the biases of our history."

(c/o *Photo Marketing* magazine, July 2000)

— William J. (Bill) McCurry
  Consultant, CEO
  Jackson, MI
  www.youachieve.com

"I consider my computer and its image-based software to be nothing more than an *additional tool* to assist in my photographic endeavors.

"It is very conceivable that digital photography will eventually surpass conventional photography in numbers of practitioners. In fact, with the emphasis on environmental issues, chemical photography may become something of a pariah."

—Terry Staler
www.onlinephotography.com

"Digital images are very easy to alter. Our readers know that this can happen, so why should they believe what they see? They should be ABLE to believe that what they see is indeed a 'photographic record' of what was actually there, because of the credibility of the SOURCE of the information. The photographer, therefore, has a huge burden of responsibility to maintain the credibility of his images, and the employer (publisher) in turn has a burden or responsibility to the photographer as well as the reader to do the same. Readers should be able to believe our product because of the SOURCE. We need to achieve our own level of excellence and, personally, be leaders to maintain the credibility of our profession. This must be done by each individual. Once the SOURCE cannot be believed, photojournalism is dead."

—Brian Masck
Technology Coordinator
Muskegon Chronicle, MI
www.zonezero.com

"A synonym for ambition for me is belief. Belief in myself, belief in photography, belief in the joining together of those two things to create a kind of a destiny

"Ambition won't get you that far. You'll shift gears. You'll see something that's shinier. But if you believe...then you're the long-distance runner.

"I knew what in my mind were the great photographs and I wanted to live up to them. And that's detached from the words career and ambition. That's belief.

"I wanted the chance to live the life. And not be a businessman, and not be a family man, but really be devoted to living the life. And as proof of the goodness and rightness of that dream, I wanted great work, interesting work that lasted, done in a way that it wouldn't get dated.

"But what's interesting has changed.

"You have to make pictures that people aren't indifferent about. You keep them in mind; it's not just you. But if it's interesting to you, it's probably going to be interesting to them."

—Sam Abell
Photographer
National Geographic
Washington, DC
www.photoprojects.org

"The responsibility for guaranteeing the integrity of the information is with the publication, not with the medium. All pictures, such as with text, are confirmed from several different sources when in doubt; otherwise, it's the photographer's responsibility to deliver an image with integrity towards the events, which in turn will be constantly monitored. We understand that integrity is not a matter of how the picture was made, but what it's supposed to communicate. Just as we don't oversee if the writers do so by hand or type on a computer, our photographers are free to use any tool they want. The veracity of an image is not dependent on how it was produced, any more than a text is credible because no corrections were done to it.

"The nice thing about this approach is that it makes sense. It frees photography from the burden of trying to accomplish something for which it is ill suited."

—Brian Masck
Technology Coordinator
Muskegon Chronicle
Muskegon , MI
www.zonezero.com

"It's impossible to avoid the impenetrable focus on methods and techniques that are thrust at us the moment we seek to educate ourselves in art.

"What makes us individual and unique becomes the basis for what we call our 'style.'

"Style is out there in many forms. Free for the taking. Anyone who understands the nature of the copyright law knows that 'style' cannot be protected, and with this free invitation in hand, the pitifully weak, bereft of creative vision, fall into the trap of plagiarism and borrowing. Plucking style as though it were the rose upon the bush is found to be a simple task for those without the personal vision and creative imagination, as well as maturity, to nurture it and bring it to life.

"For anyone seeking a place in the realm of illustration, the elusive commodity we call style is already there, waiting to spring forth, if only we place all of our efforts toward the most significant and telling opportunity that sits before us...that of becoming a visual creator with a passion for telling stories.

"Style. It's your personality all wrapped up in the little intuitive things you do, and have always done, except when you've done them long enough, these little intuitive things, such as the way you use color, the way you utilize the symbiotic relationship between words and images, or the way you create your characters with depth and originality, you are inseparable from them. They are the very essence of who you are, and always have been. And it is this very singular quality that can make you a first-class teller of stories. Not just the story… *your* story.

"Achieving a style is a lifelong process. Transformation is constant, and, quite honestly, should be."

— David A. Niles
Rhode Island
School of Design
Providence, RI

"The pursuit of technical quality is like a club, and it can be both democratic and elitist in its membership policies.

"Exhibiting knowledge is a generous act, but it is also a way to prove, by using a tangible system of measurement, how good you are. Sometimes the strict allegiance to technical quality is a way to keep the club small and exclusive. The irony is that the people inside the gates are typically the ones who lose. A rejection of a 'low-quality' technique may also be a lost opportunity. If measurement becomes the focus of the quest, it can inhibit one's ability to wander into uncharted territory.

"Stieglitz emphasized skills that were not technical, but instead were grounded in looking. He described patience, studying lines and lighting and finding the moment when everything is in balance and 'satisfies your eye.'

"In order to satisfy the eye, technique has to serve the work.

"Maintaining balance requires diligent attention. Professionals and amateurs alike, in any of the arts, sometimes wander off their intended path and get tangled in the thicket of technique, whether it's the pursuit or the rejection of technical quality. When the work is in balance, technique is neither the hero nor the enemy."

—Wendy Richmond
Harvard University
Cambridge, MA

"The Internet is our best hope for realizing new creative capabilities in a meaningful way. With the Internet we find information, solve problems, get inspiration for new ideas, and share what we have accomplished. We can go directly to a manufacturer for product information that vendors rarely have on hand, or we can buy direct online. For inspiration we can visit sites displaying ideas and current projects which in the past would have taken years to be published. And above all, we can network in forums and chat rooms: we can look for others who confronted a similar puzzle and found a solution. We can ask questions of someone within a like-minded community of people who enjoy sharing their knowledge.

"The two areas, traditional and digital, are living side by side, and they represent the dichotomy between traditions and a future that has already started.

"In other words, the entire darkroom has in essence become the equivalent to a printer. Granted, you also need a computer and some software. But where I can use my computer for many other things, i.e. to write this story, which I can print out on that same printer, I can hardly do anything other than enlarge pictures with my darkroom equipment. I own a wonderful darkroom, which I have not used now in over ten years, and yet I have never made better prints, or been more productive than now, when it became all digital."

—Pedro Meyer
ZoneZero, Inc.
Coyoacán, Mexico
www.zonezero.com

VIEWPOINTS:

ILLUSTRATORS AGAINST STOCK

"Illustrators have to recognize that stock vendors aren't our benevolent friends. Rather, they're our competition. Just ask the illustrators who lost a commissioned job to stock.

"Don't give up the rights or control of your work. To maintain the value of your commissioned work, you must maintain the value of your own secondary art sales. Take pride and better interest in your hard work. Before any client can respect your work, you must treat it with respect yourself.

"What many illustrators are trying to do is maintain our industry so the artwork we sell today affords us the opportunity to create more art tomorrow."

— C. F. Payne
Editorial Illustrator
Cincinnati, OH

"Creatives who sell images as clip or royalty-free art are condemning themselves and their peers to years of toil with only the possibility of limited, short-term gains. But simply refusing to supply images to royalty-free publishers may no longer be enough; many of these companies contract photographers and illustrators on a work-for-hire basis to provide the inventory needed to fill their catalogs. We need to provide buyers with a better choice—one that provides quality, exclusivity, and a fair price, but still acknowledges authorship."

— Paul Basista
Former Executive Director
Graphic Artists Guild
New York, NY

"If stock art has such high value (low risk, immediacy, and assumed quality) and the product itself is 'comparable' to assignment work, why does it demand a much lower fee? Great question; in my opinion, all purveyors of illustrations are to blame. Yes, illustrators, artist's representatives, stock agencies, and (most importantly) royalty-free companies. They have all bought into the philosophy that because art is now digital it should be considered software, and be marketed as such. They have been disillusioned into believing that if it doesn't cost anything (time, money, energy) to produce, reproduce or deliver, it shouldn't cost very much either. This is a huge mistake. We all must realize that art has a much more narrow market opportunity than software and should not be treated similarly. More importantly, illustration should be priced by what it is worth in relative terms, not what it costs us to produce or reproduce. Software is, by design, 'created' for mass markets whereas illustration is personalized with an artist's unique style, idea, composition, and the project's purpose in mind."

— Richard Askew
Founder
Stockart.com
Fort Collins, CO

STOCK IMAGES 14

Stock images (photographic or illustrative) already exist, either as outtakes from previous creative projects, or made specifically for sale as stock. Stock images may be in traditional or digital form, and are available through a stock agency, directly from an artist's representative, or sometimes directly from the artist (where a previous relationship exists).

INTERNET INSPIRES IMAGE PROLIFERATION

Allowing an additional medium to be an image management tool, stock image agencies:

▶ **have collections digitally available** and searchable online, allowing an expanding library of possible choices.

▶ **allow more images to be available to more people**— the selection expands greatly as more images are added. Sorting devices, user interfaces, a good editing mechanism, high quality images, and an easy order system become critical in retaining loyal customers. Speed and selection will continue to grow.

▶ **plummet image prices** due to availability and distribution, making money on a higher volume of customers and usages with wider distribution channels.

▶ **make images instantly available**—images are quickly available, either overnight from an agency or instantly when transacted and downloaded from the Internet.

▶ **inspire designers to use images** as a palette when skilled in photo-manipulating programs. A world of affordable images is immediately accessible for experimentation and low-cost visual solutions.

▶ **limit selections by databases** and search engines, which continue to improve in key words, quality choices, and ease of finding the desired subject.

▶ **can find images overused** and seen in more than one venue or industry (check the usage agreement for possible sector or usage overlaps). Images can lose their desirability if seen in other places. Achieving distinction can be difficult without buying unlimited rights. (See chapter 29, page 336.)

ORGANIZATIONS
(see also chapter 10)

www.promaster.com
Photographic Research Organization Incorporated

www.pacaoffice.org
Picture Agency Council of America (PACA)

Many designers photocopy or scan images from a magazine or stock catalog to include in their comprehensive presentation materials when showing a design concept to a client. The designer should ask the owner for permission. Generally, if used for presentation with intent to obtain an image from the owner if the concept is chosen, the need for permission is lenient. Never should an artist be hired to imitate another's work. Also, permission may be implied by the format of the source material. When in doubt about when to obtain permission for presentation, the designer should always contact the image's owner or agent.

CREATIVE POSSIBILITIES

With the triple threats of fast turnarounds, tight budgets, and demanding clients, using stock can be the best answer to a communication challenge. When budgets need to do more with less, stock has grown into being a viable and affordable option.

However, stock images can't take the place of commissioned ones (see chapter 13, page 158). If a communication challenge needs to be extremely unique, such as in high visibility or wide-audience campaigns, budgets should be allotted for commissioning original work. On large-quantity print projects, original images are a tiny percentage of the total budget and can make the difference between recognizability and obscurity.

DETERMINE IMAGE SOURCES

Always ask where any supplied images come from. There are key issues that determine ownership and permissions for use (not always determined by where an image came from):

▶ **The creator** may own and retain the copyright to the work which limits its use by others (by market, time, sector, quantities, media, or scope). Ownership does not mean the same as possession.

▶ **A stock agency** may own:

▷ *the copyright* if agreed to by the creator, backed up in written form. The creator signs all rights to use the image in any form (see chapter 31, page 346).

▷ *the digital file or physical image*, but not the copyright if the copyright is held by the artist or a foundation.

▷ *the digital file or physical image*, but work is also in the public domain (see next page).

▶ **The owner** may sell the copyright or may grant permission to use the image under specified conditions. Fees will vary by terms of the agreement, which may specify

▷ *intended use* (relative prominence, as on a brochure cover, magazine advertisement, or Internet homepage, etc.).

▷ *intended audience* or business sector.

▷ *quantity* of distribution (amounts printed or which media are used)—identifies scope of audience and how many people will see and benefit from use of image

▷ *time period* for use (single time, series, or unlimited)

▷ *degree of exclusivity*. (See chapter 29, page 330, and chapter 31, page 346, for more on ownerships.)

Kodak took an active role to protect the photographer's intellectual property by embedding an encryption code that locks images at low resolution. Similar to watermarking an image, it can be viewed, but not reproduced.

STOCK IS BEST USED:

> when it exemplifies or is the core feature of the concept

> when altered or combined to give uniqueness to it and distinctiveness for design solutions

> under deadline or budget constraints

> for adding visual enhancements economically and adding visual dimension to subjects

> by the designer when trying out several different ideas easily and econcomically.

USAGE: ASK WHEN IN DOUBT

Do not use images unless the source is known, especially when its use promotes organizations or makes a significant profit. When there is no money at stake, there is rarely cause for legal conflict. Be careful when obtaining images from elsewhere. The need for permission to use an image depends on use.

▶ **Permission is needed** when the image is:

▷ *used as a reference* by another artist in developing a new image. This is true even when the resulting image may not *appear* to be derived from the original.

▷ *included in presentation materials*—such as comprehensives to show the client. Most stock agencies release images in low resolution with an embedded ©. The designer can use these in presentation comps without permission. Rights to the high resolution image must be purchased before used in the final art.

▷ *altered for content*—ask the stock agency when an image is used in a PhotoShop collage or as a detail image amongst a range of other images.

▷ *a launching point to create new images*—either as a variation or re-creation based on the image used for a presentation. The creator of the presented image should be hired to execute the new images. (If the creator is not available or unaffordable, the concept should be changed.)

▷ *scanned into a computer*—usage fees are cheaper than legal fees.

▷ *downloaded from the Internet*—these will be low resolution and dangerous to use commercially. Encryption coding grows in sophistication.

▷ *reproduced in whole or in part* (see chapter 31, page 346) for any purpose including printing, publishing, presentation, or online.

▶ **Permission is not needed** when the image:

▷ *is maintained for future reference* in the creative group files for inspiration or future consideration.

▷ *is reproduced from a public domain source*—book, magazine, or collection. (See chapter 14, page 174.) Generally, images published more than 75 years ago have lapsed from copyright protection. However, this is not always true. An older image may be protected as part of an estate. Care should be taken when using an older image. However, there are public domain collections and compendia available commercially that can become part of a library of images. (See chapter 15, page 184.)

RESPONSIBLE PHOTO MONTAGES

When designers alter or combine images, they need to be very careful about ownerships. Unless from copyright-free CDs, the designer should get permission for every image used.

SELLING POINTS FOR COMMISSIONED PHOTOGRAPHY

"1. The client, in using original photography, makes their own unique photographic or marketing statement rather than trying to fit an advertising or marketing message to an existing photograph that may have been used in another context.

2. There often can be a better price break in using an original photograph, especially in a multiple-use situation, because the rep and photographer are trying to establish a client-talent relationship for the long run as opposed to the stock situation where they can only make money by charging for every aspect of the photograph. Something like this can mean that a stock shot can cost almost as much as, or more than, an individual shoot.

3. Often art directors and creative directors prefer an original photograph over a stock shot, because of the greater control they have over the situation (casting, settings, props, lighting, etc.). Morale can also be an issue: a happy work force (e.g., the art director or creative director) is always a plus and can only result in the best outcome for the client."

—Barbara Gordon
Artist and Photographic Representative
Barbara Gordon & Associates
New York, NY

When scanning or taking data from the client organization's systems, some experimentation is almost always needed. Versions and platforms vary so widely, it is necessary to plan time for conversions into the project plan.

▷ *has tangible artwork and high resolution files* that are owned by a stock agency, even though the image may be in the public domain. Once purchased from a stock agency, if it is a public domain image, it may be used any number of times without further payment.

▷ *is part of a CD collection* sold with unlimited rights (see chapter 31, Copyrights, page 346) and thus becomes part of the new owner's library. (See Libraries, chapter 15, page 184.)

MAJOR DECISIONS

set project foundation
plan rough project skeleton
assemble category outline
solicit proposals
choose creative group
finalize project plan
select design theme
approve design
review writing outline
edit first draft content
review revised content
review final content
approve visual components

SELECTING QUALITY TRANSPARENCIES

"These tips will help ensure that every photo you choose is a good one.

1. COMPOSITION. If there is enough room on the transparency to fill your layout space—both height and width? Enlarge or crop a horizontal photo into a vertical, and vice versa. But if you enlarge too much, it will look grainy and out of focus.

2. DETAIL. Is the photo in pin-sharp focus? Use color-corrected lighting for final selection. To minimize distraction, out-of-focus areas should look like broad swathes of pure color, with no discernible detail.

3. COLOR. Has the photo been exposed properly? Does it have an overall cast due to lighting problems or film biases? If you separate a photo with a color cast, be prepared for color correction, multiple proofs and possibly even re-scans. These will probably be at your expense."

Constance J. Sidles
Print Production Manager
Seattle, WA

CLIENT GROUP

To illuminate publications, the client may wish to utilize preexisting images often to fit budget or time restrictions. These are chosen as part of the design concept (see chapter 10, page 105).

SOURCES OF STOCK IMAGES

With a great proliferation of digital images available, the quality and reliability of the source can partner with the creative group for finding the best resources. The client delegates the finding of images to the art director who:

▶ **researches and locates public domain images**—verifies legal usability by contacting publishers and obtaining correct authorizations and permissions.

▶ **investigates resources**—taps any organizational archives as a good source for original material. Provide any photographs, logos, or existing artwork to the creative group in digital file formats that can be used by them in production. This may take some experimentation if bridging different file systems. (Small incompatibilities can easily cause big translation problems).

▶ **has the creative group obtain images**—usually after a proposed design direction is chosen. The creative group

▷ *is responsible for researching* and finding images to embellish a specific design direction.

▷ *may charge for search and purchase fees.*

▷ *can work with a variety of sources* and shortcut client involvement.

▷ *may use its own resources*—such as original photography from staff or other proprietary preexisting images. The client may also wish to build its own library of images for cross-purposing uses.

▷ *can maintain collections* of images available for use in designs and for purchase with defined usage rights. (See chapter 31, page 346.) Most organizations of any size develop a library of resources geared to publication campaigns (see Libraries, chapter 15, page 184) or the specific nature of their publication work. Proprietary images can be worth a lot to the identity of an organization and bear strategic applications to an organization's marketing health.

► **works directly with stock agency** and provides chosen images to the creative group. This means undertaking all the research and agency negotiations (see below).

CREATIVE GROUP

As part of the design concept, stock images are often shown during presentations. (See chapter 10, page 105.) They may be provided by the client or found upon creative group research. Usually they are used within the comps or prototypes to convey the design concept and solution for discussion and approval.

DEALING WITH STOCK AGENCIES

If directed by the client to locate existing images (usually when approving designs), the creative group most often deals with stock agencies. The creative group

► **gives instructions for visual direction** and key words in the search and selection of images. The Internet has transformed the research process. Limited only by search engines, key words, and the ease of site use, research includes both online and hard copy resources. Catalogs and print promotions balance in use with the ease and spontaneity for a designer to research an image online.

► **negotiates uses and fees**—usually requesting a quotation that is incorporated into the proposed budget. (See chapter 5, page 50.)

  ▷ *arrange for the transference of original or high-resolution images.* Low-resolution might be downloaded already as part of presentation materials.

  ▷ *keep track of all originals* and return after converting to digital formats for publication use according to the agencies' practices.

  ▷ *attach © symbol information to any original image used*—investigate encryption software and tracking.

► **is responsible for** how images are to be used and

  ▷ *receiving images in usable formats.* Any conversions necessary may need to be negotiated. When working with client-provided files, experimentation may be required to get workable formats. Clients need to

Knowing how to find specific subjects on the Web takes a lot of skill and experience. Dealing with the limitations of search engines and the use of key words and links between sites can reveal tremendous options.

supply native files for digital images like logos or photographs which remain in the creative group's archives to use as needed.

▷ *financial agreements* for usage or any usage changes.

▷ *care and condition of original materials* while they have possession or use of them. It is easy to lose track of original slides or artwork. Therefore, any transfer of such images from one place to another should be documented in writing by the creative group. If converted to digital files, images can be more accessible. After such conversion, originals are returned to their owner.

▷ *good computer backup systems*—which includes both the designer and the stock agency. Lost material due to erratic or unorganized file management can lead to major losses in time and money. Remember that any system can go down at any time—especially when the system has grown complacent! Have a searchable archival system, as not being able to find a backup is like not having it at all. Also, keep an off-premise copy of important documents as extra safekeeping in case of theft or fire.

▷ *financial reimbursement* to the stock agency if the image is damaged or lost. (See chapter 28, page 322.) Be sure to have all agreements in writing. If uses expand, so do fees to cover the wider audience.

► **communicates to the client**:

▷ *all usages and related fees.* (See chapter 31, page 346.)

▷ *costs for the entire stock package* or packages, such as collections, books, or disks from which any image is used. Often this will require reimbursement from the client. Who keeps the collection as part of their library needs to be negotiated. It may become part of an image management system. (See chapter 15, page 185.)

▷ *an explanation of the difference in cost between stock versus commissioned images*—generally stock is less expensive—but this can vary widely on the caliber and originality of photographs. Photo montage, favored by many designers, is generally in the category of original art. Care must be taken when combining found images to make sure they are copyright free before using. So much is now available as ingredients for photo montages that they are quite popular as visual and unique expressions of an organization's message.

► **receives duplicate digital files** from the stock agency, making sure that they keep the original scans or native files.

> *uses encrypted files* as low-resolution images for comps to show the concept to the client. (See design development, chapters 10 and 11, pages 100 and 126, respectively.)

> *purchases high-resolution images* upon agreement with the client, then receives them from the owner of the images—with promise of returning once digitized.

> *puts all agreements in writing* with the stock owners and is sure all terms are understood.

PRODUCTION GROUP

The stock agency is part of the production group, as they provide final art to the project. The art director is usually the buyer of stock services:

► **production specifications are outlined** as part of the project plan (see chapter 7, page 62, and chapter 8, page 70), yielding specific images needed.

► **the agency must guarantee ownership** and originality of images within their collection and be willing to transfer rights where appropriate.

► **fees can be for research**—each agency should publish any fees to access collections. The more specific an image that is needed, the more there may be a fee to locate appropriate options.

► **provides high-resolution images** ready for production processes.

The production group receives digital files of images with the rest of the publishing project for the assembly and finishing. (See chapters 16 and 17 on Web and print development, pages 196 and 208, respectively.)

RESPONSIBLE IMAGE HANDLING

The production group is responsible for maintaining any original or digital images received from the client or creative group in good condition and

► **if an original image is damaged or lost**, unless by situations for which the print group has no control (see chapter 28, page 326)—the production group reimburses the owner. (This usually is not a problem working digitally.)

► **must return all original material** (artwork, disks, samples, comps, mock-ups, and prototypes) to the buyer at the conclusion of the project. Often it is wise to return all project materials with each proof so the creative group has direct access to the same files for changes, thus minimizing possible confusion and having all references handy.

THE STOCK PHOTOGRAPHY DELIVERY

> "CATALOG. We believe that the print catalog is here to stay. There's simply no better way to showcase and view the higher-end, most artistic of stock images. The layout and design of the images on the pages of a print catalog can truly create a mood and set the tone for a client's opinion of the agency.

> CD-ROMs. In terms of production, CDs can contain three to four times the number of images of a 400-page catalog. Most importantly, through keywording images on CDs, the searching/brainstorming process has been transformed to new heights. In moments, clients can find appropriate images for concepts which, in a print catalog, might be found in diverse categories or not so easily found. With the ability to export images immediately into layouts, the designer can prepare presentations immediately.

> INTERNET. Digital delivery of images vs. receiving a transparency is becoming increasingly important to clients. With tight deadlines, it's helpful to receive the desired image in hi-res form in order to expedite the production process. Having the [stock house's] library online is the ultimate step to a fully fluid library. This means immediate access to clients worldwide, without time-consuming shipping/customs activities and replication costs [for catalogs, discs, and transparencies]. There will also be substantial savings in the reduction of paperwork and image liability. To protect the integrity of content, this sort of electronic pipeline will certainly need to be monitored for copyright protection."

—Gail Mazin
Marketing Manager
The Stock Market
New York, NY

"Nothing can replace the use of custom photography that provides an intimate look at a client's company environment and culture or that presents products with heroic imagery. However, circumstances often preclude custom photography for a variety of reasons—time/schedule constraints, geographic limitations, budget concerns, or off-season issues. In such instances, the creative process is often put to the test to identify suitable images. This is a task that is increasingly easier in fulfilling the creative objectives of the finished piece. Stock images have become a staple of the creative process because the industry has matured into a serious art, offering a broad variety of subjects, subject treatments and diversity of options. The broad talent resources that are presented deliver a total spectrum of quality and innovative photographic or illustrative options, so much so that stock images become a consideration early in the creative process. In fact, the convenience of image accessibility through the Internet has encouraged the use of stock images in the initial stages of design. No longer are schedule and budget constraints the motivating determination for using stock images. Stock images are now a legitimate component in the armament of the initial design phase."

—John Racila
Paragon Design
International
Chicago, IL

"I don't really have the patience to search for a stock visual on the Web. Yet, I can spend quite a bit of time looking through a print catalog. Somehow, paper seems to make it all real and relevant...."

—Amar Shekdat
Texas A&M University
College Station, TX

"Speaking as a graphic artist and photographer, I find that there is a multitude of high quality, royalty free images available. I can't even begin to shoot these 'general' images myself for the same price. There is still a huge need, though, for custom, local photography."

—Christopher Waldron
Fort Wayne, IN

"For advertising, stock is a good fit. Clients want projects completed in a matter of hours or days. Assignment photography can be a big production and seems to be more appropriate for big national campaigns or special ads whereas the stock photo is perfect for day-to-day client needs."

—Matt Orser
Lois EJL Advertising Agency

"The emergence of digital technology has forced designers to work at an accelerated pace. Clients want concepts, layouts and printed materials ASAP. Stock agencies and royalty-free stock enable us as designers to keep pace with client demands by eliminating the need for a photographer (unfortunately). We can utilize comp images to present a concept and ultimately complete the assignment that much quicker."

—William Kelly
Hartford, CT

"As clients continue to force us to find less expensive ways to produce our concepts, we are turning toward more stock photography. On really tight budgets we're looking at royalty free photography as the first option. The quality of stock photography has improved significantly in the last few years to the point where we are actually pleased with being 'forced' to go with less expensive resources. When we find the right photo that helps communicate the message and save the client money, we come out as heroes."

—Stuart Bran
Stuart Bran Advertising
Park Ridge, NJ

"The pros of royalty-free stock: a cheaper alternative for my customers gives them the option to have photography where before they could not afford it. The cons: Everyone else has access to the same photo, and it is not as creative. Clients settle for a photo that may not be in my opinion the best photo for the piece. I am designing for them, but since they have the royalty-free photo at their disposal they use it. I call it 'Forcing the Photo.'"

—Renae Benson
Principal Financial
Group
Des Moines, IA

"There are times when the stock photo is just the right price, mood, and subject matter. If I can fit a stock item into the project I will. It avoids having to locate photos, lab fees, photographer fees. Sometimes you have to really get creative and perhaps digitally alter the stock photo."

—Vincent Sperancio
Macvision Graphics
New Rochelle, NY

"Stock is both convenient and ready-to-use so there is no waiting involved. You can always download comps free, so if customers don't like what they see you have not spent any money on photography. Choice, content, and quality have definitely improved in leaps and bounds. There are almost too many choices."

—Christy Kelley
Cole Parmer
Instrument Company
Vernon Hills, IL

"The online search process can be very convenient if—and that's the key word—the search engine is flexible and bug-free with lots of brain power attached to its creation. Some search programs don't even recognize plurals of a word. For example, only a few weeks ago we put in 'trees' and got no listings but under 'tree' we were given hundreds of listings. It seems so basic to us, the users, but they must use non-users (in a true industry sense) to program them."

—Carol Levy
Worthington Levy
Creative
San Jose, CA

"When you're in a service-oriented field (such as graphic design), you're constantly in a position where you need to provide numerous ways to convey a message to an audience. In order to keep up, you need every possible resource to be very flexible. I have searched and purchased high resolution images from the internet. Having that resource gives the designer great power, leverage and peace of mind."

—Richard Santiago
RS Graphic Design

"We shouldn't establish price by what it costs us but by what it is worth to the market. The only difference between a stock product and a commissioned piece is that the archived image has probably been used before at some level and is not obviously 'original.' This is its only weakness and can be easily overcome if we keep accurate and meticulous records of each and every usage. Therefore, stock illustration can have the same proprietary effect that an assigned image could create. The secret here is strict control and full disclosure. If a client wants 'exclusive' rights to a stock image they will first want to know the history of the image and then need to trust that it won't be marketed again while the license is active. Obviously, this situation would demand a higher fee, as greater importance should demand a greater fee. If quality artwork floods the market without due control and the proper value placed on each specific usage, we'll continue to see an erosion of 'fair' pricing for 'quality' illustration."

—Richard Askew
Founder
Stockart.com
Fort Collins, CO

"The overload of stock photography has contributed to the dilution of the overall design quality of print communication. Principled art directors and buyers of photography need to continue to have assignment photographers shoot what they want and need—an image that was meant to be used for that assignment—not the photographic equivalent of clip art—one size fits all. It doesn't and clip art (and stock photography) often looks exactly like what it is—a generic, cheap solution to a challenge that requires much more."

—Marc Daniloff
MD2/Marketing Design
Saukville, WI

"Stock images are suitable but, frequently, the picture I need is more specific and less staged than what they provide. Stock photos are normally stiff and boring/conservative. I especially detest the science and technology sections of the catalogs; there seem to be the same photos in every book. There exists a great need for more original photography concepts and more cutting edge compositions and subjects."

—Chris Terkuile
Berry Brown Advertising
Dallas, TX

"Online resources are really useful. They allow me to show the project managers comps on the fly and to refine the search (and results) based on their feedback. As long as we have Web access we can speed up the approval process."

—Jeff Breitenstein
King County
Solid Waste Division
Seattle, WA

"One problem is that you often see a royalty-free photo in many places. They tend to be trendy—they're hot for a while and you see them everywhere. Then years later when you see it again, the piece looks dated. Hot stock photos become the fashion of our medium."

—Travis Woods
Devoted Press Design
San Francisco, CA

"There are several advantages of using royalty free stock photo CDs. Firstly, they save time. There is free comping so you save time in negotiations of fees/usage. There is a lower risk of loading images as well as lower costs because less administration is needed. Also, there is no need to go back to renegotiate if the client ever wants to reuse an image for another application. The sacrifices are lower quality, color, and fidelity without intervention, less uniqueness, and less flexibility with the size of the image."

—Daniel Ko
Patrick Soo Hoo
Torrance, CA

"Sometimes you sacrifice creativity with royalty-free collections. Photo CDs have gotten better as far as the quality of images and types of images…. Also, you tend to see the same images in everyone's work, so uniqueness of images is limited. The advantages are, of course, reduced cost, as well as production time. Another advantage is being able to create a new piece of art by compositing and manipulating to create a whole new image."

—Jeff Stachula
Lighton Coleman
Chicago, IL

"An advantage is that royalty free stock photos can be manipulated in any way imaginable. Therefore, my design will be unique and have an original quality to it. The negative to this is that you can sometimes end up spending too much time looking for a particular shot or subject matter, which can distract."

—Lisa Ponder
Ponder Design
North Hills, CA

LIBRARIES 15

The world becomes more visual through technology. Electronic documents are the norm, and, as a result, the obtaining and use of images has become easier and cheaper. Methods for defining, sorting, and storing are needed to make organization out of image handling. Digital technology can integrate documents with text, images, and databases that need management: to be cataloged, referred to, updated, altered, analyzed, and incorporated into communications systems. This is the era to translate from analog to digital and to utilize digital tools to make new kinds of documents. The technology will continue to accelerate the delivery of information for the technologically astute and growing community of users. Because communications are so visual, communicators need to build up libraries of images, data, and other elements that can be cataloged and accessible for cross-media applications.

FORMS OF LIBRARIES AND SOURCES

There is quite a range of places to find images, methods of collection, and ways of accessing additional material:

▶ **found images**—old books, magazines, and other sources, including images downloaded from the Internet. Any found image should not be used unless it is part of the public domain. (See chapter 14, page 173.)

▶ **databank companies and agencies**—these provide facts, information, images, and searches for requested subjects.

  ▷ *Use of key words* to find specific subjects through browsers or through subject categories in online catalogs.

  ▷ *Specialized areas of interest* are served by subject-specific stock companies or portals. Fees vary greatly.

▶ **collections of public domain images**—there are many libraries that make CD collections downloadable from the online catalog. Print versions help designers in planning.

▶ **subscription services**, distributed online or via CD-ROM. Many stock photography and illustration agencies sell collections of copyright-free images that can be used in unlimited ways. Selection is growing.

▶ **the Library of Congress** accepts publications digitally and this vast resource of information continues to grow and be accessible through the Internet.

ORGANIZATIONS

http://salesdoctors.com/directory/dircos/3106a02.htm
Association for Information & Image Management (AIIM)

www.documentary.org
International Documentary Association

www.xplor.org
Xplor International
Electronic Document Systems Association

MAGAZINES

www.transformmag.com
Transform Magazine

Library of Congress
Washington, DC 20559
General Information:
202-707-9100
www.loc.gov/copyright

> EXPENSE—both in converting
non-digital images to digital,
converting file formats when
obtaining from other sources.

> HARDWARE CAPABILITIES and In-
ternet connection capacities.

> FORMAT AND SOFTWARE STAN-
DARDS—there are many com-
peting standards being developed and proposed by a variety
of organizations.

> GOVERNMENT REGULATIONS for
telecommunications—including noncompetitive controls and
rate structures.

> COMPRESSION OF DATA—image
file sizes and transfer through
hardware limitations. Delivery
gets faster and the Internet
more visually capable.

> SPEED OF TRANSFER METHODS—
cellular is slow compared to
cable but cable installations are
limited. Much research and development is spent to speed up
devices.

Accessing non-original created information, text, or images requires
professional responsibility. To avoid
digital liability (see chapter 29, page
330, on ownership of materials):

> DO NOT SCAN IMAGES into the
computer and use them in recognizable ways.

> DO RECEIVE PERMISSION from the
copyright holder if you do want
to use an image (see chapter
31, page 346).

> DO INQUIRE ABOUT THE SOURCE
of employee- or supplier-pro-
vided images.

> DO PROTECT YOUR OWN PROPRI-
ETARY IMAGES through policies
of use, encryptions, and employee training.

▶ **private libraries** built by purchases, permissions, or commissioned originals and made available to a network of
users. This is most common in a department or company
which can be:

▷ *format- and subject-specific*—accessible through an
asset management system (see below).

▷ *configured with locks and security to control accessibility* per appropriate user (chapter 18, page 227).

▶ **personal libraries** for specific workers who develop their
own images and then categorize for later use. This intellectual property is worth a lot to the owner, as originals
provide a lot of creative freedom and options. Many illustrators and photographers earn major portions of their
incomes from the sale of original images.

For anyone involved with the management and organization
of visual images, system maintenance is always ongoing.

ASSET MANAGEMENT SYSTEMS

For libraries containing components for cross-media publishing, many asset management software packages are available scaled to the size of the workgroup, numbers of images
housed, and configuration of users.

▶ **Software capabilities**—most of the packages:

▷ *retrieve archived images* for reference and use—including logos, photographs, illustrations, and charts.

▷ *have the ability to combine images with databases* and
textual content. Information is modular and tagged
with various tasks and levels of importance.

▷ *sort components* to flow into templates (Web, presentation, print)—template capabilities vary between
packages. The most customized scales pages to individual users.

▷ *have security lockouts* to control who has access.

▷ *have image tracking*—to keep control of not only the
image, but all the background, source, version, and
color-treatment information. This information can stay
with the image no matter where in the system it goes.

▷ *encode ownership information* into images so anyone
who "borrows" will have access to copyright information and the address for permissions.

▷ *encrypt original visuals* with © symbol so that if someone else uses without permission, the owner's copyright makes clear any disputes. Tracking methods grow
with technological capabilities—and could be considered an important intellectual property investment.

- ▷ *achieve smooth work flow* when connecting to network, both to coworkers as well as to suppliers.
- ► **Benefits of implementing digital system**—key ingredients for the communication banquet:
  - ▷ *saves on time searching for images*—but invest wisely so that setup is a onetime expense and not an ongoing drain on resources.
  - ▷ *allows a smooth work flow* and shared database for fluid publishing processes between team members.
  - ▷ *provides instant availability*—Intranet allows library to be accessible to users at any time, fitting unusual work schedules and project demands.
  - ▷ *allows corporate guidelines to be maintained*—with the correct versions and usage instructions available to all users, consistency between marketing efforts can be reached. Visual investment is maximized
  - ▷ *utilizes technology to advantage*—speeds up preparation processes and responds to new opportunities—the upgrades force system advancements and are necessary to remain part of a seamless work flow.
  - ▷ *can be a source of revenue*, as images and data are the currency of an information society. Content is protected intellectual property. (See chapter 31, page 346.) The more designers can utilize past images, techniques, and inspirational resources, the more visual wealth there is to draw from for future projects.

CLIENT GROUP

Maximizing business communications through a digital library can enable a sharing of information and resources between creative team members. Digital networks make it possible to find, sort, update, store, transmit, and manage documents—but only through effective organization and library software.

SET UP AND HANDLE A DOCUMENT LIBRARY

- ► **Maintain a low-resolution version of image** (thumbnail), indexing the original with lockouts and security codes to determine depth of access.
- ► **Convert documents to digital distribution**—such as directories, schedules, and employee lists, catalogs, and calendars—documents that tend to change quickly—that were traditionally handled by print. Tie into image database for photographs and logos.
- ► **Use keyword indexing** and organize search categories.

MAJOR DECISIONS

- set project foundation
- plan rough project skeleton
- assemble category outline
- solicit proposals
- choose creative group
- finalize project plan
- select design theme
- approve design
- review writing outline
- edit first draft content
- review revised content
- review final content
- approve visual components

Investigate carefully the structure of asset management systems. This area must be the most user-friendly.

► **Have an easy method for adding new material**. No system can be useful if it isn't flexible enough to be easily maintained.

► **Establish file name conventions**—make sure the system stays consistent and that all users comply. It requires policing or people get sloppy.

► **Have one person responsible** for library maintenance— a proprietary guardian protects the collection from becoming disorganized or out of date. Possessiveness is a good trait to cultivate in this role.

► **Provide a database of training** and resource information to assist workers—as part of an intranet, having help material can fit the users' time frame and flexibility.

► **Utilize e-mail or a bulletin board** to communicate time-sensitive information to constituents. This is best built from a list of those who've responded to queries (see Listserv, page 71).

► **Utilize assistance from suppliers** who specialize in library systems. Many have services that will manage client resources for a fee (see chapter 25 and chapter 21, page 259).

The biggest challenge to having an effective digital library is to be able to find the images to use when needed. (See description of asset management systems on page 185.) Much time can be lost looking for images. Although the initial setup is an investment, hopefully it will save money for searches within the first year of operation.

CREATIVE GROUP

In the past, an artist's palette was filled with globs of various colors to be mixed into a world of hues. Today, the artist's palette is filled with the world of images that can be combined in infinite ways, possibility only held back by imagination.

Working with personal databases of images can become a designer, illustrator, or photographer's "bag of tricks" or even signature collection of elements. Building a protecting approach and a style is the concern of every creative person. Protecting that individuality takes the creative professional into the realm of both self-protection and self-fortification.

RESPONSIBLE USAGE:

When utilizing images within documents, presentations, or publications, responsible usage means working ethically and without risk. Here are guidelines to help when using images:

> USE ORIGINAL MATERIAL whenever possible.

> TRACK DOWN THE SOURCE when using a found image. If you can't negotiate with the originator, don't use it.

> NEGOTIATE, NEGOTIATE—everything is subject to situation.

> GET PERMISSION IN WRITING— put in your correspondence: 'Unless I hear from you otherwise ...'

> DON'T ASSUME you can change a found image beyond recognition.

> when in doubt, GET PERMISSION.

> remember PERMISSION FEES COST

LESS THAN LEGAL FEES— don't hope you won't get caught.

> BE CAUTIOUS when using celebrity images—many older images, such as Einstein and Mickey Mouse, are *not* in the public domain and are aggressively watch-dogged by their owners. Many violators have lost in the courts.

> BE CLEAR ON FAIR USE—if you can make money by using a

found image, do not skip on the ownership check!

> FINE ART of any age: get permission from the owner, whether a museum or an individual.

> GOVERNMENT-PRODUCED IMAGE, check with the National Archives, Library of Congress, or the White House Press Office for permission. There should be no fee.

WAYS TO BUILD AND MAINTAIN A LIBRARY

Anticipating the number of users and the kind of structure a library needs is as individual as its owner. Considerations for construction and building:

► **create original images** that can be used over and over in various applications.

  ▷ *Digital cameras*—generate original photographs to be used in publications. The traditional media of slides or transparencies have to be scanned, adding time and expense. The quality of digital cameras continues to grow. Be sure to get model releases where appropriate. (See sidebar, chapter 13, page 165.)

  ▷ *Past projects*—all illustrators and photographers have collections of their work, and most make them available through stock. (See chapter 14, page 172.)

  ▷ *Commissioned original work*—is used as foundational for branding or image identity.

► **attain collections via CD-ROM**—are copyright-free and allow unlimited use. They are very reasonable, but unless manipulating or reworking them into a new concept, these images may be seen frequently, used by others in publications and online. Many become trendy.

► **receive online from stock agencies** and other library sources—downloading from other collections is the most expedient. Stock agencies can conduct searches when needing a specific image. Check for research fees when requesting.

Many designers use several images in combination by altering them in photo manipulation programs. There is a craft in making composite images, much like collages used to be done with cut-out printed photos. Because these electronic collages are made with recognizable stock images, to achieve distinction, the designer may alter images to where they become something else—where the whole becomes greater than its parts, when done with talent and restraint. The good news with collections that contain copyright-free images: this can be done without limits. The bad news is that just about anyone can do it, and there is a tremendous proliferation of this kind of work. It becomes less visually distinctive.

"**IMAGE MANAGEMENT** is now a problem for everyone from digital camera newbies to seasoned prepress pros. A look under the hood of any asset-management application reveals the software's roots as a highly specialized database engine, usually a modified piece of an Oracle, Sybase, or Microsoft application. It's important to note that digital-asset managers don't store the files themselves; rather, they meticulously point the way to a file's location either on your hard disk or on a centralized server.

"Asset management is not about how many assets you need to manage but how many people need to manage them. In the evolving world of applications, software species can be differentiated by how many people they serve:

> INDIVIDUALS—image catalogers are directed at the solo user with limited asset-management needs—workgroups from two person design teams to departments with dozens of employees. Workgroup products offer various security features that restrict access to certain assets.

> ENTERPRISES—multiple departments within a large corporation, and digital service providers links (which offer company-to-company asset management). A high degree of customization is a prerequisite.

"Asset management tends to be an all-or-nothing proposition: once you start, the muck of media management only gets deeper.

[Reprinted with permission from *Publish* Magazine]

"Digital service providers promise all the advantages of digital-asset management with none of the risks involved in equipment overhead or obsolescence. The question of entrusting your assets to someone else's care is something that needs careful consideration. The big advantages to going with a DSP include no hardware to buy, no software to install (or reinstall), and no network to maintain. As a strategic capital investment, asset management can plant you firmly ahead of such emerging industry developments as variable-data printing and dynamic Web publishing. You need to be prepared to feed the beast. And to keep feeding it."

—Mark Nichoson
www.publish.com

▶ **scan found images** (which must be in the public domain— see chapter 14, page 173). Be very careful with this practice. Just because it is easy to do does not mean it is right to do. Always check sources—*especially* when using images for profit. The more money involved, the more owners get tough about protecting their intellectual property.

PRODUCTION GROUP: TECH

Images that are collected and cataloged often must interface seamlessly with databases and word processing, and be Web compatible. Maintenance of the image database could be the tech group's purview, especially for retail Web sites where a lot of images and descriptions are stored. Most libraries managed by tech groups are under the direction of the creative group. Together they design a management system that works for their applications.

PRODUCTION GROUP: PRINT

Print use evolves with the development of new media libraries and digital communication. Prepress and print professionals must expand capabilities to handle a range of file formats and storage, working with image databases and tagged color instructions.

WORKING WITH LIBRARIES

Many factors influence how the print group interfaces as part of the workgroup in the handling of libraries:

▶ **color images are used more** because they grow more economical. Document file sizes grow larger and require the print group to

▷ *handle various resolutions* at different phases in the process. Often the designer works with low-resolution images in page composition to keep file size down and speed maximized. Then the print group replaces the low-resolution images with high-resolution images as part of the prepress process. (See chapter 23, page 266.)

▷ *automate color separations* made more streamlined through digital originals—this is where costs have drastically fallen through the years. Digital images shortcut the conversion processes of color separation.

▶ **online transfer of data and images** grows in prevalence, especially as projects are done quickly. Conversion and speed problems grow less as networks improve hardware. Wider distribution channels bring down the price of Internet entry devices, making it accessible to more people.

BENEFITS OF IMPLEMENTING MEDIA ASSET MANAGEMENT

> LOWER DIRECT COSTS PER MEDIA PROJECT

> FASTER PRODUCT CYCLE TIMES

> REDUCED LABOR COSTS on complex projects

> IMPROVED SYSTEMIZATION OF WORK FLOWS

> ADDITIONAL LICENSING REVENUE STREAMS

> NEW MEDIA PRODUCTION REVENUES

> INCREASED PROFITS PER EMPLOYEE

> INCREASED COMPANY VALUATION of 5–15 times.

— Eric J. Adams
Executive Director
MediaBank Communications
Petaluma, CA

► **distribution of images with global considerations**—particularly when library images are used in both print and on the Internet:

▻ *interface with off-site proofing*—send artwork online to several sources at once for approval. This can be very useful in sending proofs to more than one publication at a time, thus expanding advertising reach.

▻ *distribute images through remote printing*—if printer is elsewhere or if printing in more than one site. On-demand personalized digital systems can allow for regional variations, special interest groups, and fast delivery of customized printed or transmitted messages.

NEW AREAS OF BUSINESS

Providers of print services can expand to offer a range of new capabilities to their clients, thus remaining viable as the technology affects print volume and use. New services to help buyers:

► **cataloging and processing of various documents** that were printed, inputting them into an asset managed format.

► **high-resolution replacement services** for low-resolution library images, storing the larger high-resolution files so the buyer doesn't have to.

► **digital printing**, for both long and short runs, that ties right into asset management systems as part of the work flow. This allows for more precise quantities and just-in-time delivery. Asset management systems can seamlessly flow directly into prepress preparation and printing.

► **asset management system** for buyer (see page 185)

▻ *as a onetime arrangement* and the buyer takes over when the system is set up. Extra help can be used in the beginning to key word images, add textual content, proof information for accuracy, and form all the segments into the structure of the library, adjusting the user interfaces and ensuring that the system is easy to use and update.

▻ *maintain the system* once it is set up through a contract or regular fee arrangement—services can include cataloging, storing, and retrieving, as well as regular updating of new material, correcting data, and keeping elements current. The expense of upgrading the system regularly should be part of the initial plan, and if the library is to service more than one project, it needs to have its own budget for maintenance and development. It is a project in itself.

"Ours is a ten-person networked shop. We're concerned with our clip art files: keeping versions straight and up-to-date, controlling what happens to the file, and getting the file back where it belongs. We have an employee who spends 50% of her time monitoring what is going on, identifying potholes, and coming up with solutions. We try to find out how to make these files accessible, test new versions of software, and to understand the features. She makes mini-training materials for others so we can cut down on training time."

—Scott Hazel
Design Manager
System Solutions
Washington, DC

"Pro—As we purchase more royalty free CDs we have instant access to visuals for tight deadlines. Cons—Seen it before. Sometimes a wide range of choice is needed to find the right image. It is not always as interesting or as unique as traditional stock."

—Susan Giovanetti
Imedia Ink

"I think it is an excellent way of communicating. The images are already scanned with enough definition that they are sufficient to use for layout purposes—this saves a tremendous amount of time. You can view so many images and try so many in a short time frame. Plus, designers can show PDF files to clients over the Internet without having to go to hard copy. This saves time, paper resources, and changes can be made easily."

—Benita Cassa Torreggiani
New York, NY

"With the abundance of photo CD libraries on the market, I do tend to incorporate 'lifestyle' type photos into my design more often. It's given me a new, cost-effective 'tog' to play with when I'm creating designs. For specific products and situations, however, I find that custom photography is the best option. I could look through thousands of stock photos and spend all day and still not have just the 'right' one. I'd love to be able to just download hi-res images on the day I need them."

—Jean Wood
Design Source
Jamestown, NC

"The Web, widespread use of e-mail, reasonably high band-width, search engines, Webcams, online-forms process . . . an astonishing array of technologies that make long distances seem piddly. We have been assembling an arsenal of tools that place us at the cusp of a global shift in publishing which will make continental separation almost entirely immaterial.

"Asset management will help companies manage content—in whatever form it takes, and wherever on the planet it happens to reside.

"The operating system needs to understand that my images are scattered hither and yon, and be able to intelligently gather them (preferably at the proper resolution and with the appropriate color profile for whatever output device I'm using).

"Let's say my pictures are on a server in London, Photoshop is running on a server in Madrid, and the page production department uses QuarkXPress in Paris. I can press a button in Los Angeles that will scale, crop, sharpen, save, import, format, caption, and lay out a page across the ocean before Jean-Pierre even gets his morning café. Suddenly, the whole of the Internet is my Local Area Network."

—David Blatner
Graphic Arts Consultant
Seattle, WA
www.creativepro.com

"The move to deploy a content management system is being driven by a few key applications: e-commerce, one-to-one marketing, and custom publishing.

"Increasingly, publishers are realizing their need for systems that can streamline the production, management, and distribution of all types of digital content—financial and alpha-numeric data as well as graphics, images, audio and video. Once in place, these systems can help publishers gain new efficiencies throughout the media lifecycle, from conception and growth to ongoing management of existing media. These systems streamline media production work flows, manage multiple file formats and file conversion, provide greater access to media assets throughout companies and between them, and streamline brand building, work flow, and media distribution.

"Companies that don't implement a content management solution early will lose out to their competitors, who will ultimately gain a strategic advantage.

"Ultimately, the payoffs also include lower costs, better quality, faster production cycles and incrementally increased revenue through reuse, repurposing, re-expression, and redistribution of intellectual property. In addition, content management systems can help publishers create new revenue streams through a fuller exploitation of their content across a range of media, both traditional and new."

[Reprinted with permission from *Publish* Magazine]

—Lorin David Kalisky
Writer and Consultant
www.publish.com

"Because of the way digital technology has changed the entire process of using photography, we use more photography in our design work than in the past. It is quicker, easier and cheaper than ever before. What has changed the most is that we are using individual stock photos from stock catalogs less and buying stock photo CDs much more. For the cost of renting and using a single image from a stock catalog for onetime use, I can now purchase a complete collection of a hundred or more images .... While there will always be those assignments that require a particular special image that can only be located from a stock agency, those cases are less and less."

— Brian Marquis
Marquis Graphic Design Associates
Alexandria, VA

"Knowledge management is the act of identifying, capturing, cataloging, storing, retrieving, sharing and measuring a company's information, content, know-how, and expertise. That's a mouthful, but understanding what it is and how it works will boost productivity and your company's bottom line.

"Knowledge is information given context. Knowledge is information with value, judgment, ideas, and insights included in the mix.

"There are three basic goals of knowledge management—improved learning, refined information, and improved access to information. Knowledge management improves learning by providing access to pertinent information as people need it. Knowledge management doesn't just give the facts, but it also imparts the wisdom of the company's previous experience.

"Adding a knowledge management system can be difficult. The biggest obstacle is often convincing people that sharing knowledge will ultimately make them more valuable to the company and the company more valuable as a whole. Because knowledge is power, employees are often hesitant to tell their colleagues what they know because they think they will end up missing out on a job or a promotion."

[Reprinted with permission from *Publish* Magazine]

— Lorin David Kalisky
Writer and Consultant
www.publish.com

"Royalty-free photos and illustrations on reasonably priced discs have been a great advance in the graphic design business. We use 2, 5, 10 photos on a project now, where we would have been restricted to one high priced photo. Traditional stock houses should see the writing on the emulsion."

— Frank Palazzolo
Design For Business
New York, NY

PRODUCTION

FINAL PAGE PREPARATION

Great concepts can be ruined by poor execution. Carrying through from a well-positioned direction and a well-crafted project plan, how the final work comes together will mean the project's success or failure. The process of final preparation begins with the approval of visual directions (see chapter 10, page 100). It is the process of converting those concepts into reality. Any project that involves print such as brochures, books, magazines, newsletters, posters, print advertisements, packaging, or point-of-purchase follows similar phases.

PROJECT PARTICIPANTS

Most projects involve a collaboration between defined functions, segmented to these major roles. One person may play more than one role. (For further description about roles, see chapter 1, page 2.):

► **Project director** supervises the entire process as the client representative to ensure the project continues to meet parameters. Authority to make production decisions and availability are key in effectiveness.

► **Art director** leads the process including schedule and proof stages, and secures approvals. Visual elements such as illustration or photography are assembled according to design strategy. (See chapter 3, page 22, and chapter 10, page 100.) Bringing together all segments of the project as the creative group liaison to the client, the art director is responsible for all quality control.

► **Designers** have a range of involvements in page assembly. (See chapter 10, page 100, for design process and designer capabilities.) It is wise to have the designer monitor the project as it develops—checking on proofs and adjusting components as needed. Designers

　▷ *develop conceptual design and work with the page composer* or prototype maker who executes the design.

　▷ *oversee a team* where each individual handles the construction of one portion, then pulls all portions together.

　▷ *may handle the entire production process*—one de-

ORGANIZATIONS
www.colormarketing.org
Color Marketing Group

http://stc.org
Society for Technical Communication

MAGAZINES
www.publish.com
Publish Magazine

WEB RESOURCES
http://graphicdesignabout.com

http://desktoppub.about.com/compute/
desktoppub/
About.com

signer works from rough designs to finished pages, ready for posting on the Web or sending to the print group.

▶ **Page composer** compiles all visual and textual elements, fulfilling the design specifications. Larger projects can require teams, splitting the project into segments. The page composer is proficient in page layout software, file formats, and converting from various image sources and platforms.

▶ **Prototype maker** sets up mechanical art for dies and constructs models to give fabricators to follow for specifications.

▶ **Project manager** works closely with the art director and the project director to ensure that the project plan is followed, monitors finances, and supervises suppliers.

DEVELOPING THE PROJECT

Final production is the creation of documents that contain all text and graphic elements assembled into page layouts or models. Ongoing concerns in project development are:

▶ **differences in end-product**—comparing the process for print versus Internet page construction (see chapter 17, page 208), each has a unique language, element creation, and mechanical assembly processes.

▷ *Print is high-resolution* and its precision lies in crisp, clear images, reflected color, and physical handling.

▷ *The Web relies on low-resolution files*, many small electronic pieces, and complicated file systems, and is received via the computer monitor transmissive color.

▶ **techniques and skills grow**—in traditional print methods, pages are keylined or manually pasted up. Page layout software allows the process to be digital, promoting levels of production skills matched to platforms and packages.

▷ *Roles shift*—as the software tools evolve, more control is given to the creators of documents. This continues to redefine roles, favoring more capabilities into the hands of the designer.

▷ *Roles become obsolete*—such as typesetting or keylining. Learning new skills remains necessary to stay in the business, no matter what level of experience is obtained.

▷ *Automation speeds processes*, making the need for quality management particularly crucial through defining and working against a good project plan. (See chapter 3, Strategy, page 22.)

REVIEWING DEVELOPMENT PROOFS

With page composition software, a number of intermediate proofs may be created. These proofs provide the creative group and client with the flexibility in the development of final documents by being able to see the progression of content—allowing elements to be verified and perfected for publication.

Working proofs (low-resolution) that precede final document phases include:

▶ **rough proofs** (laser or low-resolution color)—text and images are formatted by the page composer and worked into pages. These pages show how the elements fit. This can be rough, for many items might be unfinished. The client should read the text thoroughly for any needed adjustments. The earlier changes are made in the process, the less they will cost. This is the best time to refine materials. Any editing or content adjustment is flushed out as the pieces come together. These proofs should receive full team attention. Once in presentable shape by the creative group, rough proofs are submitted to the client group for review and then approval.

▶ **corrected client proof**—made by the page composer from the client's review that incorporates feedback and then refines into the final document. Every team member should:

　▷ *review thoroughly*—this may take several rounds until all content is correct.

　▷ *obtain all necessary approvals*—this is the best point to have finalized. Any credit information should be added.

　▷ *have the designer review pages* prior to final refinement in case elements have to be adjusted to accommodate changes or additional content.

▶ **final proof** (laser or color)—prepared by the page composer as the last step before the files go into prepress or die-making. The client needs to proofread and verify the correctness of all material one final time. However, additional content or significant changes in material now become costly, and should have been done earlier in the process. The later changes are made, the more they cost. (See formula of escalation, page 312.)

Once the final laser proof is approved, higher resolution proofs will be required (see chapter 24, page 280) from the prepress process, generated by more expensive and more accurate equipment, usually out of the purview of the creative group. All designers do is check the color balance of the work through a higher resolution (usually generated by the imaging center).

CLIENT GROUP

For most projects, the creative group carries out the page composition and other production processes per the project plans and according to selected design. This is often where designs are perfected and variations of a visual structure are tested. There are many issues that can arise for the client to review, for as content adjusts to its format, emphasis, hierarchy of information, and exceptions, all demand decisiveness from the project director.

SUPPORT GOALS THROUGH DECISIONS

The best time for the client to specify any needed adjustments in the writing, statistics, content, or design is before final production begins. To make such changes once production begins must be for a very important reason. (See pages 248 and 249 for valid and invalid reasons for making changes.) It is assumed that the content is verified and during the assembly:

► **the final text and content must be approved** by the client. All decision-makers should initial their comments on the corrected client proof or mock-ups. The client makes clear any necessary changes during finishing processes.

► **the proofs for key points** are verified by proofs during development as specified in the project plan. Deadlines are met during page construction according to this structure or framework.

► **the project director supervises** the phases of development with the creative group. As page composition and prototype work is completed, the project director (from the client group):

▷ *receives final proofs* to review and sign before the project proceeds. Availability is important for any timely factors that may affect the project development. The project director needs to

> check all the elements, proofreading, etc., very carefully, perhaps by getting fresh eyes to review. All facts and content need to be correct before proceeding. It is worth taking the time to do this thoroughly.

> initial and date the corrected client proof or mock-up to signify approval of work completed (or with indicated changes).

> know that the later changes are made, the more costly they become. Ask for how much any changes will cost rather than wait for this information to be offered. Often projects are moving too fast to slow down enough to evaluate. See this as a danger-signal.

Perhaps the clients' most important job is the day they review the final proof of a print project. This is the moment of commitment, where errors multiply with each copy printed. Be sure to take the time to review each element of the project before final fabrication.

Soliciting fresh eyes (those who've not seen the project before) is a good idea to pursue several times in a project development. They can point out something obvious that no one on the team has seen, such as a typo in a headline.

> ▷ *may receive digital files* from the creative group, rather than page proofs. This occasionally happens when the client

>> > has in-house production and the designer provides publication templates (digital page grids with style sheets for typography). Instruction may need to be worked into the plan and economics.

>> > has contracted with a publishing house or other supplier to carry out finishing processes.

> ▷ *becomes the buyer for final output*, fabrication, or printing. This is most effective when the creative group is available for consultation.

The client's role is most efficient when devoted to a review of the various proofs—to check elements versus to find the proofs an opportunity to make a lot of changes. All other aspects of project development, such as gathering content, authoritative critiques, and financial agreements, are completed as the production process completes and the project is ready for the production group. (See chapter 23, page 266.)

CREATIVE GROUP

A number of approval checkpoints and proof presentations to the client are needed in the development of documents. With the art director and project manager as the core project supervisors, communication should be a smooth process because a momentum is already established in the design development. (See chapter 10, page 100.)

ROLES AND RESPONSIBILITIES

Whether the creative group is one individual or a large team of professionals, each of these roles and responsibilities must be covered. Several individuals could wear more than one hat, but all projects require these activities:

▶ The **art director** or **project manager**:

> ▷ *makes sure that the client has approved* all the elements of the design and component parts of the project before proceeding to develop final artwork. This becomes even more important when under deadline.

> ▷ *receives finalized text* from the writer or client with any needed changes indicated (see Writing, chapter 12, page 148), usually collaborating on content from the project beginning. (See chapter 3, page 22.)

> ▷ *supervises the preparation of visual resources*—photography or illustration. (See chapter 13, page 158, for more on image development.) This includes:

A project schedule may start out compressed or become compressed for a number of reasons. As the page composer finishes, there may be last-minute changes, but there may not be time in the schedule to stop for estimating those changes before doing them. The creative group is faced with either jeopardizing the client's deadline or with financial risk by making changes without the client's approval for additional expenses. To avoid this common trap, the creative group must clearly communicate with the client that extra fees may be added for any changes made during final production. (Of course, it is necessary to communicate with the client what steps are considered part of production). A client has approval over how they spend their budget.

Many production decisions may influence page composition and output preparation. Every press has:

> different tolerances for color registration and film densities.

> different papers handled effectively.

> halftone dot structures and ink consistency characteristics.

Note: color of inks can vary greatly from the way color looks on a monitor. (See chapter 25, page 288, for more on press capabilities.)

Creative groups have found they can save a lot of time and money by "partnering" with the print group long before page composition begins. If the print group sees the concepts, they may suggest cost- and time-saving ways to produce the ideas. This collaboration should continue throughout the project.

> selection of artist. (See chapter 13, page 158 for how to find the right talent for the right project.)

> negotiation of terms based on usages and ownerships, see chapter 13, page 158, for factors in fees and chapter 31, page 346, for copyright information.

> review of concepts—usually preliminary sketches or storyboards. Preliminaries help to guide the refinement of concepts to best express the client's message.

> location or studio supervision for photography sessions.

▷ *supervises the page composer or prototype maker* to make sure page composition and prototypes adhere to the design and proposal specifications (see below).

▷ *is responsible for quality control of elements*, including completeness, budget, and schedule.

▷ *provides estimates to the client concerning any changes* made that alter parameters.

> The art director must be sure to discuss any extra fees for changes.

> If the creative group does not discuss extra fees with the client when the client is available, the creative group undertakes changes at their own economic risk, for clients want the choice of what changes are worth before extra fees incurred.

> If the client is unavailable for approval and the deadline is in jeopardy, then the client accepts the best judgment of the creative group, unless the client agrees in advance to hold up production until approval is received for key points.

> Any changes that do not alter project parameters are provided at no fee. This may include design refinements, incorrect images, or redoing work due to misunderstood directions.

▷ *checks over the final proofs* and secures client approval before release of files to prepress, printer, or fabricator.

► The **page composer** or **prototype maker** actualizes the concept into the final form. It is best if they've had previous input on the production issues of various proposed designs. They may give valuable input on the most efficient way for designers to prepare materials. They

▷ *work directly with the art director and designer* to produce pages or prototypes according to the approved design.

▷ *receive instructions and materials*, which include the approved designs and comps, final text, and images.

▷ *understand budget parameters and deadlines*, before proceeding with any work.

▷ *prepare the digital project files*—format all text, place text into document structure (possibly a template) combine all text and graphic elements, and build pages.

▷ *are conversant in the highest quality, lowest cost production methods* and can recommend which steps to complete in-house and which to assign to outside suppliers. Every team member needs to know capability limitations and have resources to complement.

▷ *interface with the production group* and are aware of all requirements, specifications, and technical procedures. They meet with the production group to determine the best way to prepare documents prior to providing finished files. Sometimes deadlines don't allow for this, but it's a good idea to do whenever possible.

▷ *place date and file names on all output*—versions can easily become confusing and the date can be the distinguishing factor when deciding which is correct.

▷ *prepare preliminary proofs* first for the art director and designer, then for the client. After the client reviews, the page composer prepares a corrected client proof and all final proofs for the art director, designer, and client to review.

▷ *prepare files for prepress*—how much is handled by the designer versus the prepress specialist varies from project to project and from organization to organization. More pressure is placed upon the designer to shoulder prepress, driven by software advances.

▷ *handle color separations*, where appropriate. If the page composer creates the separations, the page composer is responsible for how those separations will function for the printer. The page composer becomes responsible for screen angles, resolution, color balance, trapping, and all file preparation. Background and training must partner with the software tools to be effective.

▷ *provide all final digital files to the print group*—include complete instructions (such as system configuration, software versions, font requirements, and output specifications), final proofs, any fonts the print group doesn't have (to be used only on that document: see chapter 23, page 266), separated proof pages or composite proofs for any multicolored specifications, and any other required components.

SEYBOLD TOPICS:
(also see quotation on page 279)

The Seybold Conference is a barometer for developments in the graphic arts industry. Information about its conferences can be obtained from www.seyboldreports.com. These are industry forefronts:

> MEDIA ASSET MANAGEMENT from work flow to digital asset to color. Managing content and images between workgroups and for cross-platform publishing seems to be key in the content.

> FILE TRANSFER systems allow more and more companies to look to remote proofing solutions and communication via the Internet.

> COLOR management solutions are a big part of the digital proofing and printing work flow.

> WORK FLOW MANAGEMENT capabilities, whether managing a project through completion or distributing time-sensitive information to core workgroups, identify the importance of data management and collaboration in both the content development side and the production side.

> PREPRESS AND PRINT PRODUCTION technologies are hot with announcements from proofers to RIPs. Without prepress solutions, you don't have a final product.

> DIGITAL PRINTING and all it encompasses (variable data printing, short-run printing, etc.) has quickly become "the way to go."

> STOCK IMAGES have become an integral part of many designers' creations.

> PRINTERS. Both wide-format and desktop printers are gaining in development.

▷ *keep backup files* of all digital and proof documents in case of later possible client changes. The page composer should receive final disks back from the production group. But timely backups are an essential component in business insurance. Have backup conventions that all staff can adhere to. This makes project documents accessible to anyone who needs to work on them.

▷ *are available by phone* to answer any questions the print group may have, especially when new processes are being used or deadlines are particularly tight.

▷ *set up communication between production coordinator and printer* so project requirements are reviewed and understood by all participants. Follow up on this.

PRODUCTION GROUP: PRINT

Page composition and printing is always a collaborative process, and the digital environment encourages and inspires a smooth work flow from designer to page composer to prepress specialist to printer.

SUPPORTING DEVELOPMENT

The production coordinator, the prepress specialist, and the printer should be available to answer questions throughout the process. (See page 2 for definitions of specific roles.) They serve as resources to the page composer. In this way, they work in partnership with the creative group. They can

► **give feedback on page construction** ensuring that the pages will go through the imaging process efficiently. (See Prepress, chapter 23, page 266.) The most common output problems reported by imaging centers are

▷ *incomplete instructions*

▷ *pages with unnecessary elements*

▷ *font incompatibilities*—such problems can be avoided if the imaging center advises the page composer on appropriate preparation and receives a test page on large projects or to test new processes before page composition is completed.

► **suggest and provide samples of techniques** and materials for prototyping including paper samples, examples of printing techniques, or samples of experimental processes.

► **can offer additional image management services.** (See Libraries, chapter 15, page 184.)

► **can serve as partners** in planning updates, reprints, or regular publication schedule and methods.

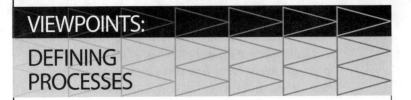

"A big part of the ingenuity of design is coming up with ways to get things produced.

"The client-designer relationship should allocate ample time and consideration to this topic. It can take as much (or more) time as conceptual phases to sift through the nuts and bolts of how something is going to get to its finished form.

"The process can be frustrating. Finding ways to produce things that are inherently different is not an easy task.

"Keep in mind that it is every bit as creative and every bit as important."

—Richardson or Richardson
Hopper Paper Company

"File management is like housework. Do a little bit every day, and you keep ahead of things. Let it go for a few days, and it starts piling up. Eventually it gets bad enough that priorities must shift. Something must be done.

"The same is true with computer output and files. When you get to the point where output is everywhere, and you see dozens of names on your hard disk that you recognize but can't associate with faces, it's time for action.

"Keeping track of output and data is really no difficult task. It's just a matter of a little organization and a little time each day to adhere to that organization."

—John Ivory
Writer
Chicago, IL

DON'T BE A CONTROL FREAK

"A few years ago, when many designers sought ultimate control, the assumption was, 'If the computer is capable, then, by gosh, I must do it!' Not anymore. Now there seems to be a level of acceptance that a designer doesn't have to do it all—it's OK to farm out processes like trapping and color correction. Maybe it's acknowledgment that you can be a jack-of-all-trades but then you're also a master of none. In most cases, outsourcing is a smart idea. Trapping requires knowledge of the press, so it's usually far easier for a vendor to handle the task."

—Debbie Johnson
Writer
Cincinnati, OH

"In the rapidly changing world of publishing technology, strong vendor relationships are paramount to success. Vendors must stay current with a barrage of technological advances, or be buried with the past. We users must draw upon their specialized knowledge to build our future and direct our businesses to survive and excel in this constantly evolving business climate. The most successful vendors stay ahead of technology acting as R&D centers to help their clients stay ahead."

—James Hicks
  Director of Print Services
  Valentine Radford
    Communications
  Kansas City, KS

"Your corporate responsibility goes beyond choosing the proper supplier. In no instance, whether selecting a specialist or generalist, should you release control or influence over the process that leads to the proper answer. You cannot simply describe your problems and then remove yourself from the stages leading to solutions. Feedback and direction must be provided so that progress remains on a true course or is repositioned before time and effort are wasted. You must develop the original objectives, review the initial concepts, eliminate and refine, then when you finally judge the outcome, the rationale and ultimate solution should tie into those original objectives."

—Anthony A. Parisi
  Director of Corporate Design
  General Foods Corporation

"In today's business climate many of the functions provided by specialized vendors have been combined into the desktop operator's job function. These continually changing responsibilities have left operators to fill gaps previously provided by the vendor. Before each specialist contributed knowledge and experience to the finished piece. By shortening the process we have deleted many vital steps. These functions and others insure that our agreements with vendors facilitate accuracy in the final product and are best done through establishing a relationship. We must agree on where the responsibility for errors between disk output and the final product will be eliminated. We must also agree on what proofs are to be seen and in what form. These agreements must be made at the beginning of projects, not during or after. Agreements guarantee consistency and accountability that can be sustained into the future."

—James Hicks
  Director of Print Services
  Valentine Radford
  Communications
  Kansas City, KS

"Printers often clean up files and don't tell the designers or page composers what is wrong with the file. Not telling them does not help the situation. But printers fear losing their job if they share their knowledge."

—Clint Funk
White & Associates
Northbrook, IL

"Lithography is not trivial. It is insulting that designers think all they need to do is buy software and they can do a printer's job. Remember: a journeyman's skills require years of training; on-the-job training is the best teacher; only experience yields expertise; only expertise yields results. Similarly, it takes more than drawing software to make a good designer. Remember: good design is a highly developed talent; design is visual, not mechanical; design requires talent, not template. Both sides must recognize their limitations *and* their opportunities. Both should openly pursue a synergistic relation-ship with the other."

—Herb Paynter
President
ImageXpress
Lawrenceville, GA

"There are two kinds of graphic designers: One is primarily production oriented, the other primarily idea oriented. Although the two are not mutually exclusive, one by-product of the digital revolution is a clearer distinction between those with skill and those with imagination.

"The computer has opened the realm of creative possibility further than ever. The designer need not be a detached participant in an assembly line, but rather a major player in a total production.

"An authorpreneur provides rather than merely interprets content.

"While there will always be those more proficient at editing than producing, or designing than writing, today's graphic designer should not be content only to design the book, magazine, or Web site. Either collaboratively or individually the designer must be totally vested in a project and product."

—Steven Heller
Art Director, Author, Cochair
School of the Visual Arts
MFA/Design Program
New York, NY

"Service is still the key. Give me a human being who cares, who shows they care and can be a partner in helping you achieve your end. Who can be honest by suggesting cheaper ways or simply being professional enough to suggest a better way to achieve your goal. Certainly, quality is highly valued—but in a digital age, I feel the standards have been raised and more easily achieved with less cost."

—Jim Lange
Jim Lange Design Associates
Chicago, IL

— Kurt Klein
Customer Support
Services
Graphic Arts Center
Portland, OR

"Designers should do it all. Printers *love* change orders. (Laughs) Actually, we recommend that our clients do not do pre-press. But if you are going to do it, be aware of the responsibility you assume."

"Technology is forcing turnaround time on jobs to collapse at an impossible pace. There's going to be a point where we can't do things any faster. Clients don't want to hear this! I also see technology creating a greater appreciation for handmade processes—interesting bindings and papers and printing methods—and a willingness on the clients' part to pay for these."

— Jessica Spring
Springtide Press
Chicago, IL

"I see PDF work flows enhancing the way we produce collateral. Also, 'Direct to Plate' offset technology will reduce costs (no film) and cycle time. I think this is wonderful for all involved with the creative process. Better reproduction means better business."

— Frank Gallucci
Dow Jones & Co.
Princeton, NJ

"Digital technology has shaped our industry over the past 15 years in a big way. This has been great for business in several ways. Speed: we can produce more work due to the technology. Quality: a lot of the design today is at a high quality level. Don't get me wrong. The technology needs to be in the hands of an educated designer; otherwise, it's like giving a monkey a handgun. Viva technology!"

— Dean Timmons
Cactus Communications
Denver, CO

"Of course, the personal computer changed the entire industry, making it easier in some respects and more difficult in others. As software becomes more advanced and versatile, designers are able to create visually appealing art more simply. However, we've gotten to the point of needing our technology to be bigger, faster, more complicated. How many Photoshop filters, special fonts, and must-have design programs do we need? Not to mention the fact that anyone with a computer and a word processing program with built-in clip art thinks they're automatically a graphic designer. Basically, technology has brought out the very best—and very worst—in design."

— Elise Korn
Ray V. Anthony & Company
Knoxville, TN

WEB SITE CONSTRUCTION

Because designers are required to wear many hats between the demands of various media, how sites are constructed, and by whom, is unique to each. Most designers handle a blend of print and online media, so well-rounded knowledge and experience pays off.

FACTORS OF WEB SITE PROJECTS

Setting up an effective Web structure is essential for maintenance and for having more than one person to be able to update it. Flexibility built in the Web plan should be part of the creative process. Web projects require a knowledge of

► **differentiating print versus Web-based projects** by

  ▷ *scope and global reach.*

  ▷ *goals and expectations of site*—retail, membership, news, information, and how the two media can work together.

  ▷ *concept of interactivity* (see below).

  ▷ *composition of creative team:*

    > content dictates appropriate contributors.

    > size and scope determine the needed technological team.

► **how to integrate with other business systems** such as

  ▷ *sales and marketing databases.*

  ▷ *publication mailing lists.*

  ▷ *product availability and inventory systems.*

  ▷ *membership services.*

► **nature of design concept** possibilities and requirements.

► **concerns all Web site projects have in common**—design, management, and construction. Sites have attributes that make them very different from other media:

  ▷ *interactive media communicate with a single viewer at a time* only, not a mass audience like broadcast media. Presentations such as educational work, absorbing sales descriptions and playing games and other entertainment all engage viewers through participation,

ORGANIZATIONS

www.siggraph.org
Special Interest Group on Computer Graphics (SIGGRAPH)

MAGAZINES

www.bigpicture.net
The Big Picture

http://cgw.pennnet.com/home.cfm
Computer Graphics World

http://ep.pennnet.com/home.cfm
Electronic Publishing

www.internet.com/newmedia
Newmedia

www.media.sbexpos.com
Seybold Reports

www.webtechniques.com
Web Techniques Magazine

www.wired.com/wired/current.html
Wired

WEB RESOURCES

www.library.nwu.edu/media/resources
Northwestern University listings of industry resources

www.ruku.com
PixArt

KNOW PRIORITIES

Constant new developments in the software require constant learning. The busy worker must have ways of coping with change. Each individual must recognize that they will never be caught up on:

> MAGAZINE READING

> ALL UPGRADES

> ITEMS ON THE "TO-DO LIST"

> CORRESPONDENCE

> PHONE CALLS

> READING E-MAILS

> CONFERENCES TO ATTEND

> SOFTWARE IDIOSYNCRASIES TO COPE WITH.

The best thing to do is plod along with trying to be ahead of everything, have lots of friends, and do what most needs to be done. Priorities are the best navigation tools.

which can be shared with others through new channels and communities. A distinction is drawn between interactive presentation media (see chapter 19, page 236), and the Internet because of the relationship-building nature of the latter.

▷ *the viewer is engaged in a dialog*—allowing for two-way communication. Many of the visual tools between presentation media, interactive media, and online media are similar and the cross-purposing of content can be efficient if planned as part of the design process. (See chapter 11, page 126.)

▷ *interactive media can blend with other media*—combine several communicative vehicles into a campaign where a conceptual design theme is applied to print, interactive media, the Internet, broadcast, and display. Clients gain from such consolidation, both in market consistency as well as economically. Elements are shared between different uses. Market recognition will grow with the visual consistency or branding. (See strategy, chapter 8, page 70.)

▶ **good planning and positioning** (see chapter 8, page 70). A well-strategized project plan is scaled to subject. Know the ladder is against the right wall, and then there can be no shortcut for strong organization and beginning the project with the necessary pieces. Web sites are more specific, often narrower subjects than the businesses which support them.

▶ **a dedicated Webmaster** (see page 2 for description of role) for sites that need to communicate, sell products, or provide services. No site is too small to have a Webmaster if it has any amount of building activity.

▶ **content that will attract viewers** (see Web Site Planning, chapter 8, page 70), as opposed to print which is linear.

KEY ROLES DEFINE RESPONSIBILITY

The process of developing a Web site blends the roles of many professionals, each contributing a portion through utilizing their talents and specialties.

▶ **The Webmaster** is responsible for the existence of the site—from its content, business focus, appearance, functionality, and profitability. The Webmaster needs to balance both the creative and the analytic sides of the project (above). Clarity and ease for the user must rule all content,

MAXIMIZE SKILLS

Whether the skills reside in one person or several, the production of a Web site requires two different sides of the development brain:

▶ **the creative** to

▷ HAVE A DEVELOPED SITE PLAN—identify all components with status, contributors, and ownerships. (See chapter 8, page 70, and chapter 11, page 126 for planning and design.)

▷ CONCEPTUALIZE DESIGN—understand its underlying visual philosophy and structure. Know the condition of templates and flexible elements. Be sure to comprehend the stylistic language set up by design.

▷ APPLY THE DESIGN TO THE VARIOUS CATEGORIES AND SEGMENTS defined in the site plan. (See chapter 11, page 126, for developing content.) This is the true test of design, as extremes in content, amount of variable information, and ease of use become paramount.

▷ APPLY THE DESIGN TO ADDITIONAL MATERIALS as it is added to explore possible combinations and scenarios, including special attention to sections that may update.

design application, and data configuration decisions. The Webmaster supervises the rest of the **development team**, which includes:

▷ *content developers*—may be a team where each contributes a unique talent or segment of information. Or, a single designer might research and write the content, as well as handle the visual expression.

▷ *the designers* who conceptualize and present the strategy and visuals. (See chapter 11, page 126.)

▷ *the programmers* (tech group) who convert all data to designed format, develop site infrastructure, complete and post pages, and handle updates through templates developed as the site structure.

▷ *the project manager*, for larger sites where the Webmaster assigns segments to individuals. The project manager watches the schedule and budget, implementing the plan and reporting any variations. Many exhibition and publishing sites are segmented this way with various teams responsible for individual segments.

▶ **The client (project director)** —authorizes, approves, and initializes the project. The client team (or sometimes it is

the analytic skills to

▷ SET UP THE FILES—contribute to and follow file management conventions so that more than one person can work on the site, especially for updates which must keep to schedule to be viable.

▷ CREATE ALL THE PIECES—translate design components into page-ready elements for structure. This can be a major part of the setup work.

▷ ASSEMBLE THE ELEMENTS WITH AUTHORING SKILLS AND TOOLS—program segments and any scripting to achieve defined objectives. Interactive programming requires specialized focus that can be a major portion, along with design, of the setup onetime investment.

▷ KEEP TRACK OF VERSIONS AND ACCESS—set up asset management system that allows multiuser development (see chapter 15, page 185). This might not seem important at the beginning, but if the design foundation is set up modularly, it will accommodate greater seamless digital databases of content.

▷ HANDLE ALL UPDATES—timely segments receive special design attention to allow both for visual consistency with the site as well as ease of changing content. This is one of the most important aspects of the site to test and streamline.

"BEFORE ANY WORK BEGINS

> A COMMUNICATION STRATEGY is the single most important element guiding a project from its initial stages through final refinements.

> initial research should include an AUDIT OF YOUR COMPETITORS' AND YOUR COMPANY'S CURRENT COMMUNICATIONS. In trying to establish a distinct position for your company or one of its products or services, you don't want to mimic a competitor's work or contradict a message your company just sent out.

> ASSIGN ONE COMPANY STAFF PERSON as the decision-maker and key contact for graphic design.

> A WRITTEN CONTRACT covering project parameters and responsibilities.

> MONEY MATTERS such as estimates and billing.

> A PROJECT TIMETABLE."

— *The Graphic Design Handbook for Business*
American Institute of Graphic Arts
Chicago Chapter
Carl Wohlt, Publications Chair
Tim Hartford, Concept and Design
Chuck Carlson, Writing
www.aigachicago.org

just one individual up to dealing with a committee) commissions content (see chapter 11, page 126), and supervises and approves each key step within the project. With the assigned role of the project director, the client team has assurance that the project will:

▷ *meet communication goals*—every decision along the project's development becomes easier to make against an agreed-upon structure. A good plan creates a site that is easy to use or enhances the content from the readers' point of view in the updates.

▷ *satisfy budget requirements*—with any additional costs presented to the buyer for choice in how to spend resources. There are always opportunities to expand the site—either through reach and links, through advertising opportunities, and through building a Listserv. (See chapter 8, page 71.)

These roles, combined with those of the creative group (see below) may be handled by distinct individuals or by blending roles into fewer individuals. No matter how many workers are involved, all these roles, with their associated responsibilities, need to function smoothly together for the greatest economy.

With the design team in place and designs approved, the development of the site itself can proceed.

ADVANTAGES OF THE ONLINE MEDIUM

> THE FORMAT IS PRIMARILY VISUAL—don't neglect either form or function, as a design plan is meant to blend the two.

> EXPANDS THE USE OF EXISTING IMAGES through the repurposing of content and develop a brand. Be careful of ownerships (see chapter 29, page 304) if using nonoriginal material.

> BRINGS IN OTHER MEDIA—such as still pictures, animation, and sound. These should be used strategically and with constraint or they can become overwhelming or annoying.

> INVOLVES THE USER THROUGH NAVIGATION FLEXIBILITY—lead the viewer to segments that are necessary for the conveying of the message. Everyone working on the project needs to understand what is emphasized and the hierarchy of information.

> MAKES COMPLICATED INFORMATION EASIER TO GRASP and be a more complete information source than any print media. The team should know how various media work together and be able to suggest efficiencies of technology. Much information that is timely is better placed online.

> INVESTS IN STRUCTURE—a site can be inexpensive in delivery, though costly to create. Most of the fees are in the setup, but the ongoing budget needs to be established for additions and updating.

> INSPIRES ORGANIZATIONS TO DEVELOP NEW BUSINESS OUTLETS AND MARKETS—Web sites need to always evolve to respond to new opportunities and links with others.

> PROVIDES OPPORTUNITY FOR TWO-WAY DIALOG—this becomes more social than individual as a marketing database of visitors builds.

DISADVANTAGES OF THE ONLINE MEDIUM

> DEPENDS ON VIEWING PLATFORMS AND SEARCH ENGINES that can limit visual impact. Testing designs in the most commonly used platforms is essential. Know the audience.

> LIMITED SPEED, FILE SIZE, AND TYPOGRAPHIC CONTROL—these all effect the efficiency and functionality of the design. Know these limits as a parameter designers will always push and techies will always advance.

> ALLOWS ONLY ONE USER AT A TIME TO EXPERIENCE—both an advantage and disadvantage, the tone and personalization can be ignored or used very wisely.

> IS EXPENSIVE IN DESIGN AND DEVELOPMENT—many sites are like setting up separate businesses, requiring overhead and staffing, even though their existence is essentially virtual.

> CONTENT CAN REQUIRE PERMISSIONS—which can be complicated when negotiating with several industries for audio, graphics, video, and text. In an environment where copying is second nature, it is hard to be a good citizen and dangerous not to be one.

> IS QUICK TO UPDATE AND TO RESPOND TO REQUESTS—this means significant maintenance time and attention—an ongoing commitment. Assign responsibility of answering inquiries.

> REQUIRES RELENTLESS UPDATING—having a plan is important to sustain a successful and vital site (see chapter 11, page 118) with new content and experiments, new free offerings, and always new viewer benefits. Make viewer-attraction and retention marketing priorities.

CLIENT GROUP

The client tends not to know very much about the technical side of developing and maintaining Web sites, unless the Internet is a major focus of their business. So having a good match with the creative group is especially important to a site's success—much like having the right doctor makes a huge difference in long-term health. Having rapport and confidence in the development team is the only way to end up with a strong site. Guidance is more important than control in being the client of a development team. It is important to keep control of the development process while at the same time encourage experimentation and creativity. The only way for a Web site to possess vitality is by being infused with it during its creation.

MAJOR DECISIONS: WEB

- set project foundation
- plan rough project skeleton
- assemble category outline
- solicit proposals
- choose creative group
- finalize project plan
- select theme and site plan
- approve design scheme
- refine site outline and data
- review navigation design
- approve visual components
- review prototype pages
- review first testing phase
- unveil site to reviewers
- final testing phase

PREPARING FOR CONCEPT REALIZATION

Although the client does not *need* to know a lot about site development, it is wise for them to understand the most important strategic and organizational aspects. This will keep all communication and implementation in line with business plans.

► **Know a few key factors** to make sure that a Web project is heading in the right direction:

  ▷ *what to expect from the site*—both in obtaining visitor information and in potential profits. How the site generates interest or makes money has to be one of the top goals. Few organizations can afford to completely subsidize their Web sites—the reality of finance plays a viable part.

  ▷ *the purpose and goals*—how the site ties in strategically with the other organization's publications.

  ▷ *how the site affects business*—respond to new opportunities for reaching potential customers. Blend with print promotions to complete a marketing campaign.

  ▷ *how the site integrates with various other systems* and departments, utilizing a range of content procurements.

  ▷ *how to utilize available resources*—emphasize budgets and staffing. This should be established as part of the project plan but needs attention through monitoring the attainment of key dates within the project plan. (STOP! Go no further without a project plan! Go at once to chapter 8, page 70.)

► **Reinforce the project director to supervise development**—the project director stays closely involved with the day-to-day aspects while the site is being constructed:

  ▷ *compares site materials to business goals*—makes sure that they continue to fit expectations.

  ▷ *watches for sites that should link* or have content that may be applicable.

  ▷ *develops potential information sources* and Web business relationships for Web promotions and listings.

  ▷ *provides appropriate update information* and investigates all copyrights if using any found materials. (See chapter 29, page 330.)

  ▷ *knows the form of viewer participation*—their technology and how to better understand audience characteristics and interests.

▷ *allows time for testing*

> sets up helpers who can give constructive feedback and are open to refinements—especially to increase ease of use. The site needs to be flexible and continually adjust with feedback. It should never be static, but always evolving.

> refines interactivity and gives incentives to new viewers. Never forget to maximize participation from the viewer. The more they can be part of the site, the more they will return, and potentially join, buy, or respond.

▷ *knows when enough is enough*—as long as the major aspects of a site are working, details can be refined. Plan a continuous evolution of improvement. Make sure it is always under development. If production stops, that is a danger sign of site stagnation. It's hard to win viewers back if once lost.

CREATIVE GROUP

When the production phases of the project begin, the creative group's experience begins to shine. Many young groups need to become versed in the vocabulary and conventions of the various media to know how to get the most out of them creatively. Setting up the construction phases of a project requires the blend of the right skills and experience brought to the assignment. At this point in the process, the design for the site has been selected and approved, and all background components assembled. (See chapter 11, page 126.)

COLLABORATION RULES

There can't be a smooth production process without the meshing of the various roles of those assembling a publication, no matter how many members are on the creative team.

▶ **The art director** supervises the construction of both the overall site and the individual pages, working in collaboration with the designer and tech group, led by the Webmaster.

▶ **The designer** is most effective in communication strategy when conversant in both print and Web design. It is important to note the many differences in Web site production from print production. (See chapter 16, page 196, for

MAJOR DECISIONS: WEB

- set project foundation
- plan rough project skeleton
- assemble category outline
- solicit proposals
- choose creative group
- finalize the project plan
- select the theme and site plan
- approve the design scheme
- refine site outline and data
- review navigation design
- approve visual components
- **review prototype pages**
- **review first testing phase**
- **unveil site to reviewers**
- **final testing phase**
- **review updating plan**
- **secure maintenance plan**

the print processes which require a different set of skills.) The Web designer must:

▷ *operate in a different color space*—reflected color from the monitor (RGB) is complicated by the system of Web-compatible colors. The gamut of possible colors is also totally different from the 4-color process of print.

▷ *use low-resolution images*—broken up into small individual file pieces. Naming conventions grow in importance.

▷ *set up structure of site*—plan construction and formats, form guidelines and a detailed site map to allow more than one person to update it. Have backup staffing.

▷ *understand file management system of local and remote servers*—utilize the way each works—include a consistent archival policy.

▷ *develop a programming background*—know limitations of the medium and exploit tools available. Decide how much technical knowledge is needed to express ideas. To be effective as a Web designer:

> use authoring programs—and understand the difference between them in cases of translation.

> learn site construction techniques—be able to solve content treatments where needed or to come up with an alternative around a technical snag

> explore interactive scripting capabilities—understanding the interface and what elements make the user experience enticing. Know the range of possibilities to address those communication challenges.

> integrate animation appropriately—use to emphasize the most important content. (See sidebars, page 133.)

> add audio and video elements to enhance content—but do not exclude appropriate audience if technology is too advanced. Additional professionals can be added to the team with various specialties.

> perfect updating skills by finding new ideas and techniques. Always be anxious to improve core competencies and find new solutions for keeping viewer interest. Keep files of inspiring ideas and examples.

> utilize maintenance and management software—stay aware of capabilities and advancements, watch

THE STAGES OF CREATIVE WORK

> CONCEPT DEVELOPMENT. This is an exciting process, exploring various options and weighing their merits against the communication strategy.

> REFINEMENT STAGE—begins once the concept has been established. Along the way, you see the project evolve, each time becoming more refined.

> AT THE END of the concept refinement stage, the graphic designer will usually present a final comprehensive layout or mock-up to the person at your company who has final approval authority. He or she should be satisfied with everything that will go into the final product, including typography, photography, copywriting, paper and colors."

— *The Graphic Design Handbook for Business*
American Institute of Graphic Arts/ Chicago Chapter: Publisher
Carl Wohlt, Publications Chair
Tim Hartford, Concept and Design
Chuck Carlson, Writing
www.aiga.org

Using Web-compatible colors is necessary for a fast site. Users do not like to wait for image loading. There are many excellent books and online tutorials explaining the best use of color on the Web.

for new versions and online demos. Be on the lookout for what new approaches might inspire improvements.

► **The Webmaster**—needs to consider several factors during creative construction:

▷ *match design structure to each segment*—identify extreme content treatment (such as long or short lists) and the elements that stay consistent for navigation. Always encourage solutions that will make the site easiest to use for the viewer.

▷ *attend to interactivity* of the site and encourage ways to improve participation, as well as response. Be willing to change directions and try new ideas.

▷ *can correct and update site*—it always needs to be changing. Fortunately, flaws (once identified) can be addressed and changed. Unlike in print, where a mistake is magnified by the number of impressions, on the Web, an error can be corrected very quickly once found. A constantly fluid medium, it should never stay the same.

BEFORE CONSTRUCTION BEGINS

Planning the site might take as long, possibly longer, than constructing it. Once a good approach is figured out, the actual creation of the pieces and their assembly can be very efficient, but only if the project

► **has clear client goals and expectations** (see chapter 3, page 22).

► **uses as much original material as possible** to avoid the time-consuming task of obtaining permissions.

► **secures all necessary permissions** for using found or existing material. Understand image ownerships and legalities. (See chapter 13, page 158, chapter 15, page 184, and chapter 31, page 346.)

► **has content experts** who understand resources, libraries, and how to keep the information fresh and inviting to viewers. Rapport in building two-way communication begins with the developers creating an environment to house this relationship. Site creators need to be most sensitive to a structure that inspires this kind of interactivity.

► **maintains a team approach** towards its segments and defines the role of the designer versus the programmer, indicating how they work together. A team determines who will construct the various pieces, set up file naming conventions, and the composition of initial structure. (Do not assume any of these factors are known unless the team has worked together before. Roles can be fuzzy and others can easily assume a task belongs to someone else.)

BUILDING THE SITE

Although every Web site will have a unique structure and marketing position (some organizations may even be primarily virtual with little physical existence) all of them have similar concerns in finding an efficient development. Tips for development:

► **Utilize the simplest software early** in the project while assembling components—then progress to the more complicated software later, keeping file size flexible and small while content is coming together.

► **Follow design parameters** as set up during the design phase. (See chapter 10, page 100.) Understand the purpose, language, and philosophy so everyone on the team can implement:

  ▷ *begin with site map, flow chart, outline, or script*—use tree-formatting layout to make sure that the viewer does not miss any essential points in the possible choices.

  ▷ *communicate design strategy* to the technology group. Get suggestions on how the two groups can best interface.

  ▷ *determine hardware, software, and training needs*— both in general (for core competencies) and specialty (per project needs). Be able to add to capabilities with other professionals or services, as needed.

  ▷ *work on overall elements that carry through the site*— such as pace, choices, integration, visual screens, symbols, color combinations, etc.

  ▷ *establish solid navigation design*—know what mechanisms help the viewer get easily from one page to the next without missing anything essential. Study what needs to be consistent and what needs to vary, utilizing hypertext and rollover activity to heighten the communication message and best utilize the dimensions of the page. Test for consistency and clarity, committing to one treatment carried through all the pages.

Scheduling periodic design and content reviews of the site is a good idea. Many effective project plans have checkpoints built into the schedule. A site that doesn't evolve is not going to remain interesting to viewers. With the amount of competition on the Web and design differentiation, keeping a site fresh and interesting can result in returning visitors and referrals and links.

Consistency must be established to help the viewer stay within the environment so carefully created.

▷ *scan or convert all graphics* and visual materials to needed formats. Define who handles what and determine what formats are accepted—publish, if appropriate.

▷ *work on various segments*—apply the design to visual templates and structures. Identify various possibilities and utilize templates that can accommodate the variety of needed content. Evaluate as much of the content (or types of information) that the site needs to accommodate in advance. Setting up both a flexible structure and one that gives distinction to content should really only be done once. During site construction, the design is tested. The designer should be consulted to solve any exceptions that may come up when working with the tech group.

▷ *add video or audio segments* and work into scripting. Add plug-ins or other special effects. As the site gets closer to release, use the higher level programming and special effects. Keep speed as a goal, choosing the simplest solution for every communication challenge.

► **Method for receiving update information**—develop a smooth system for how new material can integrate into the structure and design.

▷ *Identify where content will come from* and who is submitting what material. Establish a schedule and follow-up method for accountability. Establish digital compatibilities.

▷ *Form guidelines*—set up the processes that have to be applied to make provided material Web-ready. Know about any conversion problems in advance and set up guidelines for those submitting.

▷ *Determine any exceptions* to content that vary from the page templates and include designer review in the solution.

PRODUCTION GROUP: TECH

Constructing a site has to be a collaborative process. The tech group makes the site a reality, putting all the elements together and breathing life into it. As always, the better the plan, the more accurate and on budget the results. The collaboration with the creative group is led by the art director.

The W3C (World Wide Web Consortium) is a standards body involved with establishing a graphic system for platform development and compatibilities. Its code validation service is available to anyone developing Web sites. "W3C develops interoperable technologies (specifications, guidelines, software, and tools) to lead the Web to its full potential as a forum for information, commerce, communication, and collective understanding."

www.w3c.org.

AGREEMENTS BEFORE WORK BEGINS

Even with creative and tech groups that know each other, are in the same company, or are working pro bono together, it is essential to have agreement documents. Don't start without defining:

▶ **budget**—best as part of the proposal and project planning (see chapter 8, page 70). When the design phase is completed, the budget should be checked against the tech quotation used. Often, time has passed between the quotation and getting together the actual project, so much can change in the meantime. If the design deviates from the parameters, negotiation for new parameters and fees should be completed before work begins. (See chapter 5, Fees, page 50.)

▶ **deadlines**—as part of the project plan (see chapter 11, page 126), a schedule is agreed upon before page-building commences. Key dates for monitoring are built into the calendar. For most projects, fast progress is a key consideration. The only way to produce a Web site quickly is to have a strong concept and plan and to keep focused.

DEFINING ROLES AND PROCESSES

The scope of the site will most determine the kind of team and construction needed. From a small business site to a large commercial site, the most important quality of a successful site is user-rapport. This can only be achieved with a strong concept, marketing vision, inviting design, quickly communicated concepts, and a smoothly operating site construction. The creative team that put together the design continues to work with the tech group on its realization.

▶ **The Webmaster** directs every aspect of a site's development and interacts with every phase and worker. The Webmaster:

▷ *has input in the design development process*—(see chapter 11, page 126) since the Webmaster is to carry out the design concept, understanding it to implement is essential. There is no better way to do that than to have involvement during development.

▷ *controls dissemination of time and budgets* to be sure site development stays on track and there are no surprises. The better the plan, the better the early-warning system to identify variations or deviations. Setting up a follow-up process will hold those creating segments to an accountability. It also can make possible incremental reporting.

▷ *bridges any gaps in concerns between the creative and the tech groups*—always using clarity and appropriateness as guides in decision-making. Be able to look past personalities and keep the team focused on the goals of the project.

▷ *works closely with the art director* on keeping the site in tune with an organization's entire graphic identity.

▷ *makes sure the plans are followed* and that the client will receive the expected benefits from the site. Responsible for quality at every point and possessing organizational skills is paramount.

▷ *communicates and reports progress* to the client, sometimes as a formal recording structure, sometimes casually through a quick phone call or e-mail. (E-mail can be especially valuable for conveying financial deviations.) This reporting should be as early as when deviations are discovered. Costs needed should always be included in the report.

▷ *works to help the site grow in use*, popularity, visibility, watches for marketing ideas, and opportunities for ads and links, and has leadership in enhancing the site, making it an important business resource for clients.

▷ *establishes momentum for the evolution and growth* of the site, making sure that it keeps within the chosen design strategy.

► **programmer**—structures the site and constructs the pages:

▷ *receives data in usable formats* from the creative group. If any conversions are needed, communicate related fees and schedule needs to the Webmaster. If there are compatibilities to work out, experimental time should be allotted.

▷ *builds the graphic structure* according to approved designs and template:

> combines all images provided by the designer with incoming text and data; constructs code or scripts for all the pages.

> adds information and data to templates according to design and styles. Sometimes this is set up formally and the design is carried cross-media. (See chapter 8, page 70, chapter 12, page 148, and chapter 19, page 236.) Sometimes it is a matter of simply dupli-

cating some styles. The importance of new copy should determine its treatment.

> understands parameters of design structure and knows when to follow the template and when to communicate new design requirements back to the creative group for them to design new treatments.

> integrates as part of blended media strategy (see chapter 8, page 70, chapter 11, page 134, and chapter 19, page 236), incorporates print materials, packaging, and delivery methods.

> works with the designer on file naming conventions and the form for transferring files.

> uses authoring programs to allow for a smooth and efficient transition of designer-to-production files.

> debugs and perfects the navigation, flow, and sequence of the content, based on client and designer feedback.

▷ *arranges for testing* and gathers input of users—the tech group communicates new design requirements to the creative group and helps redefine solutions for areas that need improvement or expansion.

▷ *has setup for handling fulfillment and feedback methods*—responsibility for handling site-generated inquiries should be well established and defined under a single individual's purview so responsibility is taken.

PRODUCTION GROUP: PRINT

Most printers aren't involved with their clients' Web sites, but do utilize the Web in their business and production operations to:

▶ **provide customer services for quoting and tracking projects**—even up-to-the-minute checking on progress.

▶ **transfer and receive transmitted project files**. (See sidebar, page 231, for issues concerning responsibilities in the transmittal of data.) Many companies are adding archiving or asset management service to their profit centers. (See chapter 15, page 184.)

▶ **promote to the buyer community**, giving company information and raising visibility of company capabilities and geographic reach.

▶ **offer new services** such as interfacing with image libraries (see chapter 15, page 184, and chapter 25, page 301) or cross-media publishing. (See chapter 8, page 75, and chapter 11, page 134.)

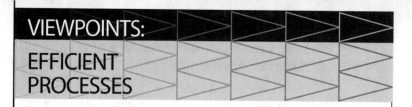

VIEWPOINTS:

EFFICIENT PROCESSES

"Recruitment advertising is growing in new ways as the digital revolution continues. The greatest change I foresee will be the increased use of PDF files electronically submitted to publications. E-mail and electronic markets like online job sites will impact the graphic designer and force new directions for creative professionals."

—Karl Kofoed
Orenstein Advertising
Philadelphia, PA

"Besides all the new technology, new media design is about teamwork. Gone are the days of a single designer being able to create and produce projects alone. Now you're part of a team of specialists, from network engineers to programmers to multimedia and animation artists. Get used to it. You'll find it's fun to create with others."

—Shawn Freeman
focus2
Dallas, TX

"We have come to a strange point of crystallization. The paradigm shift in our profession is not so much the emergence of new tools, but what we have to do to tackle that emergence; this is the generation of designers who have to constantly reinvent themselves. Previously designers received education, entered the workforce, and spent the rest of their lives perfecting their art. Now, a couple of years after we graduate, our level of technical knowledge is already outdated. The software changes, newer technologies emerge, and we scramble to update ourselves."

—Vineet Thapar
Oen & Associates
Memphis, TN

"When the user experience is determined by the screen size of a cellular phone and by push-button navigation, design will undergo yet another transformation."

—Pamela Pfiffner
General Manager and
Editor-in-Chief
www.creativepro.com
Portland, OR

"Designers should expect to see more outsourcing. Therefore, we should expect integrated communications to move to the forefront of design as technology catches up with our ideas. It is becoming easier and less expensive to mix media and thus to convey our ideas directly from our mind's eye to paper or computer or film or all of the above. Time also is quickly becoming less of a factor as these new technologies allow the designers to compress the required time frame, to produce more in the same time."

—Dave Romeo
Wet Ink
Bronx, NY

"Since the approval process may involve more than one client representative, expect changes at each decision-making point. It is important, however, that the client's key contact person keeps track of and agrees to all changes before the designer makes them."

—The Graphic Design
Handbook for Business
AIGA/Chicago Chapter
Carl Wohlt
Tim Hartford
Chuck Carlson

"New visitors might now wait for a lengthy download on a high bandwidth front page. Even the entertaining free stuff will irritate serious, repeat visitors if alternate points of entry or direct links to the core of the site aren't easily accessible. The best sites hook an audience before the audience even knows it.

"In entry tunnels, it's no longer practical to ask people to register. Some entry tunnels say, 'Register here free!' Who wants to register for free? If you really want people to register, you'll have to give them something major in return. Registration is a barrier to entry—be sure you need it before putting it in.

"New approaches to registration will replace today's pleas for marketing information from surfers The Web is advertising-driven. The advertisers will figure out a way to get what they want to know. In the future, your browser will be able to keep a fairly rich profile on you and automatically tell sites much more about you than they do today—subject to your approval, of course."

—David Siegel
Chairman
Studio Verso
San Francisco, CA
www.killersites.com

"Designers know that if people don't consume their content, it doesn't matter how well structured it is. Designers put design above structure. Designers have taken the structuralists' HTML and bent it horribly to suit their visual needs. I believe design drives the user's experience of the content; it is the designer's responsibility to present content appropriately. Who cares how powerful your database is if people can't use the interface? Who cares how great your content is if people aren't attracted to it or don't find it pleasurable to read? But as I predicted, we can now swing the pendulum back toward structure and layout, whereas before the choice was structure or layout."

—David Siegel
Chairman
Studio Verso
San Francisco, CA
www.killersites.com

"Hate it or not, the availability of information online and the speed with which it can be retrieved will probably shape (but not necessarily overthrow) much of the print world. Multi-platform formats (like Acrobat or programs that can deal with PostScript files) will probably see a bubble of activity. Of course, the nature of computerized documents, which has seen a geometric increase in compatibility between Mac and PCs, will probably change to something we can't even recognize within five years. So the best we can do is wait and cross our fingers."

—Lloyd Swift
This! No! That?
Houston, TX

"There's just too much information out there, and while it's easy to search for specific words or phrases, it's relatively difficult to search for meaning.

"Once you have assigned meaning to content, you can specify rules for how that content should appear, depending on the medium you're using.

"Turn your page-layout program into just another stop along the way to wherever your content is going. The key is separating your content and the form it takes.

"By tagging text with XML, you'll be able to tell how to translate meaning into form. All your content can be tagged, and those tags can move from one program to another .

"Content and form are fully separate, and your documents can shape-shift depending on the medium and the message."

—David Blatner
Graphic Arts Consultant
Seattle, WA
www.creativepro.com

"You need to develop a level of respect between teams. Engineers believe they can design a user interface when they have little understanding of the design process, and designers believe they can program Web sites when their knowledge of engineering is limited."

—Audrey Witters
Director of Engineering
iSyndicate

"Print production is very linear—you finish writing and hand it off to a designer, who hands it off to prepress, and so on. But on the Web, there is no direct handoff. It's a far more fluid process, with the early participants having a greater stake in later phases of the design."

—Maria Guidice
Creative Director
HOT Studio
San Francisco, CA
www.creativepro.com

"We need to remove barriers between designers and developers. Designers and coders need to talk every step of the way. The best Web sites are those where teams take the effort to understand each other's worlds. Graphic designers need to understand engineering limitations, and engineers need to understand why color and typography are so important."

—Aaron Oppenheimer
Design Lead
Art Technology Group
Boston, MA
www.creativepro.com

"Designers are beholden to coders and programmers to realize their vision. Designers need to forecast potential usability as part of their complete creative vision."

—Lisa Beaudoin
Art Director
America Online

"You have to make the leap from the computer as a tool to the computer as a medium.

"Blurring, the integration of design and production, is absolutely a good thing. It allows designers to have a much closer link to their craft. Yet I think there is also a danger of becoming good at all things and excellent at none. The designer becomes a jack-of-all-trades and is no longer a specialist.

"Not knowing how to hand-code is like not learning how to kern type. For some designers, I'm sure the mentality is, 'I'm not a programmer.' Well, hand-coding is not programming; it's like kerning. It's about craft."

—Clement Mok
Designer
Sapient Corporation
San Francisco, CA

WORKGROUP COLLABORATION

As communications audiences grow, so must the ability to target and customize. Develop a broad background in the project components to build in flexibilities and best service expanding challenges. Put together a team with balanced skills, supported by software tools, to speed development processes and make production more efficient.

DEFINITION OF WORKGROUP PUBLISHING

When more than one professional is networked in a work environment, aided by commonly used software, efficiencies and connections can be realized between groups. A fluid process can enhance collaboration and quality of the product. A workgroup

► **shares files** where libraries and active documents are disseminated to a network of users via a server. On the low end, there may only be a few users connected. On the high end, it could be entire departments of editors and designers, or a channel of project progress from designer through to the printer or Web site.

► **allows simultaneous work** on one document by more than one worker through interactive software. This can connect many workers in a variety of locations, and be in actual time anywhere globally.

COLLABORATION TOOLS

Working interactively with other workers is a blend of technology, project parameters, and peoples' habits. Although roles can blur, the process itself is very definable. It can be placed in a software format that brings together many workers over a network.

► **Workgroup publishing** is more than just transferring documents from one computer to another; it:

  ▷ *allows simultaneity*—several people can work on the same document at the same time from different locations.

  ▷ *uses communication speed* in project development— from e-mail to sophisticated authoring and page composition systems, automation equals less development time.

- *implements a system with version control*—to eliminate confusion caused by "versionitis" and determining who is permitted to change what content.

- *uses customized lockout* with role definitions for editing and for security. Only certain workers are permitted access to certain content and documents.

- *tracks progress and monitors process completion*—if kept current, anyone authorized can check the system for up-to-the-minute information.

- *coordinates project and work flow management* across departments and locations. Almost all systems need customization. Follow development to keep on track and on budget.

- *shares information* through databases and messaging functions. Take advantage of new channels that continually become available as more compatible users become linked.

- *uses the interactivity* and collaboration with the audience towards a common goal.

- ► **People make the system work**—the essence of groupware is dealing with work relationships: owner to user, requester to fulfiller, coordinator to members, assigner to doer, proposer to approver, etc. Groupware is abstract and those using it don't need to worry about the system underpinnings. Multiple workgroups can also interlink and overlap, coexisting in a webwork of productivity.

CLIENT GROUP

What the client wishes to accomplish through a workgroup will determine the equipment and system used. When adopting a new system, cultural changes are necessary. Workgroup tools can't be assimilated into an organization as simply a new way to do what is already *being* done. These tools will change both the nature of the work itself and the way the work is being done.

THE CHANGES WORKGROUPS INSPIRE

To transform the way that work is done, the system must concern itself more with the process and the way that people work than the development of individual projects. Getting the momentum and connectivity of the workgroup can't help but dictate the way that work is organized. Changes include:

- ► **role redefining and adjusting**—because workgroups are linked digitally, who sets up files and how they are built, managed, and updated becomes an efficiency concern of the entire group, in addition to who is responsible for each segment. Assignments and accountabilities should be set up as part of the project plan (see chapter 8, page 70).

► **work flow transformations**—the flow is not linear, but can move in many directions simultaneously. Identify a critical path that will help keep clarity and the project moving in a consistent direction.

► **collaboration between workers** in new ways as quick decisions are made. Relationships must evolve with new development and team experiences.

► **cross-skill sets** to transfer easily from one function to another. The more techniques are learned, the easier they are to learn. Staff should both collaborate and back each other up.

► **determining ways to troubleshoot** and constantly improve and evolve the system. It should never become static.

► **setting guidelines and page standards** that fit an organization's unique way of working. This is critical in groups: whether a formal workgroup system or just file sharing, it

  ▷ *enables a fluid production between projects*—efficiencies gained from one project applied to another, and building consistency on file handling.

  ▷ *allows more than one person to work on a project*—workers can help each other and they all use the same conventions.

  ▷ *alterations and changes are most cost-effective* with good and consistent file construction—updating can be most efficient when the design plan incorporates an updating scheme.

► **setting up appropriate time frames and work expectations**—workgroups can easily spin off track by either developing the wrong aspects, or spending too much time in too few areas, neglecting others. Keep a balanced development.

► **scheduling work sessions with participants**—be sure the decision-making team is complete before embarking on a session, as anyone missing may cause much work to be redone later, once the missing person rejoins.

CREATIVE GROUP

The software can promote a greater collaboration between the writers, editors, designers, and the producers of documents. Small organizations may work in tandem with other small groups or directly with their clients. The originator of the file, so important in copyright definition (see chapter 30, page 340) becomes harder to determine.

HANDLING STRUCTURED WORK PRACTICES

Most documents will originate with the client; each participant adds their component to the project. The creative group:

► **defines what each participant contributes** including the balance of expertise and skill sets. Know who owns what material before using. (See chapter 30, page 340.)

► **creates images and text** in their own files for exportation, only releasing unencapsulated files to others on the team. This guards intellectual property.

► **devises design strategy** that allows collaborative working, breaking large projects into definable segments.

► **communicates limits of participation and capabilities**—if the client, or other sources, are contributing content and staff talent, be sure who is responsible for what by conveying the scope of creative group activities and outlining what is expected from each participant.

► **helps network-mates understand their role in the process**—particularly subcontractors or freelancers who will contribute a portion of the total project. Give them context and carefully prepared instructions to help them anticipate the best ways to integrate their portions.

► **learns security software controls** to lock design elements so that the unauthorized cannot change them. There is a temptation, especially when under deadline pressure, for editors to alter graphics to fit. Consult the designer if any visual elements need to change. There may be a better visual solution to making changes unknown to an editor.

► **learns forms of version and annotation techniques** to clearly communicate through the evolution of a document, saving versions as they develop, and to know who did what in case of questions.

► **writes conventions for group to follow**—sets up guidelines, and conforms to page standard conventions of the group. No one can work effectively together if everyone works differently. Taking time to do this will save time in production when deadlines become very short.

Edit control in collaborative computing enables the document originator to control how other individuals interact with their document during a real time computer conference. One user may only be able to edit at one time, or all can edit simultaneously as at a brainstorming session. What each can change can be limited by their job function—i.e., editors can't change design elements.

PRODUCTION GROUP: TECH

When operating in the momentum of a system that originates with the creative group, connect into the project with the selection of design, where possible. This can help set up processes that can be applied to the entire implementation. It is best if the tech group is part of the project planning (chapters 7 and 8), setting up realistic budgets and deadlines.

PROPELLING MOMENTUM

Under the direction of the art director, all creative processes are worked into the work flow for brainstorming and presentations. Creative work flow for Web creation can be seamless if designers and programmers work closely on:

► **file name and organization conventions**—establish guidelines for the workgroup so

▷ *all segments are built the same way* to enable growing staff participation and backup.

▷ *new team members can review* and become part of the group quickly.

▷ *additional material can be added easily* or it will be hard to sustain long-term—the downfall of too many Web sites that start out strong.

▷ *redundant work can be avoided between groups* who recreate files because they don't work in the same way. This cost can be avoided with an efficient file system.

► **defined responsibilities** based on plan. (See chapter 17, page 208.) The tech group, charged with the design and functionality of the system, needs to be cognizant of:

▷ *ways the site is used*—be sure to understand intention and emphasis of site components.

▷ *technical capabilities of audience*—and profile of average user to scale structure and versions that may be needed for higher level audiences, members, or a select group of users. Develop versions based on profile of constituents.

▷ *ease of learning and ways to improve*—always be open to change and try techniques to intrigue viewers.

▷ *target glitches and fix bugs*—always be on the lookout for links that don't work, clumsy or inconsistent navigation, and clumsy interfaces. Finding the simplest and most direct solution to any creative challenge is always the best idea.

▷ *availability and speed*—file size limits most delivery systems. Users don't want to wait for images to download unless they are patient for a reason. It is inexcusable to have large images on a homepage that needs to be a fast portal for the experienced visitor.

▷ *training plan for staff*—once a site is set up, there needs to be an orientation for any new team members. Each site is unique in its construction, so allowing time to become familiar with its vernacular is essential for later efficiency. (See Training, chapter 9, page 88.)

In large information system technology, digital data interchange accelerates the acceptance of open systems. In the publishing world, hardware and software systems are still mainly proprietary, but that is changing. Open systems in publishing will allow a document to be opened in any software to utilize that particular software's features. Different software packages will also be more linked, allowing updates in one package to automatically update in the linked package. This level of fluidity will cause workgroups to be a lot more expansive and utilize the tools to best advantage.

The production processes will continue to compress the distance between the designer's monitor and the digital press. The printer will be part of the workgroup and will have input for printing specifications before the project is technically finished. Often a printer publishes specifications in advance, to help the creative group know how to best prepare documents.

TYING INTO NETWORK AND DIGITAL RECEIPT

Joining the digital work flow, the print group faces a process that grows more and more digital, until it will be a seamless connection from creator to device that will make specific impressions or outputs. The Web will open opportunities for remote printing. Workgroup production capabilities should include:

► **digital price estimation** through various choice scenarios of press sizes, etc. This will make the quotation process faster and more interactive for the buyers who can explore scenarios and variables in their selection process.

► **ability for buyer to check on status of project**—fluid systems, when kept current, can improve communication between buyer and printer.

► **online paper specification and availability**—any paper merchants have Web sites as tools for designers to specify their paper. Exploring materials is easier in an environment where such materials are available instantly.

► **customization of print orders**—quantities and locations of delivery can be more sector- and audience-specific. Special considerations should be communicated to the entire team. When cross-referencing with asset management systems (see chapter 15, page 184), the control of modular material should be figured out in the design process.

► **electronic transference of prepared data**—such as submitting advertisements to multiple publications, or remote proofing (see chapter 24, page 280), or printing. (See chapter 25, page 288.) Digital documents move fluidly around the network and demand control of versions, timing, and correct distribution. Campaigns can be carried out very efficiently.

► **direct interfacing with distribution mechanisms**—such as direct mail service, broadcast faxes, multimedia, and online services. Integration professionals should be brought in to bridge the various systems and meet the parameters of each's potential. Time to market is greatly reduced through work flow development.

RESPONSIBILITIES OF TRANSMITTING FILES

To send text or image data from one location to another, there are several choices:

> TRANSPORT BY PHYSICAL COURIER for original artwork, disks, CDs, etc.

> DATA SENT ONLINE:

• the sender of the data is responsible for its usability.

• hard copy should be sent as verification and follow-up. Faxes for verification are most often used when working remotely.

Care must be taken when transferring files either through modem or by disk. Different software and hardware configurations may have incompatible formatting. Technology changes quickly in this area, but it may take several attempts to get files to transfer. The client should be warned that this is not always a smooth process initially.

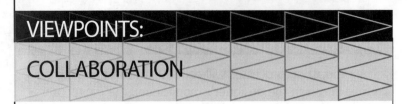
"The changing world of information technology is similar to what has happened over the years to clinical medicine. People once went meekly to their family doctors, who diagnosed their ills, prescribed a therapy and provided it—no questions asked. Today, a physician's diagnosis likely will be greeted by a patient holding two magazine articles on new therapies, who leaves the doctor's office to get a second opinion.

"Likewise, technology professionals, once seen as almost mystical information keepers whose centralized department was known as the 'glass house' in popular jargon, have lost their right to ladle out technology as they see fit—no questions asked.

"It is a profound and widespread shift in the modern business landscape, wrought in large part by technological expansion that appears to be accelerating out of any one group's control."

—Jon Van
Reporter
Chicago Tribune
Chicago, IL

"In the land of enterprise-level Web sites, there are also enterprise-level content publishing and management systems.

"These are meant for large companies or publishing operations that have content developers across states, countries, and time zones. Dynabase is very powerful in that content is stored using XML (extensible markup language) in tags and can be published in many different ways. With Palm Pilots the rage, it won't be long before developers will need to optimize sites for hand-held computers—Dynabase allows for the simple conversion of large amounts of content by changing the publishing attributes."

—John Schultz
Director
DigitalNation Interactive
Alexandria, VA

"The weakest link in your workgroup is the most inexperienced user."

— Gary Carder
President
CarPec Graphics, Inc.
Akron, OH

"When troubleshooting, your credo needs to be: 'God grant me the strength to fix the problems I can, call someone in for the ones I can't, and the analytical skills to figure out the difference.'"

— Anne-Marie Concepción
Seneca Design & Consulting
Chicago, IL

"The DP manager's creativity is often the reason for unwarranted growth. They enjoy devising and then testing new applications. Unfortunately, they rarely propose cost-effective methods of using these applications. A creative, unmonitored computer technician can incorporate new applications and programs to the point where costs exceed value. Most growing companies feel their systems are inadequate, but the root problem can usually be traced to ineffective systems planning."

— *Small Business Report*
April 1983

"Desktop publishing is often a solitary endeavor, and it's marketed that way, as if that were a good thing. But what we've lost is the team of professionals who backstopped each other, so the specialists in each field did not make idiots of themselves in areas where they lacked expertise. There were often fights, but they led to creative solutions. The intellectual conflict resulted in better, fresher pieces because two or more heads are better than one."

— Jan White
Designer, Author, Consultant
Westport, CT

"Workgroup technologies are only successful if you understand how they will improve your business process."

— George Goldsmith
President
The Human Interface Group, Inc.
Wethersfield, CT

"I believe that productivity level has greatly increased in our company because of technology, proper training methods, managing job descriptions, setting standards, making sure that everybody knows the proper end product and all the steps that are necessary to get there, and to maximize that they are part of that loop. They must know the job that is ahead of them, behind them, and how it all works together. And from an owner's standpoint I am seeing productivity increases that are reflected in the bottom line. This allows me to continue to invest in new technology, to keep people paid well, to send them to seminars, to educate them and bring them along in this new world we've created. We're hit, though, with so much new stuff, what does the production staff do first: learn or produce? Cross training is so important: knowing the jobs of two or three different areas—although you know you don't want to be a typesetter, you must learn to be a typesetter. Although you don't want to be a designer, you must learn what design means to the total project. You don't want to be a scanner operator, you must learn what the scanner operator goes through to get the flow properly into the network."

—Gary Carder
President
CarPec Graphics, Inc.
Akron, OH

"Managers tend to concentrate on the expenses of hardware, software, and training. They don't realize they need a fourth evaluation: new responsibilities and new work flow.

"We find that the technology has completely changed the services of our in-house graphics department. Many in our bank who used to depend on our services for their presentation needs are now doing it themselves. With them handling the execution of what they do, the quality has gone down and many don't realize this because they fall in love with their own work. Instead of selling or doing other things they are supposed to be doing as part of their jobs, they are playing artist. They are enthusiastic, but not skilled.

"So now the graphics department is called on to fix up and finish many of these presentations."

—Cathy Reedy
Art Director
Johnson Publishing Co.
Chicago, IL

"Designers who don't learn how to work within enterprise solutions will lose creative authority. They'll be relegated to sitting in front of a computer assembling individual elements instead of entire documents.

"Even the driest technical document can be made more effective by good publishing tools. That's why it is exciting to see companies finally providing products that accommodate both the schematic-obsessed engineer and the marketing specialist who knows how a color scheme can enhance readability.

"Designers need to get over their arrogance about design applying only to certain documents and not others. There is tremendous opportunity out there, and like it or not, it revolves around the enterprise and its voracious appetite for collaboration and compatibility."

— Gene Gable
www.creativepro.com

[Reprinted with permission from *Publish* Magazine]

"A workgroup environment has improved our productivity. We are able to communicate because of the definition of each job. I'm a stickler on procedures. I laid the work flow out precisely and defined each person's position and job. An editor would not touch a picture and change it when they are making edits to their text. The flood of information that has been hitting us has improved productivity if you improve your people and communication skills.

"It makes people take on more personal responsibility for their jobs. In the past, you always had someone else to blame. You would send your text out to be typeset and the typesetter made the mistakes. If you had someone shooting a stat for you, and they did it wrong in the darkroom, you had them redo it. But now, if you know what your role is and it is getting defined, then you have to take responsibility for it. I see it as an advantage."

— Christina Nemenic
Harris Bank
Chicago, IL

PRESENTATION GRAPHICS

Exciting venues of presentation media continually open a landscape of possibilities. From screens reaching directly to consumers in their living rooms, to kiosks in retail centers, to screens inside cars, to portable computers, multimedia infiltrates many levels of society simultaneously.

DISTINCTION BETWEEN PRESENTATION AND INTERACTIVE MEDIA

The purpose of presentation graphics is to promote information, explain products, offer services, or communicate a specific message (such as an event). The desired effect on the audience could be inspirational, impactful, or educational. Presentation graphics tend to explain a subject to an audience—to communicate one-directionally. With CD deliverables, more interactivity can be introduced so viewers can make choices in their experiences. (For two-way interactivity, see chapter 8, page 70, chapter 11, page 126, and chapter 17, page 208.)

► **Presentation graphics** encompass a static visual format that conveys information to a passive viewer through:

  ▷ *slides*

  ▷ *overheads*

  ▷ *videos*

  ▷ *documentaries*

  ▷ *CD-ROM deliverables*

► **Multimedia** has motion and combines text, graphics, video, animation, and sound. (There are those who claim that presentation, interactivity, and the Internet are all subsets of multimedia.)

► **Interactive media** is multidirectional between the presenter and the viewer:

  ▷ *control the experience through choices.* As a one-to-one medium, recipients can view whenever they want, and view at their own speed and at their chosen level of detail.

Redundancy is very necessary in interactive presentations. Unlike print (where redundancy is both annoying and wasteful), when depending on time, perceptions, and attention, repeating key information is appreciated and heightens the user experience. (See Web Design chapter 11, page 126.) Consistent use gives the viewer a sense of comfort and logic in the presentation, and the viewer will comprehend more of the content presented.

- ▷ *utilize a branching structure*—open to selection and different for each viewer. Guard that the most important information can be obtained no matter what choices are being made.

- ▷ *utilize new deliverables,* such as information kiosks, conveying localized information according to user selection.

- ▷ *adapt editorial information* to digital magazines, customized by sector and reader interest, available via subscription (CD or online).

- ▷ *explore interactive publishing via CD-ROM*—CD books grow as a market along with the ease of hardware usage. Many publishers invest in new delivery scaled to "palms" or handheld devises. The lack of platform standards holds back a full explosion of development, though many traditional book publishers are perched to pounce as soon as the hardware catches on. (How readers will adapt to pure digital delivery remains to be seen.)

- ▶ **Blended media** combine several communicative vehicles into a campaign where a conceptual design theme is applied to print, presentation, interactive media, Internet, broadcast, and display. Clients will gain from such consolidation, both in market consistency as well as economically, when elements are shared between uses.

UTILIZE PRESENTATION MEDIA

Presentation media is meant to create an enriching experience through:

- ▶ **live presentations** such as lectures, exhibits, demonstrations, or displays, aided by visual presentation media.

- ▶ **online publications** that can be a subscription service, delivered through the Internet (see above).

- ▶ **electronic books** available for handheld devices (see above).

- ▶ **manuals and directories** that can be distributed on CD-ROM (more expensive to update in print and harder to keep current).

- ▶ **instructional devices** that give "readers" choices—these are being used in self-training by many corporations, saving overhead on instructors and lecture halls.

- ▶ **augmented presentations** such as for sales or education—laptop computers have become common for salespeople to show company capabilities.

- ▶ **reference materials**—utilize digital delivery when content needs to be accessible but doesn't need to take up permanent hard drive space.
- ▶ **broadcast, newsletter, and bulletin faxes**—still considered presentation even though the delivery is more like print. In its immediacy, it is a form of presentation graphics that should be used strategically with other venues.

The growing flurry of development and activity is untempered by incompatibilities between formats and lack of standards. Manufacturers are in a race to produce the best and the most comprehensive new possibilities to reach an audience. The audience follows developers—what becomes available (if affordable) gets used. The developers lead the presentation market.

The trailblazer for making information more visual was *USA Today* in the 80s—this publication supported the reader's need for fast 'bits and bites' and appealed to an accelerating communication environment. Now, short attention spans demand the condensing and presenting of information for quick comprehension.

FROM DEVELOPMENT TO USERS

The market for presentation graphics is growing, and so is competition between the providers of development services.

Currently, there are many advantages and disadvantages to the new media:

ADVANTAGES TO PRESENTATION MEDIA:

> expands the use of existing images through the repurposing of (finding new uses for) content.

> utilizes time independence and portability.

> involves more than one sense of (sight and hearing) perception. The more senses are used in a presentation, the greater the recipient's retention.

> integrates with other media and applications, such as with still pictures, animation, sound, and video. When all materials work together, the presentation becomes the 'hood ornament' of marketing, supplying further involvement from those who show interest.

> can be inexpensive to deliver (via CD), though costly to create. (Most fees are in the setup, programming, and creation of the content.)

> makes complicated information easier to grasp by condensing and editing; presentation technology can also make content more visual.

> can be unusual, making promotions stand out from competitors, developing levels of sophistication. In a world where high school students can produce a professional-looking presentation, there is no excuse for any professional presentation to look crude.

DISADVANTAGES TO PRESENTATION MEDIA:

> hardware and software may be incompatible and dependent on the viewing device, often not portable. Presenting on location offers many challenges. If traveling with a presentation where projectors or computers are needed, plan these needs in advance and allow setup and testing time. Many speakers have been crippled when their presentations' formats didn't work with the hosting facility's.

> only one user at a time can experience, limiting the ability to have a simultaneous mass audience.

> can be expensive in design and development, with most of the costs going to setup programming and content development. An ongoing budget is needed to be maintained to update presentation materials, as many can be very time-dependent. The design should allow for ease in updating.

> permissions can be complicated when negotiating with several industries for audio, graphics, video, and text. (See chapter 31, page 346.)

> lack of standards in platforms, formats, and vernacular holds back significant development. Users will adopt what is easy and what is cost-effective. Lack of standards keeps the industry fragmented and complicated, and, therefore, slow to be adopted in use.

CLIENT GROUP

The astute client is not seduced by media technique, but is focused on the purpose of the organization's marketing efforts. To the educated manager, new media are valued if they present new distribution opportunities to further the organization's exposure. The medium does not change the purpose of the message but rather the form of the message and the way it is both delivered and received.

UTILIZE MULTIMEDIA EFFECTIVELY

Presentations are part of every organization's arsenal of marketing weaponry. It is rather like the precision of single rifle fire versus an overall bombing effort. More positively, it is focusing on specific markets or subjects versus the overall approach of advertising. To make the best use of multimedia, the client must balance:

► **business goals and expectations**—this never goes away, but must always remain foremost as the road map through choices, egos, and conflicting ideas.

► **audience characteristics**, interests, and capabilities—formal market research is not always needed. But it is true that the more the client and the creative groups understand the audience, the more effective the presentation.

► **background information** and current image resources. (See chapter 3, page 22, for a list of helpful information.)

► **potential information sources**, careful of copyright (see chapter 31, page 346). Know where provided information comes from or don't use it.

► **available content with new content**—the more original the content that is used, the better. Make sure to protect intellectual property. (See chapter 13, page 164, for encryption, and chapter 31, page 346, for copyrights and legal enforcement information.)

► **structure needed for a creative and production process**—understand basic capabilities and phases of development. Make priorities obvious and continually as the basis for all decisions. Use to navigate through personality differences and conflicts within opinions. Do not place form over function or function over form, but balance both. Stress ease if conveying information.

► **support for project director** who is assigned the day-to-day authority for developmental decisions. As a liaison between the groups, it is the project director who accepts the ultimate responsibility for a project's outcome.

There are more professionals involved in multimedia processes than in print publishing. Most designers have presentation skills, as well as print and online construction skills. (See page 66 for skills.)

BALANCING TALENT AREAS

The development team is led by the art director, and generally includes the clients' marketing professionals and project director, the writer, and the designer. The team can also include:

► **producer**—handles project management, management of components, financing, scheduling, and promotion.

► **director**—in charge of scripting, treatment of photography, editing processes and staff, and creation of product.

► **production staff**—can have many configurations from just an editor to an entire postproduction facility.

► **audio and video professionals**—talents that are project-specific such as cameramen, sound specialists, videographers, and musicians.

Presentation projects can be as small as a slide show and as big as a multimedia sales presentation. The team needed for each kind of project scales to match its scope.

PROJECT PROCESSES

Multimedia offers possibilities for new forms of visual language. Hardware and software are growing rapidly with capacity and complexity. When creating multimedia projects:

► **plan, plan, plan**—it cannot be encouraged enough. The better the plan, the more potentially the design will be on target and the smoother the production processes.

   ▷ *Know client goals and expectations* before beginning a project. Don't start without it and do share this information with the entire team.

   ▷ *Have a team approach*—no one person can be an expert in all areas of multimedia. (See strategy, chapters 7 and 8, pages 62 and 70, respectfully.)

   ▷ *Become content experts*—understand resources, libraries, ownerships, and legalities. (See chapter 29, page 330.)

   ▷ *Use original material*—always the best idea so that ownership is never an issue. Interface with asset management systems when possible. (See chapter 15, page 184.) This can be limited by budget constraints.

   ▷ *Secure all necessary permissions* for existing material—request that clients do with provided material.

   ▷ *Utilize the simplest software early* in the project while assembling components and then progress to the more complicated software and additional media later in the project. (For other ideas for development check chapter 16, page 196, for print and chapter 17, page 208, for Web sites that may have parallels.)

▷ *Lay out production calendar*—with as much detail as is possible to that the plan can target for deviations. (See chapters 7, page 62; and chapter 8, page 70, for other planning suggestions.)

► **follow design parameters** established so far in the process (as accomplished in chapter 10, page 100, and chapter 16, page 196):

▷ *begin with a flow chart*, outline, or storyboards (similar to Web development in chapter 11, page 126).

▷ *determine capabilities* in hardware, software, and training needed for project. Set up for any outside resources that will be required.

▷ *roughly score and determine flow of the content*—identify major categories or segments.

▷ *use a tree format* for interactive structure to ensure viewer does not miss any essential points in the possible choices. Make sure content repeats in strategic places.

▷ *design interface and method to inspire the viewer to interact* if a live audience or a virtual audience. Success of reaching the audience is determined by whether they respond the way the publisher desires.

▷ *scan or convert all graphics* and visual materials to needed formats. This may involve cleanup, especially for multimedia viewing, which has different visual requirements than print or Internet graphics.

▷ *work on various segments* and design pace, visual screens, consistency versus changing elements, choices, and integration of all components.

▷ *add additional media to visual files*—such as an audio voice track, working with voice talent, or adapting musical clips. Storyboard for careful planning.

▷ *synchronize animation* to the audio track, and other special effects after the essential structure is approved.

▷ *supervise tech group* in development of content applications, updates, and any variations needed. Know in advance the various uses and longevity of the project, as single presentations can be designed very differently than an evolving series or one that must be regularly updated.

"THE WEB PROVIDES TOOLS TO:

> CREATE DYNAMIC PRESENTA- TIONS through the use of sound, animation, and movement.

> PUBLISH DYNAMICALLY—not just one time. You can update the site with new (and deeper) project content at will and with relatively little expense.

> via the most basic hyperlink technology, PROVIDE VIEWERS WITH MULTIPLE AVENUES OF EXPLORA- TION through the project. This allows the creation of a multilayered, multifaceted presentation which the viewer can be free to explore in their own way depending on their interest in the project's subject matter.

> PROVIDE OPPORTUNITIES FOR VIEWERS TO CONTRIBUTE TO THE PRESENTATION ITSELF: e.g., by adding comments, reactions, and stores/images of their own. On- line forms of community (e.g., forums, chat) can easily be created around the project theme, as well."

– Sheldon Potter
Web Designer and Photographer
The Essential Connection
Denver, CO
www.essential-connect.com

PRODUCTION GROUP: TECH

Much, but not all, of the creative control lies in the hands of the creative group. Applying content to the design format can be the role of the tech group, which can include programmers for complex scripting and effects and editors (both script and film), as well as musicians, video artists, etc. Many roles overlap as each element contributed works with the others. Often, video production backgrounds overlap with those of the designer. The collaboration needs to be nurtured to avoid competition.

RESPONSIBILITIES OF THE TECH GROUP

Similar to the many professional hats a designer can wear, the techie has a great choice in focus for multimedia presentations. However, each member of the tech group

▶ **is responsible for functionality** of presentation—both hardware and software, in every appropriate location of delivery.

▶ **combines images, text, and sound, according to design template**, and works closely with the creative group to

 ▷ *develop information in a sequence* along branching structure to allow for viewer choices.

 ▷ *refine menus, navigations, and transition mechanisms*—working with the designers on the understanding and implementation of the visual strategy.

 ▷ *arrange for testing sources*—then incorporating feedback, calling upon the designer as needed, to work out exceptions to template or make additions to design.

 ▷ *debug and perfect* the flow of information.

▶ **integrates cross-media**—with print materials, packaging, and delivery methods, as part of an organization's overall marketing plan. (See chapter 3, page 22.)

Every project is different in its work flow due to the huge variety of content and formats possible in the presentation media. Be sure to collect those with the best expertise to ensure a cost-effective project and not an expensive learning experiment.

PRODUCTION GROUP: PRINT

This is not really a print group anymore, but an "output group." Many imaging centers handle not only high-resolution output for print but computer-to-video production. Many more of these companies have arisen from the video side of the industry and can provide editing services, charging by the hour. The division between the groups dissolves under collaboration, but the basic needs of planning, communication, continuous learning, and servicing new buyer need remains continuous.

"The complexity of work will likely bring overlapping media to the designer's domain; knowing more about these interweaving disciplines will be necessary. We cannot be experts in all technologies and applications, but we can begin to understand how to creatively apply the basic language of each to our work."

— Alyce Kaprow
Computer Graphics
Consultant
Newton, MA

"Multimedia authoring requires the language skills of a writer, the creativity of an artist, the management abilities and sensitivity of a film director, the determination of a project manager, and the patience of Job."

— Joel Orr
Orr Associates, Inc.
Virginia Beach, VA

"One of the things going on at WGBH is something we call multiversioning. It is the idea of producing a lot of different products in different media from one basic idea. Rather than saying 'OK, we have the television program, now let's make the book, now let's make the interactive video disk, now let's make the CD-ROM,' it's more with this idea: 'Let's think about what simultaneous media we want to create.' When you start thinking of a project that way, there's a lot to consider in terms of design and production from the start. There is the potential for the designer to have a much larger role both in terms of conceptual work and what's going to be the proper medium for his idea."

— Wendy Richmond
Codirector
WGBH-TV Design Lab
Boston, MA

"Multimedia isn't 'multi.' Multimedia is about multi*sensory* communication through a *single* medium: the digital file. A slide show accompanied by audio tape and film—*that's* 'multimedia.' Computer files with text, sound, animation, video, and hypertext links—they are multisensory *mono*medium expressions.

"It's not just the 'multi-' part with which I take issue. A medium is by definition a means—not an end. But McLuhan's dictum still holds: The medium IS the message. Multimedia today is like a talking dog; it is not what the dog says, but the fact of its talking, that commands our attention.

"In truth, multimedia is a place-holder, a temporary substitute for what we really want: direct contact, immediacy. No medium at all. Intimacy with the subject—our designs, our writing, and those of others."

—Joel Orr
Orr Associates, Inc.
Virginia Beach, VA

"An R.R. Donnelley client is making good use of on-demand services: Primus, which is part of the college division of McGraw-Hill Book Co. in New York. Professors can look through a McGraw-Hill catalog and order the journal excerpts, articles, and reviews pertinent to a particular discipline.

"Donnelley pulls chapters together from its duplicate library, renumbers the pages, and generates a table of contents and index for the custom book using proprietary software.

"The professor then gets a bound copy that includes a unique ISBN number. If the professor approves the publication, the book is produced in precise quantity and delivered directly to the college bookstore for student purchase.

"According to Gerald Harper, senior systems engineer at Donnelley: 'Everyone benefits from Primus. Students' costs of textbooks are kept down; the professor orders texts with chapters in the exact order of the course's syllabus, the publisher cuts down on warehouse inventory, and the author gets royalties.'"

—Anita Malnig and
Karen Houghton
Writers
MacWeek Magazine

"Notions of the best form for an electronic book reader vary greatly. An eBook is simply electronic text that's readable on any computer system, independent of specific hardware. Or is it a hardware device specifically tailored for easy reading of electronic documents, books, periodicals—and even Web pages? Most likely, an eBook is a combination of both. Technology developers and media publishers alike are struggling with this balance, trying to create an open definition so that a market can grow, while carefully positioning themselves in such a way that if their technology wins they'll be able to close the market and limit further innovation.

"The Open eBook Initiative (www.openebook.org) is an attempt to solve the eBook definition dilemma by creating a forum for both the technology and publishing industries that allows them to direct their efforts toward a standard.

"Publishers and booksellers are also pushing and pulling on the eBook paradigm, signing exclusive partnerships with individual hardware and software developers, while attempting to build new publicity and marketing opportunities. After all, whether the eBook revolution comes now or in another hundred years, it will come, and it will likely change the way books are authored, printed, published, and sold. And as with any multibillion-dollar industry, there's a lot of money to be gained or lost in the process.

"Clearly this whole boondoggle will depend on the consumer. And until eBooks demonstrate an obvious advantage, technically or spiritually—and certainly at an economic scale that is accessible to more than the Silicon Valley elite—they will continue to float in the limbo between the computer and publishing industries. In the end all I care about is having something thought-provoking to read, whether it appears on a computer screen, a piece of paper, or even a stone tablet."

—Hans Hansen
Developer of Rare Books
in Electronic Formats
Octavo Corporation
Oakland, CA

"Multimedia can best be defined as various combinations of text, graphics, sound, video, and animation that are controlled, coordinated, and delivered on the computer screen. Multimedia also implies interactivity where the user is actively engaged in the presentation of information, and is not just a passive observer of a fixed procession of sights and sounds."

—Syllabus
Mac/CHICAGO
Chicago, IL

If communication breakdowns are responsible for the majority of problems on projects, poor budget management is probably responsible for the rest. Good budget management is a skill that can be obtained from training and experience like any other. The bad news is that many designers are not trained or skilled in budget management. They may know enough to estimate in planning but lack the right management in its dissemination. Most designers learn about budget management through experience. It is best when these skills are taught and emphasized along with others like visual and software, for then designers can be good businesspeople and of increasing value to clients. Great ideas that don't fit budgets are not responsible ideas—and unless justified with greatness, the best ideas are those which can be done with the resources available.

PROFILE OF A BUDGET-RESPONSIBLE GROUP

Keep projects on track financially through their many phases and changes in development. This is a talent and skill that must be honed as much as the creative ones. Meeting budgets must be the responsibility of every team member. All groups that succeed financially have:

► **an established agreement** *prior* to doing billable work. Do not assume that either the client group for development or the creative group for beginning will embark without economic assurance.

▷ *Set up initial parameters as defined in the proposal* (see chapter 4, page 38). A well-prepared plan defines the scope and serves as a guide throughout the development process.

▷ *All participants understand* their roles, compensations, and procedures. These are itemized as part of planning. For larger teams, the roles should be delineated with accountability assigned. Smaller teams need to utilize each other's expertise without duplication.

▷ *One person is in charge* of monitoring the budget, usually the project manager, who acts as the liaison between the art director and the client. On many projects,

"In planning, no one can peer into a crystal ball before starting. All projects are planned with the best scenarios in mind and the best intentions to meet expectations. But no project ever goes completely according to plan. It is impossible, for reality intervenes with unexpected circumstances and delays. Reality is always different than a plan. But having a proposal to check helps everyone know when there are deviations, what are the causes, and what repercussions they have. This is where experience shines: there are ways to get snags unstuck."

—Michael Waitsman
Designer
Synthesis Concepts, Inc.
Chicago, IL

the client and the project manager may be the same person.

▶ **management and supervision** with

▷ *realistic expectations* built into the schedule. Because many projects have short time frames, a big challenge is making the schedule realistic and possible. Be sure to have input on time frames from other team members. The more complete the description of phases, the more easily the team will meet the overall goal. Any trouble can be flagged quickly along the way.

▷ *structuring the decision-makers*—know their priorities, availabilities, and responsiveness. Anticipate stumbling blocks during the approval processes, but minimize with communication and reporting avenues established. Utilize e-mail to cover all strategic decisions, including all those who are impacted by the decision.

▶ **a monitoring system**—part of the project plan:

▷ *early warning detection for deviations*—probably the most important purpose of a plan (other than defining the scope and project goal for scheduling and budgeting).

▷ *sets up input method for compiling time and project data:*

> systems should be easy to implement so staff members can efficiently keep up. Accessible data is always kept current, so each participant needs to be conscientious. The team must be comprised of people who are diligent with record-keeping and turn in project data on time. Any management system can only work well if the data is up to date. Then, reports should be quick to generate and shared with participants.

> everyone should understand the importance of how data is used to make systems effective. (Dissenters can sabotage effectiveness very quickly if the data is not current.)

> require report points and specify key date deadlines to keep project in control. Develop follow-up methods.

▷ *one person is responsible* for keeping systems up to date, researching new upgrades or additions, and teaching others any upgraded systems or methods. If it is not a particular individual's responsibility, the data will not be kept current, for such work does not mix well

LEGITIMATE REASONS FOR A PROJECT GOING OVER BUDGET.

All these reasons reflect significant change and include:

▶ DEVELOPMENTS IN THE INDUSTRY (outside the client's control) where

  ▷ *competition comes out with the same product or service first,* causing projects in the works to be reevaluated. Many industries are subject to racing products and services to market, often before they are ready. Testing and evaluating must coincide with marketing and design development.

  ▷ *a new discovery makes information dated.* Technological and health care clients are particularly vulnerable. But even in the services sector, the difference between one competitor and another's products or services may be only in their image, marketing, and customer service, for discoveries are quickly spread from one company to another.

  ▷ *change in laws or in the political scene* affects the content which must be redone. New governmental administrations can be anticipated only in that new requirements will be required, but not necessarily what those requirements will be.

▶ THE CLIENT ORGANIZATION'S GETTING SOLD or merging with another group, so projects are rescheduled, redefined, or (hopefully not, depending on the level of investment) deleted. More often than not, a new vision is ushered in with the new owners.

▶ CHANGES IN MANAGEMENT that make a project obsolete or in need of significant alterations. If a change is seen in advance on the horizon, projects should not be started until the new management is in place, or hurried to finish before it happens. But not all of these kinds of changes are possible to predict.

▶ NEW OPPORTUNITIES where another project takes priority, either putting the original one on hold, or causing it to evolve. If the new direction is more profitable than the money lost aborting a project, then it is worth it.

▶ CHANGE OF DIRECTION—often this is the product of bad planning, but it may be legitimate if a new idea enhances the content enough to make a delay worth it.

▶ THE NEED TO DO MORE RESEARCH OR DEVELOP MORE IDEAS, realizing that the market can be better reached through making changes that will lead to a more effective result. This may require kill fees. (See chapter 22, page 264.)

▶ LOSS OF FUNDING—can be hard to foresee and a surprise to everyone. But it is hard to go on without the budget to verify the expenditure of resources.

Who pays for any additional work becomes a question of who initiated the change and whether the change alters project parameters. (See chapter 26, page 310, for more about alterations and corrections.)

with other priorities. Assisting coworkers will be haphazard, casual, and distracting if not anticipated into the work of a tech specialist (see below).

  ▷ *continually streamlines reporting methods* and approval channels. Roles are well defined and appropriateness is understood by each participant. This becomes critical when time frames are short. But don't add unneeded reports that slow down progress. Utilize the technology to speed along analysis and production wherever possible.

▶ **responsive tech support** with sufficient resources to provide a dependable infrastructure:

  ▷ *makes access to data easy* and allows reports to be drawn from appropriate databases and libraries.

POOR REASONS FOR GOING OVER BUDGET

These reasons reflect poor planning or vision, and could have been anticipated:

► NOT ALL DECISION-MAKERS WERE CONSULTED and a new one has surfaced, bringing additional parameters. This almost always requires additional work. Make sure all decision-makers are chosen before proceeding with the project.

► PROJECT DOESN'T INTEGRATE WITH OTHER MARKETING MATERIALS and needs to comply. If there are guide-lines in place, any concept must honor them. If not known in advance, such guidelines (when they surface) will cause rework. Make sure everyone is familiar with any campaign, branding, or corporate identity concerns.

► DECISION-MAKERS CHANGE THEIR MINDS after making a choice, such as approving a design and then later changing to a different one. Indecisiveness can be expensive. Choosing the best design and staying with it requires thorough initial positioning. It requires leadership and vision to keep all committee or team members focused on the same goals.

► DATA IS NOT ALL AVAILABLE and the project proceeds, then needs to be altered when new data does not match what was expected. In some cases, this is unavoidable, so variable flexibility should be part of the design structure. Extreme contingencies should be explored, minimizing surprises when suddenly under deadline crunch.

► TOO MUCH INFORMATION IS MISSING—the project should wait to be started until the content is together enough that any possible changes are predictable. Flexibility can be designed into the project to handle any variables that must be completed simultaneously with development.

► MATERIAL IS ADDED LATER such as more pages, photographs, charts, reports, etc. This is another product of bad planning, usually caused by content not finalized or approved. This will always be expensive to do. The later in the project this happens, the more the expense. (See escalation formula, chapter 26, page 312.)

► TIME IS RUNNING OUT—there is always an urgency to get new ideas to market, especially with the fast speed of e-commerce. If the project begins late, expect to need extra staff, outside suppliers, and rush fees to achieve an unbreakable deadline.

► CHANGES ARE MADE THAT WON'T AFFECT THE USEFULNESS, function, or purpose of the communication, such as perfecting details that no one will notice. This is usually caused by a decision-maker's ego and/or perfectionism. Know when a project has been perfected enough.

If any of these reasons occur, it means someone, usually the client, is going to lose money. (See chapter 28, page 322.)

▷ *can handle a variety of databases*—management and updating, both relational and interactive. Content from projects can come from a huge range of sources. Any plan needs to encompass resources to handle com-patibilities and translations.

▷ *acts as a troubleshooter* to help with system problems. If it is someone on staff who earned the role de facto, it is important to give this person recognition and in-centive to keep the systems running smoothly.

▷ *encourages developing skills*—opportunities are open for professionals to learn through further training and incentives. (See chapter 9, page 88.) Take advantage of the many opportunities available—view them as a strategic investment.

▷ *maintains and upgrades system* to be on the same level as, or better than, industry standards. This means a system should always be evolving. If it stays the same too long, then the group will fall behind the industry and not receive the services and support needed to remain viable.

Watching budgets is as much a part of the project as the design or the distribution of the finished work. Often, a highly successful result makes memories dim regarding projects that go over budgets. But most projects that cost more do so needlessly.

HOW TO BEST MEET BUDGETS

Any individual involved in a communication project must take responsibility to meet budgets, whether supplied or quoted. Below is a checklist for making sure that budgets stay within control:

► **go through planning** to consider all variables (see chapter 7, page 62, and chapter 8, page 70) and get any necessary input from contributors.

► **have a graphic foundation** plan or theme for overall marketing consistency. (See chapter 10, page 100, and chapter 11, page 126.) Evaluate each project in relation to other marketing efforts.

► **organize the project** so it is not in pieces but has all (if possible) components together. Know which elements are incomplete or subject to change. It may be efficient to begin the project while some aspects are still being compiled, but allow contingencies in case new elements don't fit. Some components can be created to fit a given space or time requirement and these can be commissioned simultaneously with other progress.

► **have preliminary meeting** to outline goals. Do not work on any project unless the parameters are outlined against an economic plan. Allow budget management to be an open topic that has reporting mechanisms built into the plan. Economic secrets or manipulations rarely benefit project development. The more a project team understands the resources, the better the solution and shared investment. Do not make an exception to this rule when working with friends or relatives.

► **build in time** for approval processes and reviews. Evaluate work before moving it to the next phase of development. Be sure to have all decision-makers agree to phases before incurring any additional work.

Many clients and creative groups do pro bono projects for volunteer organizations. Even these projects should have a budgeted amount defined so that the receiving organization knows the value of the gift. If they don't know it is a gift, then future projects will be expected on the same basis. Many designers don't like to talk about money, so having a document can communicate the real value of the service. Everyone wants to be appreciated.

- ▶ **facilitate communication** with team members. Take advantage of technology and build channels, copying all team members in progress.

- ▶ **share elements** between projects to amortize expenses, as well as effectiveness.

- ▶ **allow time for the unexpected** and for testing, where needed. If the budget is inadequate for the time allowed, be sure to know this as soon as possible so remedial steps can be taken.

- ▶ **manage alterations** by anticipating the costs in advance and gaining approval before doing the work. (See chapter 20, page 246.)

CONSEQUENCES OF GOING OVER BUDGET

Sometime it is unavoidable to discontinue a project, and everyone hopes that the losses will be minimal:

- ▶ **if the project is placed on hold**

  - ▷ watch for the start-stop-start syndrome—it takes more time, inertia, and invites new or altered elements. Check the budget as soon as the project has a delay. Most often, fees are paid up to that date.

  - ▷ pay development fees and any "kill," or cancellation fees appropriate for uncompleted visual images. (See chapter 13, pages 164–165.)

  - ▷ reevaluate the budget when the project starts up again. Usually, elements change in the meantime, so it is rare that the same parameters will still apply.

- ▶ **when the project is cancelled**

  - ▷ pay all work to date and any appropriate "kill" fees (as above). Do not expect to begin the project at a later date with the same budget.

  - ▷ determine ownerships of the work to date. Although fees may have been paid for artwork, it does not mean that the client owns the artwork. (See chapter 29, page 330, for ownerships.) Buyout fees may be appropriate if the client wishes to use the elements of a cancelled project in other projects.

If a project has a short schedule, it becomes more of a challenge to watch the budget—to take the time to evaluate, especially when it seems there is no time to take. Not checking the budget is a danger that the project might be reeling into the red and no one knows it. Take the time to check.

In the planning phases, it is a good idea to build in a contingency percentage to cover possible necessary changes unforeseen at the beginning of the project but necessary to finish it (especially if the deadline is critical). Some organizations add 10 percent into their proposals (see chapter 4, page 38), and some higher. It will vary from group to group.

SIX PLACES TO EVALUATE A BUDGET

Many projects will require more monitoring steps and some less. But each of these six points should be ignored at the project's peril. Always review the budget:

► **at the beginning of the project**—review the initial budgets that are set up as part of the proposal and plan. Every project should have a projected cost to judge its viability and flexibility, without exception, even if pro bono. Any communication project has a range of what will be appropriate. If a budget is set, then it determines the design direction.

► **at the concept presentation**—the design group should know which ideas might potentially cost more and discuss these before design selection is made. Costs are always a factor in what ideas are affordable. The client will assume an idea fits the budget unless told otherwise.

► **before content is assembled**—all text is written, photographs shot or collected, illustrations delivered, and all the elements are ready to be placed into pages or artwork prepared for fabrication. Know where there is flexibility, if needed.

► **at the rough proof or mock-up stage**. After the creative group has assembled all the content into the chosen design, the roughs are presented, indicating any missing elements and explaining any further production processes needed.

► **when artwork is finalized**—but before posting or printing. Final alterations happen, which may have financial implications. Be sure the production fees are in line with estimates before completion; otherwise, surprises are invited.

► **after completion of posting, fabricating, or printing**—when the project has reached its intended audience. Determine final fees before invoicing in case any negotiation

- set project foundation
- plan rough project skeleton
- assemble category outline
- solicit proposals
- choose creative group
- finalize project plan
- select design theme
- approve design
- review writing outline
- edit first draft content
- review revised content
- review final content
- approve visual components
- approve first prototype pages
- approve pages and testing
- **approve final artwork**

is needed. After invoicing, such negotiation can cost both sides overhead. (See Payments and Disputes, page 322.) There should be no surprises if there has been effective planning and communication throughout the project.

If these checkpoints are worked into the project plan, scheduling dates for these budget checks makes them easiest to accomplish. Ask for the reports if the art director does not provide them. Keeping control of the project is easiest when establishing the key check-points at the beginning with the creative team. This secures their agreement and commitment

WAYS TO SAVE MONEY WHEN COMMISSIONING CREATIVE PROJECTS:

> PROVIDE ADEQUATE LEAD-TIMES. Sometimes the faster a project must be done, the more resources are needed.

> CHOOSE AN EXPERIENCED SOURCE. The hourly fee may be higher, but the work may be done faster and better, thus costing less.

> CHOOSE A SOURCE THAT UNDERSTANDS THE INDUSTRY SECTOR. It can take time to get a creative group up to speed on information in an industry new to them. However, when fresh approaches are paramount, having a creative team with no experience in a given sector is an advantage and can lead to greater originality. But such a creative group should still be experienced in the publishing processes to add speed.

> GET CONTENT TOGETHER before proceeding too far into production. It is very tempting to race forward before a project is ready, often to appease a board of directors, a boss, or management, or to hurry ahead of competition. Hurrying is a sign to slow down and be sure of positioning, planning, and getting the components. Often, assembling the elements together takes longer than actually doing production. Patience is a useful skill when in a hurry.

> HAVE EFFICIENT MEETINGS. One of the biggest ways that projects can vary is the number of meetings required for planning, presentations, and discussions. The best meetings are collaborative efforts between participants and have a specific agenda. It is tempting to add a lot of incremental meetings if the content is developing in spurts.

> MAKE SURE ANYONE WHO IS A RESOURCE FOR THE PROJECT KNOWS ABOUT IT, such as support staff and suppliers who will be playing roles later on. When the project is expected, it often can be finished faster.

> SEGREGATE TIME-SENSITIVE INFORMATION and have a good update plan for projects that are in series or for Web sites that change frequently. If planned in the design from the beginning, the information can most efficiently change. The budget is an anticipated expense, with an agreed upon payment plan.

> BE AVAILABLE while critical services are being performed. If there are fast decisions to be made, there will be no delays if all decision-makers can be reached.

> ALWAYS ASK HOW MUCH CHANGES WILL COST. Don't wait for staff or suppliers to report. Most changes do affect time or budget, so negotiations should be complete before work is started.

> AVOID THE START-STOP-START SYNDROME wherever possible, as this almost always is a red flag for possible costly changes.

to the deadline and increases the collaboration and reason for attaining the finish.

TYPICAL CHANGES

Most changes from the client group involve text and statistics once the design is approved. Usually, proposals do not include these kinds of changes because they are very hard to antici- pate. A percentage contingency may be included in the esti- mate. (See chapter 4, page 38.) Most creative groups charge additionally for client-initiated changes unless specified as part of the agreement. Clients must be prepared for needing to correct

► **final art**—proofs or prototypes as the project is finishing and ready to be printed, posted on the Web, or sent to a fabricator. This is probably the most important point of client approval because the client is responsible for all material that is in the project. If there is any content that is incorrect, the client is liable.

► **elements**—if any parameters change. This can happen at any time in the process and is probably the most impor- tant and most difficult time to evaluate a budget. If param- eters change, there is usually an accompanying time pres- sure, as well as economic consequences.

Try to ensure that components and decisions are as final as possible before embarking on the design phase. It is better to start a little late and save than to start premature and lose time and money later if not able to meet the time frame or budget.

CREATIVE GROUP

Watching and accommodating parameters is the art director's responsibility. So is providing budget reports to the client. Unless the client agrees to any additional fees (usually due to changes in content or project parameters) it is expected that the creative group will charge according to the approved pro- posal (see chapter 4, page 38). Any changes need to be negoti- ated if the creative group expects to be paid for them. The worst situations arise when expectations are vague.

FINANCIAL RESPONSIBILITIES

There is no one working on a project who does not have a financial impact on its profitability. Each member of the cre- ative group has financial responsibility to:

"If you aren't fiscally minded, hire a bookkeeper or accountant. Most prin- cipals are eternally optimistic and tend to ignore 'red flags.' Hire someone to keep the books and sound warning bells when your firm's finances are looking shaky—even if the person isn't a full-time employee. To estimate the number of hours you'll need to devote to bookkeeping, multiply the number of employees you have by two hours per week."

—Jenny Sullivan
Writer
Annandale, VA

► **not charge for having to redo work** that doesn't meet parameters or to correct mistakes such as:

▷ *not incorporating all specified client changes*—this is most common with very rushed projects, for the creative group has no time to check or correct their work.

▷ *underestimating how long processes take*—very easy to do for an inexperienced creative group.

▷ *mismarking or mistakes in the files* for printing or fabrication instructions. These mistakes are very common, especially in large projects, and are one reason for needing proofs. (See chapter 24, page 280.) Printing mistakes can also come from the printer where the designer only has supervisory control. They are corrected on the proofs.

▷ *risky effects and new processes* without approval and appraisal of risks involved. Usually they take more time.

▷ *missing deadlines*—if in jeopardy, the client needs to be told immediately. Solution negotiations may commence, but unless parameters change, the creative group must find the best way to meet the agreement.

► **charge for changes caused by client parameters** (see sidebars, pages 238, 239, and 255)—these can sneak in gradually to a project and need to be watched carefully.

▷ *Small changes* are going to be assumed to fit within the agreed budget, unless informed otherwise. They can add up incrementally, so keep track of them.

▷ *Large dramatic project changes*—it is usually obvious to the client that there will be extra fees. Do not assume, for this is a frequent place for disputes. (See chapter 26, page 310, and chapter 28, page 322.)

PRODUCTION GROUP

Because the budgets can be established on objective project criteria, there is a temptation to make quoting projects, setting up budgets, and monitoring progress into a science—to make it as automated as possible. But in custom manufacturing, every project is different and requires insight into those uniquenesses to be most realistic about expectations.

All publishing projects need to be prepared in the latest version of the software, as demanded by all the production processes. Printers, prepress facilities, and Web production all require team members to be working with the same elements, particularly in the case of workgroups (see chapter 18, page 226). Nothing slows a project faster than translation problems.

REEVALUATE PROJECT AS CHECK-IN

The parameters used for the estimate that was prepared for the proposal. (See chapter 4, page 38.) should be reconfirmed before any finishing work begins. Time has passed and the project has evolved. It may have added elements or use materials that were not anticipated in the beginning. Finishing projects includes:

► **posting to the Web**—finalize pages. (See chapter 17, page 208.)

► **fabrication**—displays, packaging, or products. (See chapter 19, page 236.)

► **printing**—from brochures to publications, stationery to posters. (See chapter 25, page 288.)

A binding agreement can be formed, keeping in mind the conventions of specific processes like percentage of overs and unders (see chapter 25, page 300) and what is considered acceptable quality.

Any mistakes originating within the production group will be rectified at no cost to the buyer.

It is assumed that the budget will match the estimate unless new fees are negotiated and agreed upon. (See Payments and Disputes, chapter 28, page 322.)

DEALING WITH MISSED DEADLINES

"The key to timely project execution is making sure planning covers the proper resources for meeting the milestone goals. Milestones are used to keep track of accountability. If milestones are met, projects tend to be on time.

"Some projects begin slipping as early as the first milestone. This usually occurs due to a communication breakdown between the analysis team and the client. I've seen many firms fail because they assign developers or designers to perform the role of project management in addition to production. That's a bit like asking the limo driver to pour drinks while driving. A single person needs to take responsibility for production and ensure that the client is kept in the loop on status.

"Another common reason for missed milestones is the sheer size of the milestones. Complex projects require many large milestones. Some-times these milestones need to be broken down into smaller, more manageable ones. Missing that first milestone creates an avalanche of issues.

"Even with missed milestones and delivery dates, keeping the goal in focus will help bring the project back under control and help prevent divergence among the team.

"Guidelines for 'reeling in the project: What can you do to take back control and be successful?

> DON'T PANIC. Panic leads to more undue stress and causes teams to be unproductive.

> NOTIFY THE CLIENT OF ANY DELAYS and explain the reasons for the delays, immediately. This can save a lot of grief in the long run.

> REEVALUATE THE PROJECT from a global standpoint.

> Take the time necessary to SEE WHAT ISSUES REMAIN, prioritize them, and set smaller, more attainable milestones.

> ADD ADDITIONAL STAFF only when they can be used effectively; add-ons need even more specific goals so their work can be integrated with the work of the rest of the team.

> ADJUST TO THE PROJECT'S EVOLVING ENVIRONMENT. Knowing the magnitude of the problems is half the battle.

> DON'T BE AFRAID TO SAY 'NO' TO A CLIENT. Large-scale change in specs or scope is the single biggest issue that kills projects — this change often comes from people who don't understand the production process. A 'no' now can lead to better work in the future and goodwill all around.

"Not all overextended projects turn out to be failures. The biggest failure can sometimes be losing composure and professionalism."

—John McHugh
Senior Applications Developer
DigitalNation Interactive
Alexandria, VA

"If you produce a winner, people aren't going to care if you're a little over budget. And if you produce a loser, the fact that you came in on or under budget is not going to help a bad situation. I mean, you've got graphic evidence, thousands of printed copies that just ain't so hot. But if you produce a magnificent piece and the budget exceeds whatever, most people won't care."

— Edwin Simon
President
The Pelican Group, Inc.
Hartford, CT

"It is better to be late by one day and be right than to deliver a project on time but be wrong forever."

— Patricia Maloney
Marketing Supervisor
Smith, Bucklin & Assoc.
Chicago, IL

"To maximize profitability, find out which budgets *can* be dedicated to your part of the job. Often, a creative fee is set aside and named as photography, illustration, or design. That is just one line item on a project budget. Since technology today allows you to do more of the actual work, ask about other line items in that project budget. Check on preproduction, research, retouching, postproduction and prepress, to name a few. The money is there to pay you for the extra work; it just may not be called 'creative fee.'"

—Maria Piscopo
Creative Services Consultant
Costa Mesa, CA
http://MPiscopo.com

"If you design right, you save money. I've found it most cost-effective to hire experienced designers. Use good people and you get good work."

— Grant Hoekstra
Design Manager
Trinity University
Deerfield, IL

"When your client pays half-price for creative services, they will get less. The question is, will it be less creative or less service? ... The bottom line is do not accept less money for the same work. You will damage your chance at a profitable relationship with clients. You will give your work away. Learn successful pricing techniques to sell your freelance services. Then take a deep breath and go get the return on your investment!"

—Maria Piscopo
Creative Services Consultant
Costa Mesa, CA
http://MPiscopo.com

OUTSIDE RESOURCES

During a project's development, outside materials or services are required beyond the capabilities of the creative group. Outside purchases by the creative group or production group to fulfill a client's order may include services such as photography, illustration (see chapter 13, page 158), or die-cutting, special binderies (see chapter 25, page 290), etc. Out-of-pocket expenses include materials, messenger services, deliveries, and source materials.

PURCHASING RESPONSIBILITIES

Any party who purchases "third-party" services assumes responsibility for that service which encompasses:

► **accuracy** of work performed and received

► **completeness** of materials and communications

► **timeliness** in response to client needs

► **financial clarity** through planning and monitoring

► **legal issues** such as ownerships, usages, and contracts.

The buyer (see chapter 1, page 2) pays for all acceptable work— work that meets all defined buyer parameters. The buyer doesn't pay for unacceptable work—work that fails to satisfy the defined parameters. A buyer cannot claim work as unacceptable if it is actually put into use. (See chapter 28, page 322.)

Every group is responsible for the services and materials within their purview. For example, the writer and editor as originators of the text are responsible for proofreading; the printer is responsible for paper, and each member of the production group is responsible for the quality and longevity of the media they produce.

CLIENT GROUP

No organization has all aspects of publishing in-house. Core competencies are balanced by the specialties of suppliers as part of the project team. When engaging the creative group, the client should

► **know which services the creative group will provide** and which, if any, the creative group will purchase from a third party. (See chapter 28, page 325.)

► **being billed directly**—by the third party may be an advantage. The client can save a markup fee, added by the creative group for handling the financial responsibility. A supervision fee is appropriate (see next page).

► **share purchasing responsibilities** with the creative group and in the satisfaction of work performed. Sometimes the creative group agrees to take on this responsibility fully.

► **approve all invoices** according to agreed upon fees.

CREATIVE GROUP

Every creative group offers different capabilities. From single designers working out of small offices to full-service departments, the formula of offerings varies with talents and skills. Matching the right creative team to the right project requires balancing experience with challenge.

RESPONSIBILITIES OF BEING THE BUYER

For each function the creative group performs or purchases directly, such as writing, design, text entry, proofreading, programming, prepress, buying paper, and printing, the creative group takes full responsibility—including financial—in the event of any problems, unless other agreements are made.

For handling third-party services, taking on the financial negotiations and credit burden, most creative groups are careful with how many expenses they are willing to undertake for clients. The riskiest is buying printing—especially when the printing fee equals, or is greater than, the creative fee.

DIRECTING SUCCESSFUL DEALS

Being the liaison between the client and the production groups, how the creative group handles communication to either direction is a large factor in project success. The creative group, led by the art director, needs to

Each creative group has their own markup policies when handling the financing of outside purchases. Industry standard is generally 15.75%.

► **describe its markup policies to the client**—this includes shared responsibilities when invoices go directly to the client. It is best for the creative group to review all invoices before the client receives them to be sure of agreements.

► **describe any policies for direct billing of outside services**—especially when such fees are greater than the design fees, as in printing or fabrication. When having the client billed directly, a supervision fee is appropriate and covers much of the project manager's time for supervising the various contributors.

► **put together the most appropriate production group**—hold them responsible for their roles, including quality, delivery, and budget, according to agreements. Coordinate, manage, and communicate with the production group—the client pays the creative group for this service:

▷ *through markup fees* that have been agreed upon in advance, usually outlined in the proposal (see chapter 4, page 38).

▷ *through supervision fees* if the production group bills the client directly. These are usually line items on the budget.

PRODUCTION GROUP: TECH

Similar to the responsibilities of all team members, the tech group's need for outside services may be more limited. However, they may have to purchase special services software to handle client files such as:

▶ **specific database or file management packages**—particularly when there are proprietary publishing systems to integrate. Often tech groups will

▷ *contract out for specific integration needs*—if the task is a full network upgrade, then large contracts may be awarded to outside systems integrators.

▷ *use online services* for purchasing, specifications, accounting, asset management, and many other publishing services, which can be fully networked and handled in a seamless virtual environment. When suppliers are tied into their customers' purchasing systems digitally, time and efficiencies can greatly enhance shipping and delivery.

▶ **storage media**—handle archives of past publications, and image libraries. (See chapter 15, page 184.)

▶ **file conversions**—convert past publications into modular content formats for template publishing (see pages 129 and 343) or handling the translation of print into Web. Many new services are becoming apparent as opportunities increase.

▶ **special programs or extensions**—for multimedia projects, special events, or capabilities for a specific application; often additional training and software may be needed to fulfill a specific assignment.

▶ **specialized programming expertise**, such as animation or video, when utilizing special talents and applications.

These may be billed to the client directly (see page 258), or covered by the tech group as part of their proposed fee.

PRODUCTION GROUP: PRINT

Unless otherwise agreed to in writing, the production group will

▶ **mark up any of their outside costs**—such as paper, inks, die-cutting, foil stamping, film, and other materials, which are usually part of their estimate.

▶ **define variables in markup fees**—usually as part of the quotation in the beginning of the project. (See chapter 4, page 38.) Give line items for special processes like die-cutting, foil stamping, or special bindery.

"Distribution is one of the most important but overlooked things you'll run into. If you're looking to develop a custom piece of multimedia for a single customer, you could care less about distribution. But if you're going to go out and use an existing CD on dinosaurs, add hypertext, turn it into an educational program, and get that distributed to schools, now distribution becomes a problem. You've got to know *how* to do it, what the sources are, and get to them, and it's not cheap. You have to pay up front and see the revenue downstream."

—Bob Croach
JHT Multimedia
Winter Park, FL

"To guarantee fair compensation for your work if it's lost or damaged—and to ensure that your client treats your work with necessary care and respect—you need to provide for 'liquidated damages' in your initial contract. A liquidated damages clause establishes the amount of money your client must pay you if the original work or transparencies are lost or damaged while in her possession. In a recent court decision about reimbursement for the loss of original transparencies, an appellate court said that if a liquidated damages clause was included in the contract, a prespecified amount for each lost transparency would have been awarded."

—Deborah Michelle Sanders
Publishing Lawyer
San Francisco, CA

SCHEDULE CHANGES ▮22▮

Although carefully planned from the beginning, almost every project forms some deviations—particularly in pacing and timing. The plan is to help identify deviations early so that schedule changes don't turn into budget changes. The plan needs to adjust to reality.

Projects might be put on hold or have schedule changes at any point in the process. This is the most unpredictable aspect of project planning. If a change in the organization is on the horizon, it might be wise to delay a project before beginning it. However, because most changes that happen are inconceivable in advance, knowing how to best handle them can ease the ramifications if they do.

REASONS FOR SCHEDULE CHANGES

At the beginning of a project, the client should set a deadline for the completion of published materials. The creative group generally establishes a project schedule working back from that due date. (See chapter 3, page 22.) However, the client may change the schedule during the project for many reasons (see pages 248 and 249 for legitimate and illegitimate reasons for budget changes, as many parallel schedule changes):

▶ **sudden market shifts** that cause the audience to change.

▶ **addition of new material** such as statistics, photographs, or additional pages. (See chapter 26, page 310.)

▶ **event date changes** beyond the client's control.

▶ **new rulings in the law** which affect business practices or product introductions.

▶ **management reorganization** that was not foreseen.

▶ **staffing changes** such as finding and training new staff.

▶ **funding changes**, usually accompanied by change of scope.

▶ **dissatisfaction with work performed** such as when parameters are not met and work needs to be redone.

When the client requests postponement or advancement of the deadline, the creative group and the production group may adjust fees and schedules accordingly. It is wise to evaluate the project budget when change of status occurs. (See chapter 20, page 246.)

When scheduling a project and setting the deadline, the client needs to factor in time to approve all steps. Unavailability of the decision-makers may require flexibility in the schedule and allowances in expectations for the final deadline.

CLIENT GROUP

When a client commissions a project, they approve the proposal to cover fee payments and schedule. (See chapter 4, page 38.) Variations can occur in the schedule once work is in progress.

STATUS POSSIBILITIES

Although no one may want to halt a project, it is rarely within a client's control (unless a product of bad management). The client may

► **put a project on hold temporarily**

  ▷ *renegotiate the deadlines* with the creative group.

  ▷ *add overtime or rush charges* if there are delays in approval or decisions but no change in the final deadline.

► **put a project on hold indefinitely** (without specifying a new deadline)

  ▷ *pay for time and materials* up to that date.

  ▷ *incur additional fees later* for extra start-up activities, such as meetings, implementation of a new direction, or inclusion of new material.

► **cancel a project** that has begun

  ▷ *receive accounting of work to date* from the creative group.

  ▷ *pay for time and materials* to the date of cancellation.

  ▷ *pay any "kill" fees*—there may be photography or illustration fees to cover the time for the creation of unused ideas or the lost booking opportunities. (See page 151.)

Some schedule changes may be inevitable, but they can cause less disruption if flexibility is built into the project process.

Starting and stopping a project due to delayed approvals or missing information are the most common reasons to put a project on hold. When this happens, anticipate additional fees once the project continues.

CREATIVE GROUP

The creative group accepts a project based on an agreed-upon deadline, often part of a project plan or proposal (see chapter 4, page 38).

STATUS VARIATIONS

The best thing an art director can do as soon as a project stops, whatever the reason, is to reevaluate the deadline and what must be done to accommodate it.

► **If the project parameters do not change**, the creative group

  ▷ *has the responsibility to meet the deadline.*

  ▷ *must communicate to the client*, in advance, any reasons why the deadline cannot be met, with suggestions for what can be done.

Future workload is difficult, if not impossible, to predict. When a creative professional commits to a project, there may be no way to anticipate if more projects will also join their work flow. If really busy, the original commitment should not suffer. This balancing act is particularly challenging for freelancers. So having additional resources to call when needed can make the difference between satisfied and unsatisfied clients.

- ▷ *may not charge extra fees to meet agreed-upon deadline*, even if overhead costs increase due to workload or technological changes.
- ► **If the client changes parameters**, the creative group
  - ▷ *agrees to a new deadline* before performing the work.
  - ▷ *specifies any additional fees* in advance of incurring.
- ► **If the client holds or cancels a project**, the creative group
  - ▷ *ceases all work*, compiles an accounting of work to date for time and materials, and invoices the client.
  - ▷ *returns or stores project materials.*
  - ▷ *may requote the project* when it starts up again.
  - ▷ *may require a cancellation or "kill" fee* (see page 251), most typical for photographers and illustrators.

PRODUCTION GROUP

The production group accepts a project based on agreed-upon deadlines usually specified in the quotation. (See chapter 3, page 22.) Printers and programmers cannot anticipate exact delivery dates in advance of receiving a project so usually don't commit to a delivery date until the project is in-house. Production groups handle projects in priority of client clout, finances, and committed deadlines.

FACTORS OF SCHEDULE CHANGES

- ► **If the project parameters and deadline do not change**
  - ▷ but *the buyer does not adhere to agreed production schedules* in providing materials, the production group may renegotiate final delivery dates.
  - ▷ the *production group is responsible for supplying the delivery of the finished product on time*, but does not incur any liability or penalty for delays beyond its direct control (such as outages, floods, fires, etc.).
- ► **If the buyer requires overtime** to meet the deadline, it is
  - ▷ *quoted to the buyer* at the prevailing hourly rates.
  - ▷ *approved by the buyer* in advance of the work performed. If not communicated, the buyer may not have to pay fees.
- ► **If the buyer changes parameters** by withdrawing material before completed, the production group
  - ▷ *assumes no responsibility* for the completion of the work.
  - ▷ *is entitled to compensation* for all work done and expenses incurred up to the date of cancellation, such as paper purchased for printing or conversions for data.

**WATCH FOR
START-STOP-START SYNDROME**

Any time a project is placed on hold, and then starts up again, those working on it need to reacquaint themselves with status, reassemble the process plan, make appropriate adjustments, and proceed. This is very inefficient and can cause all groups to lose time and money. Whenever a project stops and starts, it is a red flag that the budget is in jeopardy.

Many inexperienced buyers will promise to have a project to a supplier at a certain time. Printers cannot reserve press time, for if the window is missed, downtime on the press is very expensive. A printer best schedules the project when received. Programmers can't really keep a window open, hoping for a project to come in. Most groups work on a first-come, first-served basis because only so many commitments can be honored in any given time frame.

"There is no guarantee that careful scheduling will prevent all problems, but it can lessen the chance of running into an error that can't be fixed. With this in mind, preparing an accurate schedule for a job should begin with the last operation—delivery. If the project is due four weeks from now, for example, the planner would begin by subtracting the days required to deliver, bind, and print the job, leaving, if possible, some leeway at every step.

"Working backwards from the final deadline helps determine how much time the designer can profitably devote to the piece."

—Boller Coates Spadaro, Ltd.
Consolidated Papers Co.
Wisconsin Rapids, WI

"By getting into prepress functions in-house, we hope to reduce costs, speed production, and simplify processes such as accounting by reducing the number of suppliers with whom we have to work, and gain greater control over the entire production process."

—Jerry Borrell
Writer
MacWorld Magazine
San Francisco, CA

"Since production is often invisible (and mysterious) to clients, it's even more important to keep technical problems to a minimum. The best solutions are simple, internal procedures that grow to be second nature. Following procedures like supplying detailed color markups with final artwork and completing a preproduction checklist saves hours of last-minute troubleshooting."

— Sarah DeFilippis
Principal
Dialogic

"The trouble is that time, so crucial for the out-of-time creative aspects of this work, has been nibbled away so severely in recent years that the quality of that work is getting harder and harder to uphold. Efficiency and cost-cutting have become a near-religion in business. If you want to be a corporate hero, a sure way is to hammer costs down and make sure you're seen doing it. It's as though prehistoric hunters were told to maintain their mastodon quota while cutting spear use by twenty percent.

"Now, of course, there are lots of good reasons for management to cut costs, but there are also reasons to know when to stop. In a business such as ours, where immeasurable qualities really count, you can cut TO the bone and into it before you know it. I think a lot of our clients don't understand where the bone lies, at least not in the area of creativity." [reprinted with permission from *CA* Magazine, July 1998, p.188]

—Sean Kernan
Sean Kernan Studio
Stony Creek, CT
www.seankernan.com
www.thesecretbooks.com

PREPRESS 23

Technology evolves the craft of prepress as it streamlines more and more functions. But technology is only half the prepress process. A knowledge of printing tolerances and properties, finding the best solutions to each project challenge, and supervising an efficient process are not handled by software.

WHO DOES PREPRESS?

The craft of preparing projects for print blurs roles as tech capabilities evolve. Software provides shortcuts and user-friendly interfaces, moving more of the craft to migrate closer to the designer. For whom does the prepress specialist work?

► **Creative group**

  ▷ *aided by advanced software and skills*—the pressure is for those creating the files to also prepare them for print, keeping the project with one person.

  ▷ *has a prepress specialist on staff* that can work closely with the designer and page composer.

  ▷ *may need designer or page composer to undertake some functions,* collaborate with the printer for the rest.

► **Imaging center**—there are two types, though merges have yielded the complete services of:

  ▷ *service bureau*—generally grew from the typesetting industry and has a background in high-resolution file performance and systems.

  ▷ *color separator*—evolved from color handling traditional analog separators, bringing a wealth of color knowledge and understanding, far beyond the purview or capabilities of most designers.

► **Print group**

  ▷ *in-house department* of prepress specialists who receive files from the creative group. Many of these groups evolved from the traditional craft of separating, stripping negatives, and preparing imposition.

  ▷ *collaboration necessary*—when working with either an imaging center or the page composer, bringing the project elements together is a team effort. Trained in printing, the prepress specialist can set projects up specifically for their own equipment tolerances to increase quality.

ORGANIZATIONS

www.agcomm.org
Association of Graphic
  Communicators

www.colorassociation.com
Color Association of the United States

www.color.org
International Color Consortium

www.ipa.org
International Prepress Association

www.irga.com
International Reprographic
  Association

MAGAZINES

www.digitalout.com/
Digital Output

There is a common expression in electronic publishing for monitor viewing called "WYSIWYG," which means "What You See Is What You Get." When dealing with print files (see pages 271 and 273), it would be accurate to say that "What You Send Is What You Get."

As many corporations expand their in-house capabilities to include prepress methods (such as high-end workstations, color printers, or image setters), their need for outside imaging centers still remains. The technology evolves so quickly, it is not cost-effective for the corporation to invest in the latest technology. The future role of the imaging center may be to function as a research and development site, support center, and consulting center for the buyer whose capabilities and needs can vary widely.

Traditionally, keylined boards were placed into a copy camera to make the film needed to expose plates. The copy camera is a large high quality reproduction camera (similar to a stat camera) where the boards are held under pressure as they are photographed for film conversion. Keylined boards or illustrations done conventionally are also called reflective art (see Glossary). Segments of film would be stripped together to formulate pages.

Automation continually moves the skills of prepress functions towards the purview of the creative group. Although file prepress preparation is being done more by those who create the files, designers and page composers need to work closely with the prepress specialist, who understands the mechanisms and tolerances of the process in greater depth.

PROCESS OF PREPRESS

When deciding who is to do what portion of the project, it should be foremost to choose the best solution based on competency and experience.

▶ **How prepress is handled** is determined by several factors, unique to each organization:

▷ *skills of the team members*—level of knowledge and balance of talents. At least one team member should be able to research new processes.

▷ *nature of the project*—each team member should know limitations. Decisions can be made as to how far to go in one area of expertise. Two-color, for example, might be handled by the creative group in-house, but four-color process film preparation is done by an outside imaging center.

▷ *structure of the fee arrangements*—if the designer does prepress, the printer may redo much of the work. Be sure not to pay for services twice.

The prepress profession is being squeezed between the creative groups doing more functions (including the investment in training and proofing) and the printers (which scale preparation to their capabilities and equipment).

New technology can allow digital files to image directly to the plate—without film—sometimes right on the printing press. Through laser technology and advancements in the printing plate, this has become the future direction of printing. Quality will take time to develop. As with any new printing technology, it is first applied to projects where quality is not critical. But eventually, this technology will reduce costs, especially with color printing. High-resolution proofs are still needed. (See chapter 24, page 280.)

▶ **The finishing process** begins when the final low resolution proofs are approved by the client. (See chapter 17, page 208.) The creative group, led by the art director, then supervises first the prepress of the printing. (See chapter 25, page 288.)

▷ *The page composer* sends the electronic files and low-resolution proofs to the prepress specialist.

▷ *Advance planning* directly affects quality and efficiency. The better the plan, the more money saved in production. (See chapter 10, page 100, and chapter 11, page 126.)

▷ *Adequate time* for prepress activities is critical. When under unrealistic deadlines, the price goes up (rush charges) and the quality goes down (not enough time to check elements and finesse output preparation).

▶ Prepress includes

▷ *scanning color or black and white images*—digitally capturing photographs, illustrations, or line art. Most images supplied by a stock photo agency are already in digital form. (See chapter 15, page 184.)

> These files may then be incorporated into electronic pages or output separately.

> Some scanning is done at the page composition stage—most notably low-resolution scans of photographs to serve as position images to be replaced by high-resolution scans when in prepress.

▷ *preparing images, color separations, and file formats* for backgrounds, screens, and other images. These elements become incorporated into pages.

▷ *preparing color separations for multicolor projects*—color photographs or slides are converted into the four process colors (each requiring a separate piece of film). Final output is color-separated film. Many photographs are provided in a digital form already, if shot from a digital camera or furnished from a stock photo agency. The price for color separating has dropped dramatically as more images are digital. It is one of the great benefits for smaller organizations who can now afford to use more color.

▷ *creating "print files,"* which capture and encapsulate all the information needed for output, including graphics and fonts. Whoever sets up the "print files" (sometimes called PostScript files) is responsible for

> software print settings—such as screen angles and trapping.

> file and document configuration—such as font installations, and any consequential improper output. (See sidebar, page 273.)

▷ *high-resolution output*—an image setter generates final paper or film output according to the buyer's specifications.

> Paper output is needed when a combination of electronic and traditional elements are used. Paper is also chosen when the project doesn't fit the limitations of output technology (such as an oversized piece), or when last-minute changes are expected (which can be handled as strip-ins rather than requiring all new output). Film is then made from the paper output, shot like traditionally keylined boards. This practice is becoming less prevalent as content

becomes digital. Paper output is uncommon compared to film.

> Film output—used when a project is entirely digital, or with color separations, or both. Film allows higher image quality than paper output, which then must be photographed to create film—which adds another optical generation, and reduces sharpness.

▷ *identifying ownership of materials*

> when the buyer provides traditional artwork, the client owns the boards or original files, and the prepress organization owns the prepared files. The printer owns the resulting film. (See chapter 29, page 330.)

> when the buyer provides digital artwork, the client owns the film (see chapter 29, page 337) as the tangible product prior to printing. This can be important for reprints, series, or updates.

MAJOR DECISIONS:

set project foundation
plan rough project skeleton
assemble category outline
solicit proposals
choose creative group
finalize project plan
select design theme
approve design
review writing outline
edit first draft content
review revised content
review final content
approve visual components
approve first prototype pages
approve pages and testing
approve final artwork
approve prepress proof or test

CLIENT GROUP

Many clients do not know very much about prepress. Once they sign off on the final lasers of a project, they often feel that they are done with it. However, so much of publishing comes together at the prepress stage through an ever-growing sophistication in imaging technology, client attention should remain astute.

KEY CLIENT RESPONSIBILITIES

Decision-makers' availability is critical, especially for projects that are moving very quickly. Though the client is mainly concerned with the proofs that come out of the prepress process (see chapter 24, page 280), the client must:

► **approve the schedule and fees** before proceeding with prepress.

► **allow adequate time** for prepress to ensure quality. (For new methods, it is best to allow twice as much time for output as estimated.)

► **understand an overview of the process** to develop accurate expectations, especially for proofing methods. (See chapter 24, page 280.)

► **keep track of the film** and make arrangements for its receipt or storage. (See chapter 27, page 318.)

► **approve all related proofs**—reviewing printers' proofs is one of a client's most important approval points. (See chapter 24, page 284.) This is the last opportunity, albeit a quite expensive one, to make changes.

The art director communicates to the client about the prepress process. Whether capabilities are handled in-house or with an outside supplier, the art director is the supervisor of each step.

KEY CREATIVE GROUP RESPONSIBILITIES

If the creative group takes on more and more of the prepress functions, they must also

► **check references and samples** in advance when utilizing a print group for the first time.

► **make determinations of quality at key proof stages**

▷ *set up the best expectations* by using samples and descriptions—every art director has preferences for what comprises good color. Demonstrate to the buyer, using samples to describe quality expectations.

▷ *check for accuracy*—one of the most important reasons to have proofs. Read early proofs thoroughly. Check placement of colors and elements on later proofs. Be sure to have fresh eyes read copy.

▷ *alert the client if any problems* arise and recommend the best resolution along with proof appraisal.

COLLABORATIONS THAT PAY OFF

The creative group should request advice or guidance from the production group on building pages to:

► **achieve visual effects**—such as the handling of typography, backgrounds, vignettes or blends, borders, etc. Identify any technological limitations or guidelines.

► **combine several image formats and methods**—such as portions from different software packages, integrated database publishing, or statistical diagrams.

► **handle photographs** with separations, screen settings, color information, or special effects.

► **combine traditional methods with electronic methods**—for some concepts, the old ways may still be the best ways.

► **understand the printer's specifications**—such as paper and press tolerances. Make sure that the prepress specialist has access to the printer for questions and will confirm file needs and receipt. This communication will enable the prepress specialist to prepare output appropriately. If this communication is not facilitated, the creative group may have to absorb extra costs if the printer cannot utilize the output provided.

Collaboration between the creative group and the production group, before the page composer prepares final digital files, allows for a faster finishing process. Send a sample page in advance of the project if systems need to be tested. Many page composers make a mistake when they send a project with an urgent deadline to the imaging center without a previous test or input from the prepress specialist. Perhaps the most important thing to check is any incompatibilities between the two systems so that when time becomes critical, the systems will work predictably.

CRITERIA FOR SELECTING AN IMAGING CENTER:

> MATCH CAPABILITIES WITH NEEDS. If a project requires typographically related services, make sure the imaging center has experience in this area.

> FOR HIGH QUALITY PROJECTS, select an imaging center that has both traditional and electronic experience. In this way, the best techniques will be employed for each project's needs. Make sure the prepress specialist is a color craftsman.

> CHECK THE IMAGE SETTING OR COLOR SEPARATING EQUIPMENT used. Review samples of projects that most closely relate to the project needed.

> CHECK COMPATIBILITY OF EQUIPMENT between the computer system where documents are created and the system of the imaging center. Allow time for experimentation in translation if needed.

PREPARING FILES

The production coordinator from the print group should advise on the best approach for the page composer to provide files for imaging. Building a working relationship between the two makes the collaboration smooth and seamless with practice. Test files if using a new supplier. Prepare files as:

► **"live" documents** (from the originating computer that created the file as a native version). The page composer

> ▷ *helps to handle any differences between systems*—the native system (where the file was created) and the foreign system (where the file outputs). This is particularly true when using new software upgrades.

> ▷ *sends the digital files* complete with all appropriate graphics and images to the prepress specialist.

> ▷ *gives instructions* to prepare the "print files" to the printer's specifications. Unlike a "print file" (below), the document file does not necessarily carry fonts within it. The imaging center is required to own the appropriate fonts and applications for outputting "live" files or to be provided fonts from the designer. The transmission of fonts to the imaging center is generally prohibited, except in "print files."

► **"print files"** (such as Encapsulated PostScript files). The page composer

> ▷ *chooses the most appropriate settings*—such as for line screens, page orientation, percentage of enlargement or reduction, crop marks, etc.

> ▷ *sets up the pages for the desired kind of output* according to the printer's specifications and encapsulates the page into a format for the prepress specialist to make film. Capabilities between the role of the page composer and the role of the prepress specialist will vary from group to group.

> ▷ *is responsible for how the files perform*—if the files do not image appropriately, the page composer absorbs any additional costs associated with correcting errors and reimaging the project. This can be time-consuming, so it should be avoided.

> ▷ *does not require the imaging center to own the needed fonts* because the fonts are embedded in the document. These files typically can't be opened or edited by the imaging center.

> ▷ *has fewer chances of error* because they are encapsulated on the same system where they originate. These files will typically be downloaded directly to the im-

Many files that imaging centers receive have "fatal errors"—the document will not open at all. Other files may have minor errors that cost time and money for extra preparation. The page composer can avoid most of these problems and costs by discussing the document composition in advance with the production coordinator.

age setter and will image according to the original system's software and hardware settings.

▶ **and receive final output** back from the print group. When proofs are received, the page composer and the art director:

▷ *check over the quality and consistency* of the final output proofs (see chapter 24, page 280) and materials.

▷ *receive the original disk and support materials back* from the print group and return appropriate originals to the client. (See chapter 29, page 330.)

PRODUCTION GROUP: PRINT

Generally, the project manager and page composer, with supervision from the art director, are the buyers of imaging services. In some cases, the creator of the digital file turns it over to the client group, who then becomes the buyer (such as in series use or when templates). In these cases, the page composer should be available for consultation.

EFFICIENT PROCESSES

The production coordinator is responsible for all aspects of the project in the preprint process:

▶ **sets up for prepress**

▷ *works with the page composer*

> gives guidance and advice on building and preparing pages (see page 202)—and is especially helpful when seeing the design before pages are built.

> provides guidelines and forms for the page composer or buyer to fill out, with output specifications needed to ensure film and press compatibilities.

▷ *receives digital files* from the page composer (see preflight sidebar)

> evaluates project materials and readable electronic files for completeness of

• software and hardware compatibilities.

• imaging requirements and instructions.

• final low-resolution proofs, with appropriate markings for color treatment.

• inclusion of all appropriate graphics.

• availability of appropriate fonts.

• print settings.

> has no liability for being able to open or read received or provided files.

Many imaging centers have a "preflight" department that handles incoming digital files for prepress. A prepress specialist is assigned to check incoming files immediately for imaging ability. In larger facilities, there is a full-time staff whose responsibility is to check files and communicate potential problems or errors to the creative group who can handle any appropriate matters with the client.

**IMAGING ERRORS CAN
BE CATEGORIZED:**

> INDUSTRY-WIDE ERRORS: imaging problems that occur because of poor PostScript code that prevents files from imaging properly. These are very difficult to explain and are not the imaging center's or the page composer's fault. Correcting these errors is part of the imaging center's overhead of doing business. Similarly, if the page composer's computer malfunctions, it is the creative group's responsibility to repair damaged or lost files. These evolve and change as capabilities mature.

> SOFTWARE ERRORS: These errors may be inherent in a software application package. The vendors to the imaging center have a disclaimer agreement which prevents them from being held liable. It is the responsibility of the imaging center to deal with these errors, and they should report them to the application vendors who can correct them in new software releases.

> PAGE COMPOSER ERRORS: The document originator builds pages improperly. These errors can include font incompatibilities, buried images, or incorrect settings. It is the responsibility of the imaging center to request corrected files from the page composer, who must absorb the expense of correction and additional imaging time.

> IMAGING CENTER ERRORS: These errors are caused by improper equipment setup, maintenance, or not following the buyer's instructions. They are corrected by the imaging center at no cost to the buyer. All imaging centers should utilize density control devices and have quality control procedures in place.

- Van Tanner
Vertech, Inc.
Greensboro, NC

▷ *notifies the buyer immediately* and requests additional information if

  > instructions are incomplete.

  > file setup is inadequate.

  > errors are detected.

  > preparation is not according to the print group's guidelines and will require extra costs to make ready.

▷ *accepts the project at their own discretion* if there are questions and the page composer or buyer is not available to answer questions or provide needed information.

▶ **handles file preparation**—the imaging center can receive a project in one of two ways:

▷ *run the project "live,"* which means opening the files in the original software and then creating the "print files." (See chapter 23, page 271.) The imaging center

  > has the prepress specialist select the proper settings which have more control for file output performance and printing than if the files are encapsulated.

  > is responsible for the film's printing performance—having a printing background is very valuable, for the behavior of dots on paper can be anticipated when knowing the materials.

▷ *guide the page composer* in setting up the "print files" (see chapter 23, pages 268 and 271), and then utilize it to generate film. The imaging center

  > has less control of how the files perform—unless the native files accompany the encapsulated, the imaging center is limited with the changes they can make.

  > should request both native and print files so that they can access the original setup if needed. It is a rare project that translates correctly from encapsulated files the first time. There are always adjustments to make on proofs.

  > has fewer chances for errors than when the "print files" are prepared properly.

  > hold the page composer responsible for the file performance when preparing "print files" without the advice of the imaging center. Only a very experienced page composer should handle prepress, and should know limitations.

- **crafts the finishing processes**—when working on the project, the imaging center is organized by the production coordinator and fulfilled by the prepress specialist who

  - *combines text, images, and other graphic elements* to construct the document for output imaging.

  - *images the file in high-resolution*, set up either for paper output (not common) or film output (most common) or direct to plate (growing quickly).

    > generates output according to the buyer's instructions. The final output should match the final low-resolution proof in content, composition, and instructions.

    > guarantees that final output is within accepted printer specifications. The creative group does not pay for discrepancies inside the print group as long as communication is facilitated between the imaging center and the printer, and the project parameters don't change.

  - *should not perform additional work* to a file or project without approval from the buyer. Often corrections, not foreseen in advance, need to be made for compatibilities that may incur additional fees. (See chapter 28, page 322.) Be especially careful if under hot deadlines, for reporting changes can seem to take up valuable time.

    > restructuring elements to ensure printability—rescan originals, reset color information or page setup specifications, etc.

THE KEY COLOR COMPROMISES BETWEEN ORIGINALS AND IMAGES READY FOR PRINTING

"The quality of the printed result is influenced both by the selection of materials and processes, and by the trade-off decisions made during the production process. The whiteness, gloss, brightness, absorptivity, opacity, fluorescence, and texture of the substrate will directly or indirectly affect the appearance of the reproduction. The number and types of inks, and their properties (gloss, transparency, pigment type, fluorescence), will greatly influence the color gamut of the reproduction. The printing process (e.g. gravure vs. screen printing vs. lithography) and the selective use of overprint coatings will also influence the quality (or 'grade') of the final result. Economic factors drive the selection process: consider the price paid for a newspaper versus a limited edition book of fine art photography. Regardless of material selection, much can be done during the reproduction process to produce the best possible results for the grade of materials being used. The key decisions have to do with compressing the contrast range and color gamut of color transparency originals to produce the most satisfactory reproduction.

"In general, the best results are obtained by compressing the tonal range to favor the 'interest area' of the tone scale (e.g. light tones in high key photographs), and by compressing the saturation to retain the approximate perceptual differences between gamut and non-gamut colors. Trade-offs can, of course, be minimized by using such enhancements as higher grade substrates, 7-color instead of 4-color printing, and overprint varnishes. On-press changes in ink film thickness or color sequence (black should always be last) can produce minor refinements in appearance for single images (e.g. a poster).

"The use of mathematical models to make acceptable tone and gamut trade-offs will always be very limited. The quality requirements of the proof (the 'prototype' for thousands or millions of production images) are extremely demanding and the models cannot handle the many 'field of view' factors and individual color vision characteristics of the person who approves the proof."

—Gary G. Field
California Polytechnic State University
San Luis Obispo, CA

> translating files or graphics (i.e. TIFF and EPS files—see Glossary) from one form to another or fixing file formats set up improperly. This can include matching all graphics to a page layout file by finding all the pieces and a construction that will image properly.

> correcting print errors—either due to differences in the native and foreign system, software versions, bugs, or the misunderstanding of instructions.

> changing typography or content—client circumstances change, causing changes to be needed at the last minute, usually affecting budget.

▷ *should provide the buyer with an estimate* for any additional work necessary to make a file ready for imaging.

> If such costs are difficult to estimate, a guesstimate or a not-to-exceed amount should be provided. Many clients are comfortable with a not-to-exceed amount, reserving some for contingencies if under heavy deadline pressures.

> If estimates for extra work are not provided by the imaging center to the buyer in advance of work being done, the imaging center proceeds at its own risk. The buyer may choose to not pay for work done without prior approval. When under deadline, be sure to stop and apprise the alterations before proceeding.

> If any work is done to the file, the imaging center needs to have the buyer approve changes, often requiring additional prepress proofing.

▷ *proofs are made to buyer specification*, expectation, and acceptability. (See chapter 24, page 280.)

► **collaborates with the printer** to

▷ *prepare prepress instructions according to press specifications*—this should be known in advance, and not when the film arrives at the printer under a tight time frame.

▷ *ensure that the film is compatible with printing tolerances.*

> If the printer cannot utilize the film, and the prepress specialist did not follow instructions from the printer, the prepress specialist resupplies film at no charge.

> If the creative group does not ensure communication between the production coordinator and the printer, the creative group assumes responsibility for the acceptance of the film and absorbs the extra fees if the printer cannot use the film.

"Color desktop will eventually shift much (most all?) color prepress activity into the printing customer's design and production environment."

—Jacques Marchand
Printing Consultant
Marchand Marketing
San Francisco, CA

"Some of us computer nerds came from a graphics and printing background; we simply evolved with our industry and the integration of technology with design. Some of us may have traded our pens and brushes for keyboards but we haven't lost sight of the art and craft of printing."

—Laura Smith
DataQuick Information
Systems
San Diego, CA

"Color approval can be complicated by differences in perception. Leaving aside the fact that about eight percent of the male population has abnormal color vision ('color blindness'), everybody's color vision differs, to a slight degree. A further source of difference is that we all experience color vision changes (the yellowing of the eye's lens is mostly responsible) as a normal part of the aging process. Color perception is also influenced by mental processes—by the observer's bias for or against the subject matter of the picture. Differing cultural experiences will also influence the "correct" expression of color creativity.

"In practice, the use of physical samples such as printed color charts or proofs of prior jobs may help facilitate communication between those involved in the color approval process. Such samples must, of course, be representative of the production conditions for the job at hand.

"A simple recognition of the fact that vision, like any of the other senses, naturally varies from person to person, can do much to reduce conflict. Nobody, from an absolutist point of view, has 'correct' color vision; but, as a practical matter, 'correct' color vision is that of the person who has the authority to accept the proof as a satisfactory reproduction."

—Gary G. Field
California Polytechnic
State University
San Luis Obispo, CA

"Generally speaking, everyone involved in the process of photomechanical color reproduction strives to produce the best possible result. The quality of the ultimate result, however, is constrained by the economic forces that dictate the use of certain materials and procedures in the manufacturing process.

"The predictability and consistency aspects of color quality can, for example, be significantly enhanced by adopting a specified ink set, substrate, proof density target, and screen ruling. Such specifications are essential for the publications industry which has to handle color pages from many different sources.

"The needs of the design and creative community, on the other hand, include the flexibility to choose from a wide variety of substrates, inks, and other materials, in order to produce printed products that have the desired impact. A further expectation of many art directors and print buyers is the freedom to 'fine tune' the reproduction through all the production stages until the final press OK."

—Gary G. Field
California Polytechnic
State University
San Luis Obispo, CA

MOBILE COLOR PROFILING

"It's not voodoo, but it is a mix of fine art and hard science.

"The printing press is the most highly variable of all of the printing processes. Different pigments and tacks of ink, variations in fountain solutions, temperature and humidity, thickness of packing, and variations in blanket compressibility are just a few of the variables. And that doesn't even mention the ink keys that control the flow of ink. All sorts of combinations of hardware, chemicals, and substrates have been tried in the offset printing business, and while some are better than others, none is perfect. In short, the physical limitations of presses cannot possibly be fully taken into account with any proofing system.

"The best way to do it is to create a profile sheet that mimics as closely as possible the majority of work you expect to run on the press and the physical and environmental conditions that are 'normal' for you. Use your standard inks and your most popular paper stock. Then run the sheet and create your profiles at the beginning, middle, and end of what would be a typical press run. Take an average of these profiles to create a composite profile. Some printers actually run these profiles for different papers and different presses. Customers who have profiled their presses in this manner find they can save a substantial amount of make-ready time for every job they do."

—Stephen Beals
Digital Prepress Manager
Finger Lake Press
Auburn, New York

"These days, it's easy to assume that the need for the traditional service bureau is being challenged by a preponderance of do-it-yourself designers, typesetters, and digital imagers, thanks to the ever-increasing accessibility of computers and prepress software. As any self-respecting digital imager knows, prepress is not made by computer alone. To make the cut in today's dog-eat-dog digital imaging market, where service bureaus are a dime a dozen and where only a select few survive and are successful, you've got to have an edge, an advantage you can offer your customers that they might not be able to get elsewhere.

"'Computer nerds' inevitably stumble because they do not understand that the industry is based on graphic solutions. They don't have experience in fonts, color calibrations, or any of the other issues that really drive the industry. Many printers were pushed into prepress, but they simply lost money trying to keep prepress in-house. Customers rely on you for lots of service and lots of knowledge. And you've got to be prepared to buck up and keep current with your equipment. Our goal now is to become a turnkey, one-stop shop for our customers. And, of course, you've got to respond to the fast turnarounds that drive this business, and being able to deliver is what has made us successful."

—Laura Lee Smith
www.imagingmagazine.com

"At last count there were about two dozen dot-coms offering a wide range of print-centric services, from equipment and supplies procurement (PrintNation.com), to auction sites where print buyers can solicit bids for projects (PrintBid.com), to sites that let buyers order templated print materials online (iPrint.com), to sites that let you preflight files online (creativepro.com). But perhaps the largest category is application service providers (ASPs), which streamline the job production, management, and administration process between printers and their clients.

"Printers are notoriously reticent about adopting new technology and sharing their hard-earned revenues with third-party 'brokers.' Not to mention the fact that the printers themselves are perfectly capable of developing and offering their own online systems."

—Anita Dennis
Contributing Editor
www.creativepro.com

"How do you sell your best work in a time when most clients can't seem to afford it? The real question is, 'Can they really afford anything less?' Lean times force businesses to expect results from every dollar they spend. And as we all know, their fear of the unknown in design increases with the pressure to succeed. Second-guessing can replace risk-taking very quickly.

"None of us in communications, can afford to produce anything but the most effective, functional, and memorable creative work. Nor can we cover up a weak idea with expensive production techniques. Smaller production budgets put more emphasis on the basic ideas being sound."

—Mike Scricco
Designer
Keiler & Company
Farmington, CT

SEYBOLD REPORT

"The industry has become so complex that each person's microcosmic view reflects the portion of the industry he or she takes part in.

"Is Internet commerce destined to dominate the way the printing industry does business, as companies such as Noosh and Collabria would like us to believe? Will we be swayed enough by the flexibility and sophistication of Adobe InDesign to abandon our ingrained Quark habits? Can we get our customers to give up expensive proofs and accept inexpensive proofs?

"The turbulence in the market is threatening much of what we used to take for granted and creating real challenges that we must deal with if we are to survive. We call attention also to asset management, cross-media publishing, and Internet-related issues. The 'real' challenge of the future is cross-media publishing, with the winner of that battle bringing home the bacon.

"Web-based systems for buying print jobs came into focus at SSF, with exhibits from six new vendors. They all purport to facilitate the process of specifying and bidding on print jobs. They can be classified on two criteria. First, is the system open or closed? PrintBid, PrintMarket, and MediaFlex are open, in that anyone with a browser can use the system. Noosh, Impresse, and Collabria are closed: the users must have contracts and go through a database setup process before they can use the system. The other criterion is the degree to which the system manages the printer's work flow.

"Nowhere is technology advancing more rapidly than in color proofing, where image quality, color fidelity, printing speed, and sizes supported are advancing while prices decrease.

"Now the market is being flooded by machines at all price levels, with wider color gamuts, higher resolutions, multiple speed-quality trade-offs, and the ability to proof press plates of nearly any size. With the aid of the Extreme architecture, it is possible to divide a job into pages and send individual pages to different processors for faster throughput."

—Peter Dyson
Senior Editor
Seybold Conference
San Francisco, CA

Publishing can't proceed without proofs. There must be a way to check the work definitively before making multiple impressions. Proofing provides all participants in the publishing process with

► **an approximation or a preview** of how the project will look when finished—before committing to press or fabrication.

► **a final review** used to obtain approvals—and the worst time to find content errors (but the last chance). Proofs are intended to check the preparation processes rather than the accuracy of content.

► **confidence that the results will be as expected**—no one wants surprises upon delivery, but wants to find printed pieces exactly as specified.

VISUAL INSURANCE I: DESIGN PROOFS

Proofs confirm that various elements of a publishing project are visually correct before the next phase begins. Various proofs verify content, typography, image placement, production preparation, color positions and balance, and, eventually, function as a guideline for printing or fabrication. The kind of proof needed depends on the phase of development, complexity of the parameters, media needed, the delivery method, the use of technology, and the approval process.

Designers prepare a number of proofs for the client's review (see chapter 16, page 196).

► **Comprehensive presentation materials** (comps) show design and color intent. These are typically prepared digitally and often look more polished than a rough idea may intend. Ideas can be polished through the production process, and it is important to do the most experimenting and option-trying in the beginning versus later on. (See chapter 10, page 100.)

► **Low-resolution proofs** show the formatted text and rough graphic elements in position. The creative group submits this to the client for text and content refinements. This is the best time for the client to make changes or add information.

- ▶ **A corrected client proof** incorporates the changes from the client, who checks it for accuracy. There may be several of these, depending on how many corrections are needed.

- ▶ **An on-screen viewing** or "soft" proof is reviewed on the computer monitor. Monitor color can only approximate the ink on paper. Calibration software helps to adjust the monitor to be more accurate, but the viewer should be careful not to expect to match monitor color exactly onto paper. For exact color, use color swatches from the ink manufacturer or printer.

- ▶ **A final low-resolution proof** receives the signature of approval from the client, and then accompanies the digital document to the print group. They are to check

  - ▷ *typographic and statistical correctness*—the client is responsible for any errors

  - ▷ *element position, placement, and registration*

  - ▷ *color specifications*—refined as pages grow more final, but the color can be inaccurate in any proofing method when it tries to predict print. Low-resolution proofs can show elements much rougher or darker than they will be in high-resolution. Low-resolution proofs include:

    - > dot matrix printers that use toner cartridges—these are very cost-effective and getting cheaper. These printers cover a fairly high-quality low end of the market and will continue to until another technology even less expensive surfaces.

    - > thermal wax transfer printers—some of the earlier color proofers, rather expensive, and being used less and less.

    - > laser printers, both color and black and white—allow high-resolution but can be rather costly.

VISUAL INSURANCE II: IMAGING CENTER PROOFS

The print group generally provides higher resolution proofs than the creative group. Sometimes, when the color is not critical, some rough imaging center proofs may be sufficient enough for client approval, if the client understands their limitations:

- ▷ *Ink-jet proofs* or *dye-sublimation proofs*, such as Iris® or Rainbow®, are made directly from the digital file. These are often used as preliminary color proofs to show the color specifications. These proofs contain no color separation halftone dot structure that approximates printing, so blends look smoother and details much finer. They often look better than the printed piece, for

Do not assume that, because you OK'd low-resolution proofs, you don't need to carefully check printers proofs. It is wise to check the first and last line of every paragraph on all proofs to check the integrity of the content. It is not wise to actually *read* the copy after the final low-resolution proof because editorial changes sneak in like unwanted dinner guests.

The closer proofs get to the end of the production process, the more reliable they become in approximating how the printing will look. However, the closer they approximate the actual printing, the more expensive they become to produce and correct.

dyes are more vivid than printing inks and halftone dots can coarsen images. They are very commonly loved amongst designers to check color treatment, special effects, and for presentation.

▷ *Digital high-end proofs,* such as Digital Matchprint® or Approval®, are also generated electronically—not from film. They do imitate the halftone dots but may not reflect the actual screen patterns that will appear in the printed pieces.

VISUAL INSURANCE III: PRINTERS' PROOFS

The print group, prior to printing, uses the final low-resolution output, any proofs, and files provided by the prepress specialist to make the high-resolution contract or print proofs. These most approximate how the project will look when printed. (See chapter 23, page 274, for what factors affect printing predictability.) This is the client's last opportunity to give approval prior to printing. Several different proofs may be utilized.

Generally, the printer makes the prepress proofs that the client will see. But as imaging centers handle more and more prepress functions, they also create more of the proofs. Generally, the group that prepares the film for plate-preparation has the responsibility of creating proofs for the buyer to approve.

► **Final prepress proofs** are made from the film that will be exposed to the printing plate. These "composite" proofs incorporate all page elements and are high-resolution:

▷ *Dylux® proofs* (or "blue lines") are single-color approximations made on a photosensitive paper exposed to the film. These are generated for one- or two-color projects (a second color usually appears as a lighter image), and often can accompany color proofs to show format. These proofs are cut and bound to approximate the intended bindery methods. A Dylux® proof demonstrates the image registration and indicates how the pages will fold and bind, or how a prototype works with dies or folding, or demonstrates other special finishing.

▷ *Velox® proofs* are reproduction-quality and show content as black on white paper. They are accurate, inexpensive, and durable for one-color projects. Used often in advertising, they are inexpensive and highly accurate if color is not a factor.

Advertising agencies often use Veloxes® for one- and two-color advertisements. Generally, magazine publishers do not provide agencies with proofs before printing the ad. For this reason, the agency provides Veloxes as proofs for the client. On two-color ads where one of the colors is black, a Velox provides the black image and a Color Key® or Chroma-Check® overlay shows the second color. However, rarely is the overlay the exact color of the requested ink color.

▷ *Overlay proofs,* such as Color Key® or ChromaCheck® proofs, show each color on a separate acetate layer. These are an inexpensive way to check spot color, position, registration, and trapping (see Glossary). They only come in a limited number of colors, so the overlays closest to the ink colors are chosen. They are used most often to check two- and three-color projects. It is a relatively inexpensive proofing method and individual colors can be viewed separately. The disadvantage is that the color is not accurate.

▷ *Integral proofs,* such as a Cromalin® or a Matchprint® check process color built from the halftone dot structure of the film. They approximate how the printed pages will look, and do include the halftone dot, but not how it will behave on the paper. These are made from ink-matched powders and laminated material, and have some color limitations. However, they provide high-resolution, accurate registration, and are reasonably priced. Experienced printers have learned how to match the reality of ink on paper to these proofs.

▷ *Water proofs*—that take the powder/laminate technology of the integral proof and allow them to be on the substrate of choice. Some can approximate dot gain or opacity of the inks. As a film-based contract proof, it is a good guide for the pressmen to try to match. Kodak Approval ™ is on the vendor front and recently became able to handle double-sided.

▷ *Signature proofs* are actually printed pages made on a small special press. The advantage is that they show the inks on the selected paper with any dot-gain and can be printed on both sides of the sheet. The disadvantage is that they are expensive and time-consuming, and may still vary from the actual press that will be used.

▷ *Press proofs* require setting up the actual printing press and performing a preliminary short run. This is the most expensive and time-consuming proofing method. But it is by far the most accurate, for the project is printed on the actual paper and press, demonstrating how much of the ink will soak into the surface. Most press proofs allow color experts to see which colors contain problems, or how best to improve color balance. This is the only way to predict the exact way the halftone dots will behave on the specific paper and to proof specialty inks (such as metallics or varnishes). Some printers may elect to create control sheets for their most used house-stock, giving the buyers stronger predictability for a given press.

▶ **Industry evolution**—the technology of color separation and proofing is advancing rapidly, and color proofs are becoming more accurate and less expensive. As color printing becomes more accessible to more buyers, proofs help lead the color evolution. Color proofs are becoming less expensive and more accurate with each generation of proofing device. During the 80s and 90s, proofing methods have been one of the fastest changing areas of the graphic arts.

MAJOR DECISIONS:

- set project foundation
- plan rough project skeleton
- assemble category outline
- solicit proposals
- choose creative group
- finalize project plan
- select design theme
- approve design
- review writing outline
- edit first draft content
- review revised content
- review final content
- approve visual components
- approve first prototype pages
- approve pages and testing
- approve final artwork
- **approve prepress proof or test**
- **approve press proofs or reviews**

Advances in publishing technology compress the phases from concept to proof, accelerating the process. Therefore, the involvement of the client is intensified, for there are fewer stages at which to catch errors, and less time between the stages. Proofs are critical checkpoints to perfect details and verify accuracy.

PURPOSE OF PROOFS

Proofs provide the last convenient opportunity for changes before proceeding from phase to phase (see chapter 26, page 312 for escalation formula). Each reviewer should

▶ **catch typographic errors and changes**—correct on the earliest of proofs. The later errors are caught, the more expensive they are to correct. Especially proofread early proofs.

▶ **check evolution of project**—each stage of the proof is provided for a specific purpose. Do not make changes later that should have been made on earlier proofs because this jeopardizes deadlines and budgets.

▶ **verify images**—previous versions and original materials (such as photography or illustrations) with client-marked corrections should accompany the next proof for comparison.

▶ **mark changes**—any corrections the client wishes to make must be clearly indicated, and another proof requested, when changes are extensive.

CLIENT REVIEWS

The client, if remaining on top of the development process, should find no surprises in the proofs received.

▶**Initialize and date three key proofs** before the project proceeds to the next step:

▷ *final writing.*

▷ *final low-resolution proof showing all elements.*

▷ *final prepress or print (high-resolution) proofs.*

▶ **Verify accuracy**—any preexisting mistakes discovered after each initialed proof are the responsibility of the client who:

▷ *controls the ultimate correctness of the work.*

▷ *makes sure all approvals precede each phase.*

▷ *pays associated fees* for any changes and corrections they make. Proofs are the only assurance the client will have regarding the visual quality of the work prior to going into final processes. (See chapter 25, page 288.)

WHO PAYS?

Rarely is a proof made that doesn't lead to changes of one sort or another—and changes always increase costs. The question of who pays for the changes depends on where errors originate.

> If the print group did not understand an instruction, the print group absorbs the cost to make corrections.

> If the creative group needs to make an adjustment because of its own errors or changes, the creative group absorbs the cost.

> If the client makes alterations in content, text, or other changes, the client pays for these changes.

It is unrealistic to expect that there will be no additional costs, as proofs are intended to catch anything wrong. Hopefully, corrections will be minor, not causing any change in the schedule or budget.

The art director shares the responsibility for quality and accuracy of the final product with the client, although the client has the final authority over content and what constitutes acceptable work. The creative group relies on proofs to obtain all approvals before proceeding to each phase of a project.

ART DIRECTOR REVIEWS

As the art director is responsible for all project processes, this professional must be sure to

Art directors each have preferences in the judgment of what makes a good color separation. For example, some prefer a high-contrast separation while others want a more subtle effect. Some may prefer the color balance to be warmer, while others prefer cooler. It is the responsibility of the art director to communicate preferences to the prepress specialist. The more the art director's preferences are understood by the print group in advance, the greater the chance the first separations will be acceptable. Perhaps the best way to demonstrate preferences is for the art director to show examples of preferred color.

► **review and approve all proofs** before showing to the client.

▷ *instruct corrections deviating from the approved design.*

▷ *direct changes caused by the creative group's errors.* Costs for corrections will be absorbed by the creative group.

► **supervise proof categories and limitations**—describe to the client what to expect with an accurate description of the proof methods due to budget or technological constraints. (The closer the proof is to the way the project will look when printed, the greater the assurance that the client will be satisfied with the completed project.)

► **manage the status of proofs** and ensure that all parties adhere to the appropriate approval processes.

▷ *Obtain the client signature* on proofs and never give the "OK to Print" without this assurance. It makes clear the client responsibility and the point where the creative group is not liable for any previous errors.

▷ *Ensure that the designer or page composer also review* the final proofs prior to print as a check that the printed piece will match the intended specifications.

Each member of the print group has different proofing capabilities and responsibilities, which should be outlined in the estimates and quotations.

BUILD CORRECT EXPECTATIONS

Match the appropriate kind of proof to the project phase. This is a printer's specialty. Planning should specify:

► **the kind and number of proofs** to be supplied.

▷ *Color needs best dictate* what kind of proof to generate. For example, a cosmetics brochure may require more color accuracy than does a brochure on tools.

▷ *Decision-makers' ability to visualize demand proof*

level—although it may be more cost-effective to use a rougher proof, a higher-resolution proof may be chosen so the client can more closely see the intended result.

▷ *Location of proofing needs*—remote proofing allows projects to be transmitted to other sites where the proof is generated by the receiver's output device. This enables suppliers to work with clients in other locations, with multiple offices, or to transmit to several publication sources at once.

► **proof limitations** and variations from printed pieces:

▷ *range of tolerances*—different for each kind of equipment, processing, and printing. (See pages 270–273.)

▷ *reasonable color match*—between original photographs or illustrations and color separations. (Acceptability is defined by the buyer.)

▷ *removal of imperfections*—created by dust, scratches, fingerprints, etc. If an original image has imperfections, the buyer needs to be notified of the costs entailed to fix it.

▷ *behavior of ink*—different on various kinds of paper. Adjustments can be made in file settings if the imaging center is aware of the paper to be used.

▷ *other conditions*—between proofing and pressroom operations that may cause variances between proof and printed piece. Explain factors to the buyer (such as humidity, paper porosity, drying cycles).

PROOFS AS INSTRUCTIONS

The print group uses the proofs as guidelines in production, and must approximate the finish as closely as possible. The print group:

► **prepares proofs according to specifications**—often the imaging center and the printer work together to create.

► **receives the prepress proof** back from the buyer with approval. Proceeds with printing only with the signed proof.

▷ *Any corrections are indicated* on the proofs.

▷ *Request for another proof is made*—when there is a color shift due to more work being performed on a file.

▷ *Any additional charges* for corrections are accepted unless they are caused by to the buyer's changes or errors.

► **anticipates a reasonable variation in color** between proofs and the printed project. If the buyer is not available to approve a press sheet, then the best judgment of the printer defines acceptability. (See chapter 21, page 258.)

When proofs are completed and signed, the project can move on to printing or fabricating for the final processes.

Viewing conditions should be considered when comparing a proof to an original. Most printers possess a color corrected lighting facility to balance colors most objectively. The viewing conditions of the final audience should also be considered and mimicked.

Paper greatly affects the way a project will look as compared to the proof. If the paper is uncoated, there will be dot gain—the halftone dots that make up the color density expand a little as the ink soaks into porous paper. This will cause the images to be muted and a little darker than the proof. On coated paper, the dots do not sink into the paper as much, and images have a crisp look. Many ink manufacturers provide swatch books that show how their ink will look on both coated and uncoated paper. With some colors, the differences are dramatic.

"The proof OK is extremely significant because any quality concerns that are missed at this stage are reproduced by the millions during printing. A misspelled word on the proofs will always be misspelled on the printed sheets in the same way that a wrong halftone dot value or color on the proof will always be wrong on the press sheet. Such errors reflect badly on everybody, hence the inevitability of the professional inspection process and all the attendant behavioral issues that exist between buyer and seller. The opinion of the final consumer (the general public) is of limited importance because the high-stake quality decision is made at the proof stage by the 'quality gatekeeper,' the art director or buyer.

"The rich variety of materials and colors that print offers will always appeal to the creative community, and will mean that certain aspects of color quality will always remain elusive. The distance between both ends of the color quality market may become less in time, but even then the path to efficient color quality will be based upon color education. An understanding of the limitations of engineering solutions and an awareness of the attendant physical, psychophysical, behavioral, and creative factors that play key roles in the color approval process is for now our best chance of achieving nearly-predictable color."

—Gary G. Field
California Polytechnic
State University
San Luis Obispo, CA

"Remote proofing allows two or more parties at different locations to proof the same file. Documents are sent and received electronically, allowing ideas and concepts to be shared quickly and inexpensively without the need for face-to-face contact or courier services."

— Debbie Johnson
Freelance Writer
Cincinnati, OH

"In some cases, the opportunity for remote proofing can be a major advantage in turnaround, as electronic files can be printed out at the customer's site.

"The biggest area for growth in digital proofing is opening up in scatter or intermediate proofs. Any work that requires several iterations of proofing is ripe for at least some digital work. Even if distrust in digital proofs resurfaces in the final stages, it can still make sense for all art directors and publishers to approve type, position, and color breaks, from images that show up on the desktop, especially when they are cheaper and faster."

—Steve Hannaford
Journalist
Philadelphia, PA

PRINTING 25

Printing means reproducing an image (usually using ink) onto a substrate (such as paper, plastic, wood, etc.). Offset printing uses spot inks and four-color process. Printing projects range from one-color to six or more colors, on every size and kind of surface.

BEGINNING WITH THE FINISH IN MIND

The buyer should choose the printer by matching their specialty to the nature of the project. Many clients and creative groups work with four or five different printers for various kinds of projects. All those involved with printing need to consider how the digital environment affects the work flow:

► **receive the project**—it can come to the printer in one of three ways:

  ▷ *digital files*—from the imaging center to be handled by their in-house prepress. Some files may image directly to press-mounted plates.

  ▷ *film that is plate-ready*—provided by the prepress specialist, often from an imaging center or the creative group.

  ▷ *camera-ready artwork*—provided by the creative group traditionally. Few projects are done completely without some digital elements.

► **transformation of capabilities**—evolves every level of skill set, work flow, equipment, and method—technology causes continual evolution.

► **new digital forms of printing technology**:

  ▷ *high-fidelity color* (hi-fi color), that adds secondary ink colors to the cyan, magenta, yellow, and black of regular four-color offset printing. The addition of these colors increases the color gamut (the range of possible visible colors made with dots on paper).

  ▷ *stochastic screening,* which digitally breaks a regular dot pattern into a random dot pattern, allowing for

---

---

greater color resolution in printing and eliminating moiré patterns.

▷ *digital press controls* to enable specific quantities, fast turnaround times, and print-on-demand capabilities.

▷ *high-speed customized digital presses* can accept artwork directly from the digital file without the intermediate step of film. Database printing, personalized publications, and small variable quantities are more economical, opening new venues. These presses begin in the realm of "acceptable color" but are increasing in quality each year.

▷ *remote printing to off-site locations* takes the printing process out of the printing plant and places it at the receiver's site.

CUSTOM MANUFACTURING

As one of the country's largest business sectors, printers can be very large organizations and business conglomerations. From small quick-printers to large book publishers,

▶ **printing is a craft** that utilizes the chemistry of ink and paper. Skilled pressmen know:

▷ *how to best match the proofs to meet client expectations* and deliver the acceptable quantity on time. Every project is unique.

▷ *how to maintain consistent quality throughout the run.*

▷ *calibration and color correctness.*

▷ *when to communicate any potential variations.*

▶ **printed press sheets (or rolls) are sent to the bindery** for assembly. Some printers have bindery services in-house, but others send work to special bindery companies. Bindery and finishing operations may include

▷ trimming	▷ foil stamping
▷ folding	▷ numbering
▷ collating	▷ stapling
▷ die-cutting	▷ gluing
▷ embossing	▷ laminating

Creative groups are all small businesses. Buying customized manufacturing means dealing with many printers who are very large businesses. This requires that the buyer both control variables and conform to industry methods. With such small organizations on the creative end and such large organizations on the finishing end, there can be a tension between the groups with opposite business needs and philosophies.

THE WISDOM OF THE WEB

New products are intended to remove redundancies, eliminate mistakes, and otherwise streamline the print chain. This means that costs are expected to be taken out of the process. Usually, reduction of cost is a euphemism for the need to reduce price. That is one of the probable outcomes of the e-commerce push. A second effect is that further increases in manufacturing efficiency can also be a euphemism for the decreased need for both capital equipment and consumable supplies.

—Robert L. FitzPatrick
President
FitzPatrick Management Inc.
Charlotte, NC

—Stephen P. Aranoff
Founder and Principal
Arttex Associates
Redwood, CA

All those in the process must share the same goals for reaching the buyer's expectations. Ultimately, if the creative and production groups don't realize that their crafts are also services, they will not survive long in the digital environment. Collaboration between the groups can build with experience together. In all production groups, technical skills are never enough for business success. Other ingredients that all have equal importance in balancing a production-based operation:

► **managing projects**—coordinating all elements and concerns.

► **respecting organization methods**—knowing the purview of various team members.

► **handling customer interests and relations**—be knowledgeable about priorities and key project concerns.

► **diplomacy among differing views**—each team member has a role and an agenda that should determine contribution, but individual motivations can also be a factor in project progress.

► **marketing savvy for what is relevant**—connection with the marketplace and what is needed.

► **intergroup communication to facilitate the process**—throughout the project, sharing information and development expectations.

A PLAN TO OVERCOME DISTRIBUTION STAGNATION

"Graphic arts growth is inhibited by:

> *distribution viewed as a sales business rather than a service business*—trying to deliver more goods to customers. Success is measured by sales volume, not customer productivity. But service is the *real* product: time, efficiency, customer satisfaction, effectiveness in solving problems, and improved productivity. The delivery of goods may be a part of the process and a profit center, but the specific goods themselves do not define the business.

> *remaining in the offset box*—will result in shrinking markets. Imaging is not just offset printing. Imaging is exploding in new output devices. Salespeople move from market to market, backed with appropriate marketing material and technical specialists.

> *resisting or underutilizing electronic data interface (EDI)*: infrastructure has not been put in place by the manufacturers [and other parti–cipants]. The Web offers graphic arts customers greater control over procurement, as:

• it provides extraordinary convenience and choice.

• can free up sales personnel from unproductive handholding, and reduce the number of operations personnel at the manufacturer, dealer and customer level.

• can more effectively monitor inventory and lower overall inventory levels.

"Customer service, an exploration of related digital markets, a resolution of channel relationships, and an acceptance of the Internet can bring the graphic arts business into the Information Age."

—Robert L. FitzPatric
Presiden
FitzPatrick Management In
Charlotte, N

—Stephen P. Arano
Founder and Principa
Arttex Associate
Redwood, C

As the project finishes, the client is anxious to receive the printed materials. In a deadline-oriented business, often, time crunches can put tremendous pressure on the finishing phase—the buck stops with the printer who should deliver to expected schedule.

DEFINING OF THE END

The client communicates printing needs to the creative group as part of the design parameters and must provide the following before printing begins:

▶ **approval of final artwork and proofs** (see chapter 16, page 196, and chapter 24, page 280). It is important to note a common dispute: Even if an error is made by the printer, if it was undetected by the client when the proofs were approved, the client can be liable for the costs to fix the error. The art director needs to communicate the seriousness of the final proof approval. The client needs to be very careful to have all necessary approvals within their own organization before giving the OK to finish.

▶ **schedule requirements** (see chapter 3, page 24).

▶ **determination of quantities** needed:

▷ *within 10% of the specified amount* (see over- and under-runs, below).

▷ *specify fixed amounts*—sometimes there are extra fees involved to hit a quantity number exactly.

▶ **destination and delivery instructions**—time worked into the schedule. In the event of changes, fees may be renegotiated. Common changes include a request

---

DESIGN AND PRODUCTION TIPS FOR VARIABLE-DATA PRINTING

"1. IMAGES WILL BE DIFFERENT than they are for print. The resolution of supplied images could be too low, resulting in pixelated output; or they could be too high with more data than is necessary.

2. THE PAPER STOCK must work on a digital press.

3. LARGE AREAS OF SOLID COLOR or screened tints can exhibit variation in toner coverage.

4. A BLEED MUST CONSIDER THE PRINTER'S IMAGE AREA.

5. A PRODUCT CAN'T BE TRIMMED, folded or bound because the layout does not accommodate the binding. Jobs must be set up to accommodate the final format and the available cutting and binding equipment.

6. Vignettes, blends and gradients exhibit significant 'banding'.

7. Variable text is cut off because the text container is too small for the longest word in a field. Designers should also anticipate odd word breaks. Jobs that cross platforms are especially vulnerable to text reflow.

8. TOO MUCH VARIABLE INFORMATION on a page causes press or RIP errors.

9. PLAN FOR TRAPPING AND KNOCK-OUTS. Trapping is the deliberate introduction of a bit of distortion into areas of abutting colors to accommodate mechanical misregister in a conventional printing press.

10. WATCH YOUR FONTS."

—Frank Romano
Rochester Institute of Technology
Rochester, NY

- ▷ *for advanced copies* sent to the buyer, or other partial delivery arrangements.
- ▷ *to add locations* for shipping printed pieces.
- ▶ **storage instructions** for film, plates, or extra printed pieces.
- ▶ **the budget** (see chapter 5, page 50) and billing arrangements.

RESPONSIBILITIES FOR FINISH

Supervising the printing, the art director is the key liaison between the groups. Setting up the best processes, however, is a collaboration of several team members all servicing the client's goal. For a smooth running project, keep several factors foremost in planning:

- ▶ **responsibility of print quality** resides with the buyer. Acceptable quality is defined by both the buyer and the art director (as a representative of the client) and by the use of the product. Each project will have different acceptability criteria. Discussing quality expectations in advance will help satisfy both the client and creative group when appropriate quality is achieved. Who is billed determines who ultimately has the final say in quality issues.
  - ▷ *If the client is billed directly by the printer—quality*

DEFINING DIRECT MAIL MARKETING:

"A comprehensive system of media and methods designed to elicit a response from a prospect or customer in order to develop or enhance a client relationship.

> Direct mail can be envelopes, cover letter, product brochure, order form, reply envelope, or other variations: packs, card decks, self-mailers

> Direct mail can reach a single individual at their home or business.

> Direct mail is an effective way to establish a one-to-one personal relationship between a potential buyer and a seller, and it can complement your overall efforts.

> Good direct mail cuts through the clutter of other mail to stand out and be noticed.

> Direct marketing makes the sale or contact now, directly with the customer, which is controlled by the seller in time and place.

> Every response of direct mail is measured; therefore, companies know if the mailing increased sales or client base or interest within a few weeks.

> Direct mail has a much lower cost than most other forms of media, strong response rate and fast results.

> The database and mailing list are the key to best results.

> Direct mail is probably the most cost-measurable and least expensive way to research prospects.

> The quantity of direct mail is increasing faster than the population.

> Self-mailers are expected to increase as well as non-standard envelopes, to help gain more attention.

> Business-to-business mailing is expected to represent 45% of direct mail expenditures in 2008, compared to 39% in 1998.

> Consumer direct mail will decrease to 55% in 2008 from 61% in 1998

> 45% of direct mail costs represent expenditures in database marketing and list services. 55% is postage, paper and printing.

> Direct mail is a proven and a cost effective way to reach customers."

—Frank Romano
Rochester Institute of Technology
Rochester, NY

responsibility is shared by the creative group (who usually receives a supervision fee), and the client, who has approval authority. (See chapter 21, page 258, and chapter 28, page 322.)

▷ *If the creative group is billed by the printer*—the client accepts the best judgment of the creative group. There is usually an associated markup fee. Check on the creative group's policy and percentage of supervision fees. (See chapter 21, page 259.)

► **availability for fast decisions**, whenever possible but especially:

▷ *at the beginning of the press run* to approve a press sheet, comparing to the proof, and bringing color up to acceptability.

▷ *during the press run*—by phone, as needed.

▷ *throughout the press run* when color balance is critical.

RECEIVING THE PRODUCT

Once the process is complete and the print product is delivered, the buyer has the responsibility to

► **be aware of how finishing work affects the schedule and quality standards of the entire project**. Although the creative group will check samples, the client should pull pieces from each carton of delivered printing for consistency of quality. This should be done promptly, for if the project is incorrect, the print group needs to be informed within fifteen days, or they will consider the project accepted.

► **request advance copies if deadline is critical**. This can help the buyer know what they will receive and when. Although the creative group sees only preliminary samples, these should represent the entire print run. The buyer should check samples in each carton of the delivered printing to ensure consistency.

PLANNING THE FUTURE

Keep abreast of technological developments—first through trusted suppliers and second through research—factor flexibility in planning for new developments and opportunities. Keep open to:

► **nature of the buyer's needs changing**—as technology advances and new opportunities arise.

▷ *color printing grows less expensive* due to both cost-reductions in color preparation and advances in printing presses.

▷ *print-on-demand* allows for economically feasible smaller and more exact quantities, rather than larger quantities to be stored and used as needed.

► **short-run quantity options growing** with the advancement of digital devices and personalization.

► **taking advantage of personalization**. Database technology combined with electronic publishing enables marketing materials to be personalized and targeted to specific interests. This changes the way printing is ordered and used. It also changes marketing strategies by making them more targeted. And it requires careful accuracy to make sure that the right matches are made between recipients and subject matter, or the resources will be wasted. This comes from developing a strong dialog with constituents.

► **expanding printer services** (see end of chapter) into new areas, such as Internet management, leveraging print and online content against each other through asset management and cross-media publishing.

When obtaining bids from different suppliers, be very specific about the format to define the project. If you leave the break down of how you need the budgets just to quantities and art specifications, you will end up with quotations that assume different variables and have individual ways of breaking down the costs. You then need to go back to each supplier to help you define all three the same.

CREATIVE GROUP

Many projects, though designed and positioned carefully, can be ruined if that same attention isn't carried through to the printing and finishing processes. The creative group can minimize the risks and challenges through experience, diligence, and open communication. Matching the print group to the project will ensure that the right team is assembled.

SETTING UP RAPPORT

The challenge for the creative group when buying printing has a lot to do with the relationships up to this point

► **with the chain of communication**—although the art director supervises, the printing is usually managed by the project manager who has received estimates. (See chapter 4, page 38.)

► **with interfacing business needs**—the creative group has obtained quotations and input from the printer on how they want to receive projects, and prepares files accordingly.

► **with a new supplier**—receive three bids to compare different suppliers, but be sure to give each the exact same specifications (see sidebar, page 40). The more specific you can be in the beginning, the better the comparisons and time will be saved. Review samples, rapport, teamwork, and experience, as well as price. Be cautious of the lowest bidder. Ask what is not included, as the lowest bidder might actually be the most expensive.

- **with the client**—the art director, as the liaison between the client and the print group, communicates key decisions to the client, helping to build the right expectations:
  - *describes print quality tolerances.*
  - *conveys variations in schedule and fees.*
  - *reports print over- or underruns.* (See pages 277 and 300.)
- **with the print group** during production:
  - *supervises outputs.*
  - *receives proofs.*
  - *communicates with and sends all materials to the printer,* with their assistance on setup specifications.

CRAFTING THE CRAFT

The buyer has matched the project with the best printing source and has secured a quotation for project production fees. The creative group, whether the buyer or whether a supervisor, works closely with the production group:

- **The art director** has the most responsibility in the printing process. Handling all members of the development team, the art director is generally assisted by the project manager to
  - *oversee the development of the finished artwork*—facilitate proofs through prepress. (See chapter 24, page

Under deadline pressures, the collaboration between the creative group and the print group may be easy to overlook. Those working on the project race ahead to finish their segment and move it along fast to meet the schedule. Not taking time to adequately review, check budgets, and communicate variables indicates potential danger points. Careful advanced planning actually is the best way to speed up the production cycle rather than trying to do the work itself as fast as possible. Lack of planning may cause more changes and delays during prepress and printing, which can be expensive and may jeopardize the deadline.

CRITERIA FOR SELECTING A PRINTER:

> REVIEW SAMPLES of their work and experience in a project similar to the one needed.

> CONSIDER THE PROMPTNESS OF THEIR RESPONSE to the request for an estimate and completeness of their itemization.

> EVALUATE THEIR UNDERSTANDING of the project parameters and examine the way they define the process and expected results.

> ASK ABOUT THEIR CAPABILITIES for handling various digital file formats and materials.

> CONSIDER THEIR PRESSES' CAPABILITIES. For example, do they use a two-color press to print a four-color project by running it through twice? Do they have a six-color press for projects that require more than four colors?

> COMPARE THEIR COSTS to their quality and capabilities. Are their fees competitive? It can be helpful to obtain three bids before selecting a specific printer.

280). Completes process through securing all approvals and permissions.

   ▷ *facilitate proof reviews*—expediting the receipt, checking, routing for reviews, and securing correctness and accuracy. (See chapter 24, page 280.)

► **The project manager communicates progress both to the buyer** and back to the team which facilitates that the right people talk to each other at the right time. This can prevent duplication of efforts or misunderstanding of instructions.

   ▷ *Supervises the pressrun*—to ensure follow-through on specifications, corrections, and expectations. Facilitates any on-site press-checking. Attendance at the printer for large or quality-critical projects is a good idea.

   ▷ *Checks advance copies of printed samples*—checking that all bindery and finishing processes are correct.

   ▷ *Reports to client at project completion* to confirm consistency and delivery dates and locations, then communicates these to the print group. The buyer

      > conveys the delivery location and requirements to the print group, thus linking the shipper with the receiver.

      > supervises the delivery schedule, and may require advanced printed pieces to check for accuracy before the project is delivered to the client.

   ▷ *Follows through on bindery and delivery results*—confirming receipt and checking samples for quality.

   ▷ *Follows through on client or audience satisfaction and results*—confirming that decision-makers are pleased.

   ▷ *Approves the printer's invoice,* even when the printer bills the client directly as a third party. This way, the creative group can inform the client of any deviations that may affect the invoice, including under- or overruns. (See page 300.)

► **The page composer** has an integral role to:

   ▷ *collaborate with the prepress specialist* to ensure the creation and output of successful digital files. This yields documents that have efficient output time and can save the client and the creative group money by maximizing the most efficient use of resources by doing things right initially.

   ▷ *provide accurate instructions* to the printer and check format samples that serve as a bindery guide.

   ▷ *be available during printing* in case there are any questions or changes.

Although there is always a great attempt for techies to turn printing into a science (or a commodity), it always will remain a craft as long as ink behaves differently on different substrates. One of the largest industries in the country, printers are as varied as the projects they handle. There can be no hard and fast rules for buying printing services because even a single factor like deadline or number of copies will change an entire project approach.

RECEIVING THE PROJECT

The print group should already know about the project before receiving it because the plan is secured by an approved estimate. (See chapter 4, page 38.) In advance of sending the project:

► **If the print group is seeing the project for the first time**, it is usually because:

▷ *they've done it or something similar before*, and are basing the parameters on the last time.

▷ *they have an ongoing relationship with the buyer* and fees are not an issue.

▷ *the project is moving too fast*—no printer should proceed without a plan for fees and specific instructions, deadline, and delivery before accepting.

▷ *there is no time for adequate proofs or approvals*—it is a dangerous precedent to print a project without these elements defined.

► **If file preparation and prepress have progressed**—either done in-house at the creative group, imaging center, or printer (see chapter 23, page 266), input has been given by the printer to plan

▷ *timely and correct output* according to both the buyer's and the printer's specifications. (See chapter 23, page 266.)

▷ *how digital materials are transmitted* to the printer. Notify the buyer of their transference and receipt.

AUTOMATED BUT CUSTOM

A printing organization is rather like a factory in automated processes, but unlike most factories because each project is unique. The printer:

► **receives the materials** from the prepress specialist at the imaging center (whether a department within the printer or a separate organization) and

- *reviews all proofs* and sets up for tolerances preparing to print and communicates back to the prepress specialist if not acceptable or if anything is missing.
- **handles imposition preparation** for press requirements—software has transformed stripping to utilize imposition packages that assemble Postscripted pages.
- **performs pressrun to consistent quality** under supervision of art director and provides the printed project according to the buyer's quality expectations and acceptability:
  - *color and registration* must match the approved proof—within technological tolerances. Any influence of paper on ink quality should be anticipated through testing methods and materials checking. If new materials are used, time should be allowed for additional make-ready adjustments.
  - *quality must be the same throughout the run*—all sheets should match the approved press sheet, with sheets pulled frequently during run.

PRINT E-COMMERCE MARKET SEGMENTATION

"We have been amazed to see how quickly this new marketplace has become crowded with players. We list each product category:

> Buy-side
  • e-broker (such as iPrint.com)
  • auction (such as 58K)
  • solutions (such as Impresse, Noosh)
> Sell-side
  • solutions (such as Collabria, MediaFlex)
  • enabler (such as HP)
  • value-added services enabler (Impresse)
> Supply-side
  • purchase solutions (such as PrintNation, Graphics One)
  • inventory control and management (such as VieNet).

"The three further determining characteristics are: open or closed solutions; type of printing supported (offset versus digital versus large-format); and type of market (consumer, SOHO or corporate America).

"Therefore, using the above (two solutions x three products x three markets), it is possible to configure up to 18 different business models within each of the eight 'types' of businesses — for a total potential differentiation of 18 x 8, or 144 specifically different business models.

WHAT IT ALL MEANS

"These components comprise the entire print channel— that often lengthy and convoluted marketplace stretching from the print buyer to the print seller to the printer to the printer's consumable suppliers to the manufacturers."

—Robert L. FitzPatrick
President
FitzPatrick Management Inc.
Charlotte, NC

—Stephen P. Aranoff
Founder and Principal
Arttex Associates,
Redwood, CA

> ▷ *acceptability is defined by the buyer*—often present during the pressrun to approve sample sheets. If changes are made by the buyer (other than to correct printing errors) that cause a loss in press time, the buyer absorbs the extra charges.

> ▷ *any flaws or deviations must be corrected* from what was approved and down time costs must be absorbed until completed. If work is not acceptable according to buyer standards and agreed-upon quality factors, then the printer replaces faulty work at their own cost.

► **is not responsible** for:

> ▷ *errors that were previously undetected* by the client or creative group on proofs. The individual who signs

---

BUSINESS, PEOPLE, SERVICES, PRODUCTION, AND KNOWLEDGE IN THE DIGITAL ECONOMY

"Printing is e-business and convergent media. This environment has rules. Its building blocks are interactive digital services and the management of extended enterprise conducted across networks. E-printing is a 'circus' with five rings:

> RING ONE: E-BUSINESS—networks conduct all stages of the exchange with customers and suppliers. Everyone is online together. The entire graphic arts industry functions as an electronic business community. Across networks, printers, customers, and suppliers market and promote; specify products and services; estimate, quote and make offers; negotiate and agree to terms; enter orders; manage the logistics of the relationship; and accept, pay, and settle.

> RING TWO: E-PEOPLE—People power e-printing. People skilled in business, print production, digital technologies and problem solving are key resources. The right people in the right digital culture will draw in the right technology. The reverse will not happen. E-printers use technology to empower people. This infrastructure provides anytime, anywhere access to services, information and support. The fastest work flow is lean and passes through the fewest hands. Eliminate bureaucracy and reduce management layers.

> RING THREE: E-SERVICES—When everyone is online, and any service can be outsourced, e-services are the way that businesses interact on a transaction fee or subscription basis.

"Services obtainable online include: business intelligence, marketing, sales, creative services, content origination, photography, illustration and image making, design, video and animation, content and media asset management, rights management, prepress, printing, distribution and fulfillment, information services, application hosting, business services, financial management.

"The principle is to capture business transactions once, then transmit and process the data multiple times as needed.

> RING FOUR: E-PRODUCTION—An all-digital work flow is the printer's core value-added process. The five stages are:
> • *planning stage*
> • *origination stage*
> • *production stage* is the transformation of content into product masters and their replication as media products. Activity includes outsourcing, prepress and premedia, digital printing, conventional printing and finishing, digital media replication, and release of Web-based media. This stage manages masters, handles customization and dynamic composition meta-

data, and interfaces with enterprise manufacturing and production management systems, as well as supply-chain partner systems.

• *distribution stage* is concerned with delivery of media products and services. This stage manages product and service information and distribution metadata. Distribution interfaces with these systems: customer sales and marketing, customer relationship management, supply chain management, distribution partners, and suppliers, rights management and e-commerce.

• *evaluation stage*

> RING FIVE: E-KNOWLEDGE—E-knowledge is knowledge in a digital form. A networked repository contains the core of digital knowledge within any particular operating environment. It integrates information from different aspects of the printing process: for example, e-business, content origination, media production, content management, process management and external systems data.

"Sharing e-knowledge requires an information architecture that supports a complex range of data types: part database, part transaction, part program and part content or media object."

—Mills Davis
CXO and Managing Director
Digital Roadmaps Project
New York, NY

the OK to the printer's proof is the one who is responsible for accuracy.

- ▷ *content or copyright of material provided* by the buyer. (See chapter 31, page 351.)

▶ **assumes responsibility for originals** while in their possession (see chapter 29, page 330) and documents their return when the project is finished.

▶ **provides printed quantity**—according to the buyer's instructions. Industry conventions apply:

- ▷ *Overruns and underruns* are not to exceed 10% (unless another percentage is agreed upon in advance) on orders for up to 10,000 copies. For over 10,000, the acceptable percentage should be specified in advance.

- ▷ *The final invoice will be adjusted* to reflect the actual quantity delivered. (Clients should budget for the possible variance of 10 percent.)

- ▷ *If the buyer requires exact guaranteed quantities*, the printer may add an extra charge in order to have enough test ("make-ready") sheets.

▶ **supervises bindery**—often there is some bindery capabilities in-house or the printer will utilize a subcontractor.

- ▷ A sample format (called a mock-up, comp, prototype, or dummy) accompanies the project to demonstrate how the finished piece is to look.

- ▷ *The printer counts the acceptable printed pieces* before sending them to the bindery. Pre-counting is not the responsibility of the binder.

- ▷ *The printer sends press sheets* to the bindery wrapped or skidded. (Containers that hold the printing remain the property of the binder unless arrangements are made for their return to the printer.)

- ▷ *The binder is responsible* for assembling the project according to the buyer's needs for

  > accuracy—converted samples should be checked periodically for quality while running. All should match the approved sample.

  > timeliness—deadlines are almost always a priority.

  > quantity—within 10 percent of the specified amount needed (see above).

  > quality—cartons should be packed carefully to avoid damage of final work in delivery.

- ▷ *The binder is not responsible* for materials damaged due to situations beyond its control.

IT'S TIME FOR THE PRESS CHECK!

"1. Have your client review the reader sheet (one of the first sheets that comes off the press) for elements including misspelled words, broken type, missing words, miscropped photos and incorrect folio numbers while the press is being prepared and the colors are fine-tuned.

2. Focus on color: Does the color match right? Is the print contrast good? Do the PMS colors match the ink draw-downs?

3. Double-check the project's page size.

4. Although the printed page might look great in the light booth, look at the piece outside or in the lighting where the most end-users will see the project. Does it still look as good as it did in the booth?

5. If an error is found during the make-ready stage, go through the above four steps for every part of the project before shutting off the press to replace the plates. This ensures that all mistakes are fixed at the same time."

—Jesse Williamson
President
Williamson Printing
Dallas, TX

FINAL DELIVERY

The print group agrees to send completed work to receivers designated by the buyer. Many printers deliver printed materials locally as part of their services. For long-distance delivery, special carriers are usually needed.

The accuracy of all shipping and delivery instructions is the responsibility of the buyer. The print group coordinates the physical delivery of finished printing. The print group

▶ **includes delivery costs** in the quotations to the creative group. (See chapter 4, page 38.)

▶ **is responsible for safe handling** of original and printed material with a timely delivery in good condition.

▶ **secures that the carrier of the finished project will deliver it in acceptable condition**—unless storage arrangements are made. (See chapter 27, page 318.) The production group is not responsible for loss or damage that occurs in transit beyond its control.

▶ **makes arrangements to store the film** and deliver the printed pieces and originals to the buyer. These arrangements need to be made prior to delivery.

NEW OPPORTUNITIES

Expanding business options for printers as the industry evolves:

▶ **increase prepress capabilities**—either by expanding current systems or acquiring a prepress facility.

▶ **include complete fulfillment** such as mailing services.

▶ **offer consulting** and client on-site systems management.

▶ **provide training services** and enable clients to do more in-house. Though it may seem that this would demand fewer ongoing services, it actually can be the foundation of a long-term collaboration.

▶ **research and development** capabilities—commitment to stay on the "bleeding edge" so that customers don't have to invest. The customers rely upon the vendor to keep them up to date.

▶ **charge for storage and archiving maintenance** (see chapter 27, page 320). This is particularly applicable to cross-media and database publishing.

▶ **publishing on-demand** (see chapter 27, page 320), which expands imaging offerings.

▶ **asset management**—uses large memory capacity to manage client image libraries. (See chapter 15, page 184.)

▶ **multimedia production** editing and finishing operations. (See chapter 19, page 236.)

The binder may charge extra for preliminary samples or partial shipments of finished pieces. This happens most frequently when materials are needed at a specific event, and the balance used later.

Many office buildings have very specific dock procedures and delivery hours. The client group should check for permits, permissions, or access information needed for its building and inform the dock that the project will be arriving. This will ensure that the delivery truck has access and that the client will receive printed pieces on time.

"The free democratic society we enjoy in the United States owes its very existence to the power of the printed word. A free country must have an educated citizenry. Libraries have been called the memory of mankind. While other media can be helpful, education fundamentally depends on the printed form of communication.

"Ideas only have consequences if those ideas are communicated. The educated citizen must be informed about public issues and controversies if he is to make sensible and prudent decisions. The products of printing can provide him with the information and ideas he must have to play his part as a free citizen in a free society."

—John R. Walter
President and CEO
R.R. Donnelley & Sons
Chicago, IL

"If I had to describe the power of the printed word, I'd have to say it's the most effective way of communicating. The printed word can be revisited, whereas the verbal word cannot. The printed word requires less concentration and allows more speed and comprehension than computer screens. Better technology will help speed the longer lead time necessary for the printed word. Electronic media will not replace the printed word. To make the future better we have to have knowledge of the past. We gather information so we can improve."

—John Darragh
President
Standard Register
Dayton, OH

"Some print is being replaced by other media. For example, directories or instruction manuals that often change are being handled by CDs or the Internet. For printers, the transition can be a barrier or an opportunity, but there are more opportunities than barriers. The barriers are set up by us, not the technology. There is a growing use for print and for new kinds of printing. And as various segments of the industry are combining, such as media buyers creating the media and media creators doing production, there are new areas for printers to explore."

—Lou Laurent
Managing Director
Laurent & Associates
Lynnfield, MD

"The printing industry is moving from a craft-based manufacturing industry to an electronically-based service industry. Our turnaround times are shorter, the work is more customized, we will be able to only print what we need, and we will see more color content. This is not a market to push new technology, but rather is a market to pull new technology. The customers demand faster service because they are in a more competitive environment. We see a growing volume but a shrinking budget. We see a demand for more flexibility. We manufacture in hours now versus days. We need to have broader skill sets."

—Charles A. Pesko
Managing Partner
Charles A. Pesko Ventures
Marshfield, MD

"The future of graphic arts technology could mean the end of the highly mechanized processes now considered the bread and butter of the printing industry. The industry is on a one-way course toward the virtual extinction of press operators and craftsmen as we now know them and the gain in technicians—'a shift from a greater deal of physical work to more thinking.'

"I think the printing press of the future will be a non-impact machine. The printing processes as we know them will be all obsolete."

—Hank Apfelberg
Professor of Graphic
Communications
California Polytechnic
State University
San Luis Obispo, CA

"Contrary to conventional wisdom, I don't believe the market will be dominated by a small handful of very large companies and a vast array of faceless local printers. Instead, mid-size regional printers with electronic interfaces that include advanced telecommunications will provide technical support for customers whose print inventory is maintained electronically by their printing companies. These printers will use a variety of output equipment to provide on-demand and just-in-time reproduction services."

—Jacques Marchand
Printing Consultant
Marchand Marketing
San Francisco, CA

"Some print buyers don't know who their printer will be as they are planning projects. Having several regular suppliers can help with this problem. Many printers are willing to help early in the project, even if they aren't awarded it later. If it's a regular relationship, as a supplier, you know you will receive another project. The client will remember that you helped them. It's important to have a longer-term view of working together than just one project."

—Debbie Ball
Printing Salesperson
JB Printing
Kalamazoo, MI

"Print still remains a strong medium. People still love to hold an object, feel the texture and see fine detail. Some things don't reproduce well electronically—print is still the best for high-resolution reproduction."

—WM Padgett
WM Padgett Design
Erieville, NY

"Browsing, bookshelves, and magazine racks, skimming, books and print media for things that might be of interest, skipping around a publication or flipping back and forth between sections on pages is still far faster with hard copy than searching the web, loading a document then scrolling through pages..."

—Niel K. Klein
Niel K. Klein & Associates
Traverse City, MI

"This society is a long way off [from] being paperless—in fact I feel there is more paper. Our clients (the salesmen) are not ready to rely only on CD-ROMs or our Web site when specing a job. They still want the sell sheets and catalogs in paper format. So now we are producing the catalog three times!! The more clients see what we can produce in a short amount of time due to the computer's color copies they want more, and more, and more!!"

—Laura K. Narby
Moen Incorporated
North Olmsted, OH

"Clients need print more today (and more innovative ideas with print and paper) to get their messages out. If they have Web sites they need to get their Web addresses out and known... otherwise, it's like building a house without an address—no one will be able to find it without knowing an address or location—and it doesn't make any difference how great the site is—if your potential customers don't know the address. More clients are going with full form color process or more client response cards, campaigns. But deadlines are tighter and a lot shorter, requiring a lot of digital photo images and innovative ways to meet time schedules."

—Darlene Hughes
Redgraphix Design
New Berlin, WI

"I don't think a paperless society will happen. Paper usage in my professional life has not changed. I personally prefer to receive a greeting card in the mail as opposed to a virtual/cyberspace greeting card via e-mail. Paper and the images on paper are real and tangible. Cyberspace is not."

—Jean Wong
226 Design
Phoenix, AZ

"The friction for the printing industry is toward consumer choice. The consumer will be able to obtain information matched to his or her needs faster, and in the format desired. Newspapers and books will never actually go away—they are, after all, in Borges' words, 'mankind's imagination'—but the format of many will soon change from paper and print the way they once did from Sumerian clay and Egyptian papyrus scrolls.

"There are several insights we can draw from the history of printing. The first is that virtual products will result from combining numerous and diverse technology advances. For example, the new printing processes are dependent upon lasers, xerography, integrated circuits, the microprocessor, high-speed communication processes, display technology, and advances in software.

"The role of the author as a coproducer is also evident. He or she does not merely create the content but also can control the presentation of what is printed. To do that the author must not only be skilled in language but must also be computer literate. This is the new trade-off: greater control demands a corresponding extension in skill."

—William H. Davidow and
Michael S. Malone
Mohr, Davidow Ventures
Menlo Park, CA

"With the advent of television and computers, people predicted the end of printing. So far, it has only grown. Computers led to an onslaught of printed materials from manuals to magazines. Television also led to special interest magazines, newspapers and books, and of course, more printed advertising. Where would America be without its Sunday television magazine or weekly edition of *TV Guide*?

"For most, printed materials can be absorbed faster than spoken language. They are portable and can go wherever the reader goes. And they can be reviewed again."

—Joe Kirschen
Writer
Printing Impressions

"The first step on the road to predictable color quality requires that the designer be informed of the budget before the creative work commences. It is frustrating for everyone if a design is created that exceeds the capacity of the ultimate production system. There is no reason why layouts should be created (or approved) that are impossible to execute because of economic constraints. It is the responsibility of the designer to exercise creativity within the constraints that the specified conditions impose. If these constraints are ignored, the integrity of the design and the quality of the reproduction will be inevitably distorted and degraded."

—Gary G. Field
California Polytechnic
State University
San Luis Obispo, CA
www.ipa.org

"The design may be great, the proofs correct, the printing and binding superb, but if the job is delivered late or to the wrong location, all those efforts are for naught—bottom line is the printer failed. This is the black and white of being on time— that's expected. We try to have this service throughout the process so there are no surprises at the end. I believe that the printing business is more of a service business than a product one."

—Jim Madden
President
Rider Dickerson, Inc.
Chicago, IL

"In the busy schedule surrounding a print project, delivery may seem a mundane detail. Most assuredly, it is not! The best design in the world can be eclipsed by the disappointment of inaccurate shipping. It is important to determine delivery expectations early on and keep checking on them. Having the printed materials available as promised can be the final ingredient in the success of your project—particularly with time-sensitive material. Program booklets are no good if they show up to an event late. Sometimes the timing of the delivery is more important than details on the project."

—Kris Erhart
Account Manager
Smith, Bucklin & Associates
Chicago, IL

"Designers are caught between demands: the market place wants us to stretch new visual capabilities and the printing industry that wants projects to come to them as digital as possible. The more printers demand direct to plate processes, the more business they give away in prepress and alterations. The less a printer handles a file, the more they put work into hands of those creating it. Shifting budgets can be very tricky. And for a designer to keep up with the changes on both sides of the industry is a challenge that gets more complicated with technology, not less. As we look to our printers to help us handle all these new processes, those that treat projects with a service view and not a commodity view will win my business. So in a way, I stay in favor of the printer doing more, and me doing less. The industry just isn't going in that direction."

—Eleanor Mandler
Principal
Mandler Design
Chicago, IL

Managing on Automatic

"To really make color work, we need automated color control, based on three elements:

> input and output devices smart enough to calibrate and profile themselves;

> operating systems that automatically identify the color properties of each attached device, then query the user to establish typical work flows;

> applications that support color management internally and take advantage of color-management capabilities provided by the operating system.

Color management will disappear into the background, working invisibly to produce high-quality color for nonprofessionals while still providing custom controls for those who want to tweak the settings."

—Michael Kieran
President
DPA Communications
Toronto, Canada
www.creativepro.com

"A good print salesperson should provide insightful printing ideas, give you an accurate estimate, plan a production schedule to meet your deadline and organize your future printing projects.

"When planning your project, find a printer you trust and involve him on the ground floor. You'll be amazed at how much time and money a good salesperson can save you if you let him work with you from the start. Instead of using your time to shop for the lowest price, focus your attention on the print salesperson's knowledge so he can produce a more economical plan for you.

"Designers are judged by the end product of their ideas. The production of those ideas is largely affected by the tools and skills the print salesperson brings to the project. This sales professional is one of the most important resources a designer consults.

"An accurate estimate should be included in the salesperson's plans, outlining the costs for each print phase. As a designer, it's frustrating when you get a bill with additional charges tacked on after the job's been printing. These charges are often due to a lack of communication between the print rep and designer before the job was started."

—Ted Archer
Direct-Imaging Specialist
Buchanan Visual
Communications
Dallas, TX

"The Internet can be used to provide more personal service while taking costs out of the delivery of goods and services.

"With the higher technology products and services required for printing, buying habits and use habits aren't quite so simple as in other manufacturing businesses.

"Just as the Internet is changing the selling model by allowing low-cost commodity sales to prosper, it is also allowing personalization and value-added sales to prosper.

"In the printing marketplace, e-commerce is beginning to be used to inventory the use of goods at a printer's site. By managing this process, a printer is able to cope with his irregular use of supplies by automatically generating purchase orders for replacements based upon delivery cycles and use. This takes administrative cost out of the business and provides more rational use of inventory. This overall view can help manufacturers plan their businesses more efficiently and to manage inventory turns more aggressively.

—Stephen P. Aranoff
Founder and Principal
Arttex Associates
Redwood, CA

"The result: Everybody takes costs out of their business model, while working together more effectively. This gives everybody the same opportunities as the bigger players, if they take it. And the customer gets better service, too!"

—Robert L. FitzPatrick
President
FitzPatrick Management Inc.
Charlotte, NC

"HiFi (for high-fidelity) color is the umbrella term for any color separation process that relies on more than four colors as primary tints for process color printing. The original term, "HiFi," is credited to Mills Davis; since then, various vendors have developed algorithms for HiFi color separations and printing, each using a different technique and even different primary colors, but all offer a greater color gamut than is possible from four-color process.

"Hexachrome, from Pantone, is the most viable of all the HiFi solutions that emerged in the early 1990s. Hexachrome uses enhanced versions of cyan, magenta, yellow and black, plus a specific green and orange. Hexachrome is capable, says Pantone, of simulating more than 90 percent of solid Pantone Matching System Colors."

— From Pantone Description
www.imagingmagazine.com

"Pursuing HiFi color work has helped StudioZ set itself apart from its competitors. You have to invest more time, more effort into each stage of the Hexachrome process [than 4/c process]. Each stage must be planned and executed properly; you need to have your prepress done perfectly because there's no way to tweak it on press.

"The best clients for HiFi/Hexachrome are not only those that can afford [it], but those that have absolute color critical material—clothing or things that have a laminate—gives them a better print quality and the colors are much more accurate."

— Trey Simmons
Partner
StudioZ
Sheffield, AL

COMPLETION

PROJECT CHANGES 26

Alterations are unavoidable. No project goes exactly according to plan though the plans can be flexible enough to allow for deviations. Even regular periodical publications have unexpected twists and turns in their development. If they don't, then they are not responsive to their audience or reflect changes in their environment. The majority of projects require perfecting the elements as part of their process, which is hard to predict in advance.

The easiest way for a client to lose money or get in trouble on a project is to begin before truly ready. Time is rarely saved when working on a project in pieces.

MANAGING CHANGES

Each team member, responsible for the quality of a portion, will make changes on a project as it develops. If not, the project is very well planned, moving too quickly, or is not receiving enough attention. Change comes in two forms:

► **Alterations** represent a change in the scope of the project parameters and specifications.

  ▷ *The client generally pays for alterations when requesting changes that effect parameters*:

    > additional text, pages, or changes in content.

    > additional photographs, or other visuals.

    > alterations on previously approved work.

    > rescheduled deadline.

    > differences in quantity of printing or fabrication.

  ▷ *Cost estimates for alterations* must be provided to the client from the creative group or production group before doing the work, or they are at risk to absorb these costs. When the client knows the cost of alterations in advance, the client may choose the amount of the changes based on priorities and available resources.

► **Corrections** are changes undertaken to fix errors generated by the creative or production groups, not by the client.

  ▷ *Corrections are needed for*:

    > typographical errors caught in the review process that were missed by the creative group.

> page composition inconsistencies and format errors, incorrect color, position, or size of images. Most of these should be caught through quality control.

> experimental effects not tried before and resulting unpredictably. This can take time to correct.

> aspects of the design overlooked or misinterpreted. Designers are good at pushing technological limits, so sometimes the page composer can misinterpret designers' intentions.

> faulty prepress instructions—such as for color overlap and position, screen angles, tint construction, etc. (See Prepress, chapter 23, page 266.)

▷ *The client does not pay for:*

> changes that must be made because the creative or production group made errors, called corrections.

> corrections due to human errors such as misplaced or missing elements, poor file preparation, inadequate image preparation, or poor quality proofs.

MAJOR DECISIONS:

set project foundation
plan rough project skeleton
assemble category outline
solicit proposals
choose creative group
finalize project plan
select design theme
approve design
review writing outline
edit first draft content
review revised content
review final content
approve visual components
approve first prototype pages
approve pages and testing
approve final artwork
approve prepress proof or test
approve press proofs
review bindery instructions

CLIENT GROUP

Know how to evaluate, manage, and approve various project components through developing a background in the phases. Without at least an explanation, tour, and a patient creative group, an inexperienced client is at a disadvantage. This can directly affect the amounts of alterations and budget containment, for the inexperienced client doesn't know what is possible.

CONTROL FINISHING

The client is responsible for the accuracy of all materials. Though the creative group created the project and may have introduced errors, once final artwork is approved, the client is liable for all content (though the creative group will help resolve any work that needs to be corrected). Major points to know:

▶ **Alterations become increasingly expensive** the later they are made in the development process. It is important to avoid the temptation to race ahead prematurely due to deadline pressures. To avoid the most common costly alterations:

▷ *wait until all the pieces of a project are together.* (Break a large project into a few sizable groups to move progress forward as other portions are being completed.)

▷ *follow a strong project plan* (see chapters 7 and 8, pages 62 and 70, respectively), so that deviations can be anticipated and everyone involved knows what pace needs to be kept.

▷ *secure all approvals* and be sure that content is finished—nothing can derail a project faster than if the copy changes frequently. Doing so requires redesign and rework in production.

▷ *don't place deadline as more important than quality,* unless it is, and quality just needs to be "good enough." Some projects don't warrant a lot of perfection. Prioritize and know when perfection is worth paying attention to every detail or okay works when trying to meet a budget.

▷ *get an estimate of the costs before work commences.* Be sure to ask if it is not being provided. Not doing so can be inviting surprises and possible conflict at the end of the project.

▷ *understand how project alterations can escalate*—as a project progresses, alterations require more work, which affects cost. For example, consider this escalation formula:

> a typographic change made to the manuscript may cost 50¢.

> the same change made after the corrected client proof may cost $5.00.

> to make the change after the final low-resolution proof is approved may cost $50.00 (if it requires the movement of elements or the restructuring of a page).

> on the prepress proof, it may cost $500 to change (as it can affect the film of more than one color).

> when made on press, the same change may cost $5,000 because it stops a press and waits for corrections, or requires rescheduling.

► **Corrections**—the final printed product should match the text, design, images, and proofs approved by the client. Unless otherwise notified

▷ *all errors indicated should be resolved* and reflected in the next step of development. Check each new version carefully (such as paragraph beginnings and endings) because unexpected changes can creep in.

▷ *deviations or corrections from the creative group* should be undertaken at no cost to the client.

▷ *the client is responsible for accuracy prior to going to press,* and so should always approve corrected proofs.

One of the most difficult reporting tasks for the creative group is to tabulate economic status on a project and quote alterations before proceeding. Usually there is a lot of deadline pressure and little time to stop and review. It may risk the deadline. But making the deadline might shortcut communication and leave the creative group vulnerable to pay for the changes.

Most project changes can be handled within the purview of the project parameters, budget, and schedule if contingency is built into the planning. Generally around 10 percent of the budget, changes can be anticipated in the proposal (see chapter 4, page 38) or by history with the client behavior on other projects. How the creative group handles changes can make the difference between the recovery of a wayward momentum and the loss of control, money, sleep, and friendships.

SANITY UNDER PRESSURE

Because some changes are expected, most creative groups know how to get the work done, but often don't distinguish, until later, between who should pay for them. The creative group, for self-protection, must stop and evaluate the budget in the midst of deadline pressure. Handling alterations and corrections are some creative groups' biggest challenges.

► **Alterations** are received from the client after reviewing proofs. The art director and project manager:

▷ *set up communication* between the page composer and the production group for any questions that may arise during the execution of alterations, especially under tight deadlines.

▷ *secure estimates before alterations are made* and communicate these costs to the client.

▷ *are responsible for the accurate execution and reproofing*—check for the correctness of the alterations made and secure all approvals. (The client has ultimate responsibility for the correctness of materials.)

▷ *see that the client understands their responsibility for paying any extra costs*, or the creative group will be at financial risk. The client does not have to pay for any work they did not approve.

► **Corrections** will happen during the version and approval processes. Corrections might be needed due to oversights or mistakes by the creative group, and, therefore, the creative group

▷ *does not charge the client for correcting any errors caused by the creative group*—these errors may include:

> mistakes in digital file preparation.

> instructions miscommunicated to the production group.

> errors generated by experimental new processes—new methods should be tested in advance.

Within the client organization, it is most efficient to ask decision-makers to initial proofs that they have reviewed. This is important for:

> it impresses upon the decision-makers the importance of their review.

> it helps them to focus on finding any needed corrections as early in the process as possible.

> it protects the creative group if there are later questions.

The client group is ultimately responsible for the correctness of prepress proofs.

The creative group is best protected from costly errors by collaborating with the production group in digital file preparation to determine who is best suited to perform each task. Dangers arise in some of the page layout software color separation features. Color separation is a complicated discipline, requiring a great deal of knowledge and experience. The designer and page composer should ask questions of the print group and not be overwhelmed by desire for control or ego.

As the creative group reviews the prepress proofs, they may be dissatisfied with the way color is balancing, the way the pages are composed, or with a technique that didn't come out as hoped. For example, perhaps the borders around photographs are too light or too dark. Extra prepress work may need to correct these. Such corrections are not billed to the client.

▷ *reviews proofs for accuracy* and catches any problems, and has corrected them for client approval.

▷ *requests corrections*, if needed, which may be due to dissatisfaction with the results on proofs or if missing details. Earlier proofs can be too limited in the processes they approximate.

▷ *facilitates client approval* and marks up with client alterations; is responsible for being sure alterations are completed.

▷ *communicates that the client is responsible* for checking accuracy on the final prepress proof. Secure signature as okay to finish.

PRODUCTION GROUP

Ideally, the alterations or corrections made with the production group should be limited to checking preparation, cleanup, and any last-minute details. Any other forms of change can be time-consuming or costly.

FACILITATING ACCURACY

The purpose of proofs is to check elements, not to have the buyer read the copy or make substantive changes. The production group should never proceed without the client signature on a final corrected proof. Managing this process will determine if budget or timing are in jeopardy. The production group can help to minimize misunderstanding or inaccurate expectations.

▶ **Alterations** are defined by the creative group according to client request when proofs are reviewed and compared to parameters. The buyer may request alterations that are not included in the quoted fees. These alterations

▷ *must be clearly marked on the proofs* or prototypes. The creative group manages the various versions, usually by marking up proofs for production changes. Try to minimize the number of corrected versions being received from the buyer. If there are too many versions and if the page composer has to resolve contradictions between them, more alterations are inevitable.

▷ *must be estimated for the buyer* before work is undertaken. If it is hard to know how much time alterations will take, a guesstimation is better than not communicating. It is best if trust has been built between the groups that results in equitable solutions.

▶ **Corrections**. As workers can make mistakes on projects, proofs are meant to reveal those, and are not meant for editing. (See Proofs, chapter 24, page 280.) Proofs show glitches from machines (also fallible), processes, and people. For complicated projects, errors should be expected and watched for. Don't depend on previous proofs, as errors can sneak in at any time. The production group

▷ *corrects errors, without charge* to the buyer, when

> instructions are misunderstood or miscommunicated, assuming they are clearly marked.

> specifications are not followed.

> questions are not asked regarding incomplete information and work proceeds anyway.

> requested corrections are not made, though instructed by the buyer on the proofs. This might be due to unclear instructions, but the production group may also misinterpret.

> quality expectations with final printed pieces have not been met—delivery is made but the printed product was not printed or bound correctly, or there is inconsistent quality (see chapter 25, page 288).

> there are misunderstood delivery instructions, or delivery could not be made for reasons beyond the transporter's or client's control.

▷ *provides, at its own expense*

> corrected pages, images, film, or plates to replace defective prepress if not communicated prior to the press run—whatever material may need to be recreated to meet project parameters.

> correctly reprinted or refabricated pieces—for however many pieces are printed or bound incorrectly, new ones must be furnished as quickly as possible. With time-sensitive material, the damaged products may need to be °anyway, and then adjustment should be made to the final invoice. Quality control must be continually addressed.

▷ *corrects errors, with a charge to the buyer,* when receiving

> unclear instructions, marked sloppily or vaguely, including mismarked proofs.

> incomplete, inaccurate, or poorly prepared copy or data, and decisions are left to interpretation.

> incorrectly prepared files and other prepress errors when film is provided from an outside imaging center (see most common errors, sidebar, page 273)— once accepted, though ensures its usability (see above).

"Clients usually say, 'Just one more change.' Always talk through client alterations with a positive attitude—put your own 'spin' on their needs and summarize alterations in writing. You're not sacrificing your client's direction and your pieces carry the effective design and strong content your client demands. If you're working with a small budget, matching their personal aesthetics against actual project costs gets them to revisit the reasons for the changes. Include an estimate of the alteration cost—it's an effective reminder of the early parameters of the projects and shows you care about, and are keeping on top of, the entire project."

—Regina Rubino
Designer
Louey/Rubino Design Group
Santa Monica, CA

"Low-balling a job results in a chain of 'markdowns' by suppliers who want to do business with you, as well as a chain of 'extras' and 'markups' that must be added to the client's ultimate bill. This inability to adhere to the original 'low-ball' estimate naturally generates mistrust for all designers in the eyes of the client. What will become important is to just what extent these changes will alter the perception of what 'design' is—an electronic function, inexpensive, unwarranted, or a vital ingredient of corporate and product image."

—Lisa M. Bruno
Project Manager
Smith, Bucklin & Associates
Chicago, IL

"There's never enough time to do it right in the first place but always enough time to do it over when it is done wrong."

—Anonymous

"Changes become increasingly expensive as a job nears completion. At the onset of a project, management should be forewarned about the costly impact of any changes. By the same token, designers shouldn't view presswork and finishing operations as opportunities to make corrections that might have been made earlier. With a few exceptions, the earlier the designer can fix mistakes, the less costly fixing them will be, in terms of both the budget and the schedule."

—Boller Coates Spadaro, Ltd.
Consolidated Papers
Wisconsin Rapids, WI

"The more you can clarify the process and make your client a partner, the less likely they'll question you later or not understand when something is late because something else was delayed. There's a cause and effect to everything that happens and the design team tries to make that very clear. This keeps us all on the same page."

—Karen Uhl
Graphic Design Manager
Pottery Barn
San Francisco, CA

"To my regular clients, I usually provide one sketch. To new clients, rather than show the standard three sketches, I prefer to provide the completely finished piece. That way there's nothing to misinterpret or change. If the client doesn't like the first attempt, I create another. I adopted the tactic because I was fed up with clients asking for constant, unreasonable changes, which I attribute to technology. ('Is that in layers?') Whenever you give clients three ideas, you know they're going to choose the worst, or they will try to combine elements of one idea/rough with elements of another. ('How will this look if you move this from here to there?') I tell clients: 'I don't need to carry, just so you'll understand, a red sofa into a green room to know the colors will clash.'"

—Mirko Ili´c
Mirko Ili´c Corp.
New York, NY

"You can't afford to reinvent the wheel every time you do something. It needs to be as efficient and productive as possible so that we can deliver the highest quality product in the shortest time frame. If everyone writes out their processes and keeps files of their projects, when someone else works on a similar project everything is right there for them and they can learn from what that person has already experienced."

—Karen Uhl
Graphic Design Manager
Pottery Barn
San Francisco, CA

"One of the most valuable techniques to practice is to share with the client the costs of changes or problems instead of letting the cost come out of the design fee. This most often comes up when you have a client make a request that you cannot fulfill without loss of profit (don't forget, time is money). The client will ask and you will respond 'Sure, OK, fine, no problem' when, in fact, they have given you a problem, perhaps a costly one! The better response to share the cost of this request is to respond one of two ways. First is to say, 'Yes, *and* that will cost (name something specific)'; second is to respond, 'No, *but* here's what we can do (name some other option).' By the way, cost here does not only mean money. Cost on any freelance project can mean time, energy, attention, quality, or money."

—Maria Piscopo
Creative Services Consultant
Costa Mesa, CA
http://MPiscopo.com

STORAGE OF MATERIALS

There is a race between expanding media and the storage to archive and maintain that media. As document files become larger with more images and special effects, storage hardware keeps pace as must the user through expanding capacity. Digital media requires storage, as do the physical proofs, originals, and printed pieces.

BEFORE THE END

With all projects, storage and backups systems need to be part of the finishing processes. As the project is completed:

► **digital files are archived** and deleted from individual hard drives unless part of an ongoing campaign.

► **project materials are returned** to the owner of each (see chapter 29, page 330). Project materials include:

   ▷ *original artwork*—such as illustrations or photographs—whether reflective or digital.

   ▷ *elements and background* supplied by the client.

   ▷ *proofs generated by the creative group.*

   ▷ *proofs generated by the production group* (includes Prepress, chapter 23, page 266).

   ▷ *digital files* as supplied by the creative group. The production group retains any files created to prepare supplied work for fabrication. (See chapter 29, page 337.)

► **project materials are stored** whether in-house or to an outside service firm:

   ▷ *digital files used for prepress*—this becomes very important in situations of reprints. Projects like stationery or capabilities brochures should have a reprint plan and arrangement set up with the production group as part of the project plan.

   ▷ *maintenance of film and reflective artwork*—a system to house materials needs to be developed, and most production groups have a time limit. (See page 319.)

   ▷ *asset management software* should be integrated into the network to make both archives and current elements available to those who need them. Access to archives can be an important business component for many publishing organizations. (See chapter 15, page 185, for more information about asset management.)

CLIENT GROUP

Because the client owns the final camera- or plate-ready materials and digital files (depending on the copyright owner of the files; see chapter 31, page 346), these should be sent to them after production, unless the client arranges for storage with the creative group or the production group.

KNOW STATUS

Clients may assume that the creative and the production groups will store preliminary materials and film. However, this storage may only be for a limited time. The client must specify length of storage time, and requires notification before any materials are destroyed.

The client may request that the creative group supervises the storage of original files and camera- or plate-ready artwork when the project is completed. The client

▶ **retains ownership** of those specific materials it supplied or paid rights for (see chapter 29, page 330).

▶ **pays a nominal fee**—The creative group or production group may charge for the maintenance and responsibility of storing the materials.

CREATIVE GROUP

There are not a lot of conventions for handling project materials once a project is completed. Assuming all digital work is backed up, archiving is most useful in ongoing client/creative group relationships.

FACILITATE LONGEVITY

The creative group should retain digital files and elements used to create final camera- or plate-ready or digital artwork, archived by name and project.

▶ **Keep output for a minimum of one year after the project is finished**—the client should be notified before disposal.

▶ **Keep the digital portions for approximately two years**—unless the client requests otherwise.

▶ **Archiving is generally a courtesy**—but if in quantity, the creative group should negotiate a monthly fee for retention and library services (see chapter 15, page 184).

▶ **Develop or tie into asset management system**—any creative group that builds a library will need to make images and past designs easy to find and access. (See chapter 15, page 185, for more on asset management.)

Most clients prefer that the creative group retains materials. In an ongoing relationship, efficiency is gained by the creative group resourcing all past components. Unless a client needs to maintain files as part of their own library (and ownership needs to be clarified), the creative group can offer storage management as part of their continuing service.

Upon receiving all project materials, part of a production group's work flow should include policies for handling.

RESPECT FOR ELEMENTS

Return or storage of all project materials should not be left to chance or to assumptions, but clarified as part of the project planning. (See chapter 7, page 62.) The production group is responsible for any materials within their care. (See chapter 29, page 330.)

Many production groups have developed expanded revenue from offering reasonable services for storing digital project files:

▶ **archiving establishes an ongoing relationship** between buyer and supplier—allowing efficiencies based on trust. This can be important for:

▷ *updates of series* like periodicals or Web sites that may rely on templates.

▷ *inclusion in other projects later* or cross-purposing of elements, especially if part of a campaign.

▷ *reprinting* as for print-on-demand or projects that may be reordered as needed.

▶ **storage of nondigital materials**, like film, reflective art, prepress files, and document files, should be set up in advance as part of the planning phases. (See chapter 7, page 62, and chapter 8, page 70.) The production group

▷ *may store for an additional period* if requested by the buyer, and may bill the buyer for archival services to cover cataloging, upkeep, and media.

▷ *takes reasonable care to provide an appropriate storage environment.*

▷ *notifies buyer at the end of agreed-upon storage period*

> the buyer may request that materials be returned.

> the production group notifies before destroying any materials. This communication can also facilitate staying in touch with current and former clients and preserving ongoing relationships. (If buyer-owned materials are destroyed without agreement from the buyer, the production group must replace the destroyed materials at their own cost.)

▶ **at the conclusion of the project**, the production group

▷ *returns all buyer-owned materials.*

▷ *retains intermediate materials*—such as scans or films, and make-ready files, generally for one year after printing has been delivered.

"Storage is a billable service. Most printers now find that a lot of their floor space is taken up with media storage. So what they're doing is turning around and saying to their clients: 'We will house this electronic file for X amount of time in case you misplace yours because we know you don't have an image management system. So that is included in the cost of the job we do for you—we'll store it for three months or six months or whatever. At that point you have to pay for the storage medium and for housing it with us.' That is an additional service a printer can offer. If a client is going to create large files, then they have to be responsible for them at one level or another and for some reason the printing industry has a mess on their hands because they have these back rooms loaded with racks of media that are unmanageable for them."

—Sandra Kinsler
PhotoLibrary
Management Service
Ventura, CA

"Many people are banking on the future of storage media. Database and document management is a crucial consideration. As storage media changes and evolves, there may be a new profession in the future: the data archeologist, who will need to recover data from obsolete media."

—Frank Romano
Professor
Rochester Institute
of Technology
Rochester, NY

"As more media join the communications lineup of choice, handling all the various versions, ages, formats, software, and hardware will demand that the archivalist be a media specialist. Although many media choices are now digital, the storage method and equipment can make past information inaccessible. As databases talk more to each other, information will require a new kind of librarian: someone who knows how to find anything anywhere and how to keep track of intellectual property. New online services will become more enabled to help in these functions. But not one should ignore the diligence for keeping information accessible and recoverable."

—Eleanor Mandler
Principal
Mandler Design
Chicago, IL

PAYMENTS AND DISPUTES

How well the financial agreement is constructed between parties can make payments a smooth part of the process or a painful experience spiced with surprises and tempers. The communication, rapport, and momentum of the project will determine the quality of its finish. Having a project plan that includes checkpoints and budget reviews will ensure an equitable project end. Even during the most frenzied of deadline races, keeping on top of the financial status is just as important as making the deadline.

PREPARING TO NEGOTIATE

No one enters a project anticipating disagreements in the end. Hopefully, all team members agree on the same objectives and develop rapport and trust through working together. Unfortunately, it does not always happen that way, so understanding the elements of conflict prior to experiencing them can either prevent or minimize any potential damage to projects or relationships.

► **Definitions**—although parties agree on fees and payments in advance, disagreements may still arise either during or after a project. Terms used to describe financial considerations include:

  ▷ *payments*—compensation for work performed.

  ▷ *claims*—requests for price adjustments.

  ▷ *disputes*—disagreements over payment amounts that require negotiation and possible adjustment.

  ▷ *liens*—the right to hold a buyer's property until payments are made.

► **Conflicts**—it is rare for publishing projects to end up in serious complications or legal disputes, but they can. Most of the trouble is caused by

  ▷ *imprecise initial agreements and definitions of terms.*

  ▷ *work performed without approval.*

  ▷ *ownership questions.*

  ▷ *unpaid invoices.*

  ▷ *broken deadline commitments.*

  ▷ *unclear negotiations between third parties*—usually arising when the projects' decision-makers are not involved in the work or are not responsible for payment.

> ▶ **Prevention**—the best protection from lawsuits is through clear agreements. However, client, creative, and production groups may carry insurance to compensate for damage or loss of original art or printing due to factors beyond anyone's control.

CLIENT GROUP

Concerned to have project planning include budgets and defined expectations, the experienced client is consistent with requests. Wavering on instructions invites later trouble. Not knowing about the ownership of materials can also foretell possible conflict.

KNOW THE RULES

Ownership transfer of commissioned artwork occurs only when payment is completed. Before that, the creator of the materials owns them, and has the right to hold them until the client makes payment per the agreement.

If the project changes, the proposal or agreement may be amended to include

▶ **additional work** that may incur unanticipated fees.

> ▷ *Always negotiate new fees* when the parameters of the project change. (See chapter 20, page 246.) Never skip this step.

> ▷ *Gain approvals* before authorizing work to continue.

> ▷ *Pay for changes* in addition to the established fees when informed in advance of the necessity to incur.

▶ **additional use** for artwork beyond the project's parameters that may incur additional fees.

> ▷ *Inform image owners if use expands*—greater use requires extra fees, such as using print images online. (See chapter 28, page 322, and chapter 29, page 330.) Additional use should be defined, budgeted, and put in writing.

> ▷ *Do not alter images* or elements without permission—there may be extra fees based on client specifications for changes. No one should alter artwork except the copyright holder, unless permission is given or the work is in the public domain. (See page 175.)

▶ **outside services** that may incur additional fees.

> ▷ *Understand supervision fees*—the creative group may charge for supervising third party work, especially when having the client billed directly. (See chapter 21, page 258, and chapter 25, page 296.)

When printing is delivered to the client, the client should check all the boxes for quality consistency. If they store some of the boxes for later use without checking them, and they are defective, if they have not alerted the creative or print groups within two weeks of delivery, the project is considered accepted. They may not be able to receive credit towards a reprint, if reprinting is necessary.

- ▷ *Understand markup fees*—the creative group may add a fee for bearing the financial obligation to a third party. The client may avoid this markup fee by having the third party bill the client directly. (See above.)

- ▷ *Understand ownerships*—the client does not own the work until the client has paid the creative group *and* the creative group has paid any outside resources. Disputes can arise when only one party has paid another. (See chapter 29, page 330.)

- ▶ **legal fees**—the client will undertake and pay for all appropriate copyright searches to ensure the originality of created materials. (See chapter 10, page 100.)

  - ▷ *Protect the user* of the materials.

  - ▷ *Use critically for corporate identity* as it will be used for a long time and have a wide market reach.

  - ▷ *Allow the client to apply for copyright* on work created for its ownership. (See chapter 29, pages 330.)

    - > This does not extend to portions of the work copyrighted by others, such as photography and illustration, unless agreed to in writing.

    - > Even if the client does not hold the copyright for the work, the client can be named in legal disputes with a third party who contests the client's right to extend use of images without compensation. (See chapter 21, page 258, and chapter 25, page 296.)

A legal search, no matter how thorough, cannot guarantee 100% non-duplication, for there are new company identities, publications, and Web sites every day. There may also be an obscure, hard-to-find example that is similar, though it is rare that possibilities are overlooked. It is very dangerous not to search a logo or a name before using it. Web site domain names are also worth protecting. Fortunately, the online registration process only allows for names unused. Visual images present more of a challenge and require additional protection, especially when shown on the Web, which automatically has global distribution potential.

CREATIVE GROUP

The creative group specifies compensation arrangements in the quotation or proposal. Deadlines and budgets are agreed on in advance of working on a project (see chapter 4, page 38). However, interpretations, changes, shifting priorities, politics, overtime fees, and unpredictable reality all have their influence on making projects deviate. When deviations happen, conflict and disagreement can often result.

Many graphic design firms require:
> one-third in advance
> one-third upon approval of design
> one-third upon completion of the project.
(See chapter 5, page 50, for more detail on terms.)

PREVENTION THROUGH PLANNING

Set up a strong project plan in the beginning as the best way to avoid disputes. Never start a project without a document that outlines fees and ownerships. Keep in mind that a client will make their own assumptions otherwise. Be sure to be explicit about any business policies

- ▶ explain how terms vary from firm to firm. (See Fees, chapter 5, page 50, and chapter 20, page 246.)

  - ▷ *Some firms request payment in phases.*

A creative group rarely carries insurance against the possibility of simultaneity of ideas (see chapter 8, page 70), and therefore secures the client's agreement to assume such responsibility. The client, whose organization is using the published materials, is legally responsible for that use. However, it is the responsibility of the creative group to provide all information about ownership and use of each image to the client.

▷ *Many firms bill for large projects monthly.*

▷ *For very large projects, a retainer arrangement is often preferred.*

► **estimate additional costs** before incurred:

▷ *notify the client of additional estimated costs* before doing the work.

▷ *itemize additional fees* on the next invoice.

► **include ownership agreements** as part of the quotation or proposal to begin the project.

▷ *Communicate the usage terms* of any images or proprietary original content.

▷ *Explain ownership of reproduction materials*—including camera- or plate-ready output, does not transfer to the client until all fees have been paid.

▷ *Be tough when necessary*—if the creative group has performed according to terms but the client does not make payment according to terms, collection costs incurred by the creative group, including attorneys' fees and court costs, are to be paid for by the client.

► **exhibit leadership on copyright searches**—explain the necessity (above) to the client. The creative group assures, to the best of its ability, that materials created are original and meet the client's specifications.

▷ *Meets the copyright criteria*—the image is secured for use and registered to prevent others from infringing.

▷ *Does not meet parameters* for original concepts and the copyright request is denied:

> the client may request further work at no cost.

> the client may pay for acceptable portions and expenses for work-to-date and discontinue the project.

► **be explicit about parameters and goals.** (See chapter 3, page 22.) If there is a conflict of ownership, the creative group

▷ *is not responsible for any writing or imagery*, as supplied by the client, that may infringe upon anyone else's rights.

▷ *is only responsible for its own created images*, and needs to depend on legal searches to ensure originality.

▷ *is not party to the third-party suit*—if it has provided all usage and ownership information to the client. The client is the liable party and should pay all invoices of the creative group, which should include third-party expenses.

PRODUCTION GROUP

No project should proceed without estimates from the production resources. These are different agreements than with the creative group because often the quantities involve larger sums of money than the creative fees or the technical requirements are complex. The buyer should pay the production group in full according to the agreement in the estimate. (See chapter 4, page 38, and chapter 5, page 50.) Most agreements stay out of trouble with these general terms, conventions, and understandings:

▶ **general terms**

   ▷ *identify economic sources*—the buyer company that places the order, not its employees, is responsible for payment on all work performed.

   ▷ *publish terms*—the production group usually specifies that payment is due within thirty days of the invoice date.

      > Amounts not paid when due may bear interest at the maximum prevailing state rate from the due date until paid. This is tough to enforce but can be part of a legal agreement for payment and/or damages.

      > If payment is not made within agreed upon terms, the buyer is liable for collection costs incurred by the production group, including attorneys' fees and court costs.

   ▷ *include time limit on quotation*—all claims of defects, errors, shortages, loss, or damage of buyer property should be made in writing within two weeks of delivery. If not, the print group may consider that the buyer has accepted the work.

▶ **conventions**

   ▷ *releasing of materials*—possession of all work remains the property of the print group until its invoices and additional charges have been paid according to agreement, even if the product has been delivered and is used by the client.

   ▷ *compensating for losses*—the production group's liability seldom exceeds the amount to be paid by the buyer for the work in dispute. The production group should carry insurance to cover inadvertent destruction of the buyer's property. This insurance covers loss or damage to materials. Liability shall not exceed the amount recoverable from such insurance, which includes replacement of blank media or reprinting where appropriate.

Some disputes arise when the individual buyer who placed an order leaves the client company, and then the company denies payment. The best protection against this possibility is to get a written purchase order or advance payment to secure the validity of the order.

The importance of backups cannot be stressed enough. Everyone involved in a publication project needs to have at least two backups of all files—one should be stored off-site.

> handling damaged images—if the print group damages or loses original artwork, they must compensate the owner of the artwork. (Fortunately, this rarely happens in a digital environment because there are usually backups.)

> > Receiving fee for lost slide—from a stock photo agency the owner is generally paid $1500 per lost image, but this can vary per agency.

> > Receiving fee for lost illustration—the owner is paid the original creative fee unless the image is digitized from an original. Most images are scanned to be handled digitally, and the conventional methods are mainly used for high quality or crafted projects.

> > Receiving fee for lost or damaged computer files— the owner is generally not compensated because it assumes that files meant for production are backed up. The buyer should never give only-copies to anyone. There should always be duplicates. The creator of the files is responsible for maintaining backups and archiving files.

► understandings

> identify Web content strategy—know content of project, who is responsible for it, and watch for cross-purposing opportunities. Any online accuracy of information:

> > depends on the client. The production group has no responsibility in the violation of any copyrights or proprietary rights.

> > because the tech or the production group is not involved in the creation of content, cannot be held responsible for invading any person's right to privacy or other rights. The tech or print groups are only responsible for any content that they generate.

> > includes indemnification of any matter that may be libelous or scandalous, unless the production group created it.

> plan Web updates and adapt new content—set up budget and schedule, and define who needs to contribute which updated information.

> the parameters will be met before payment is made— the production group adjusts fees if the buyer's parameters, schedule, or quality standards are not met.

> if parameters are met but the client is not happy—depending on the nature of the problem, the creative group and the production groups collaborate to resolve the issue to everyone's satisfaction.

"It doesn't pay to be right and stick to your principles, if it means losing the client. They are, after all, your client and you are presumably taking large amounts of their money in exchange for your services. Principles are terrific, but empathy for clients is also necessary if you want to retain them. In most instances, these are not mutually exclusive. When they become so, you get to decide whether you are in this business for love or for money. More often than not, intelligent design serves both masters."

—Mike Hicks
Hixo Inc.
Austin, TX

WAITING TOO LONG TO START COLLECTING IS THE BIGGEST MISTAKE

"Customers whose intentions are good take your calls, don't avoid you, and often initiate trying to work something out when circumstances place them in a temporary cash-flow crunch. There are three stages leading to a debt that's never collected:

> Stage one—the debtor can't afford to pay everyone everything he owes, so he pays each a little to keep them happy.

> In stage two—the debtor can no longer pay every creditor something, so he pays something to only essential creditors required to keep the business afloat.

> Stage three—the debtor gives up, perhaps making an attempt at a payment plan with essential creditors, but soon lapses into a moribund state that signals the end.

"The best way to evade bad debt is to avoid the problem in the first place. Institute a good in-house collection policy and stick to it. Not extending credit to bad risks prevents many problems. Request customers to submit a credit application." (July 2000) [See next page]

—Larry Thall
PhotoMarketing Magazine
Jackson, MI

"Make sure you have an arbitration clause in your contract. Then your disagreement can be settled by a panel of industry peers—and you won't have to deal with the court system. If it's a money issue and a small amount is at stake—under $2,500 in most states—it's relatively easy to represent yourself in small claims court. Get together all the signed agreements, invoices, and copies of the finished project, and tell your story to the judge. Many judges will be sympathetic to an individual entrepreneur who's been taken advantage of by a larger company."

—Ellen M. Shapiro
President
Shapiro Design Associates
Irvington, NY

"The credit application should state the types of agreement and finance charges. But no business is invincible to problem debtors. A time-sensitive collection plan, based on net-30-day terms, involves a six-step process:

1. Prior to the account being past due, make a courtesy call to the customer, to determine if the merchandise was received in satisfactory condition and there are no discrepancies between what was ordered and what was received.

2. At the 30-day mark, send the first collection letter. Friendly and helpful, thank the customer for the recent purchase, reiterate the net-30-day policy, mention the record shows no payment has been received, and invite the customer to call with any questions. The envelope should also contain another copy of the statement, thus eliminating the 'I never received an invoice, could you send me another?' stalling tactic.

3. At 45 days, make the first collection call. State the balance amount and ask the customer again if there is any dissatisfaction with the product or service. If the answer is in the affirmative, fix the problem; otherwise, inform the customer of your policy of placing credit limits and COD/credit holds on all accounts that are 60 days overdue. Stress that you wouldn't want to have your customer experience any inconvenience.

4. At 60 days overdue, send the second collection letter. State payment hasn't been received and the lack of communication is disturbing. It might be a good idea to ask a collections agency to run a computer check on the debtor to determine if the company has been placed on collection status before and the outcome of the action.

5. At 75 days past due, things begin to look serious and urgent, and it's time for another phone call. There's obviously a reason why the debtor hasn't paid. Attempt to agree on a partial payments plan. Offer to work something out, but restate what each of you has agreed to. Get the debtor to sign a personal guarantee if a payment schedule is arranged.

6. Send a final demand letter at 90 days overdue. It demands immediate payment in full and states if it isn't received by a given day, the account will be turned over to a collections agency. Make another phone call to inform the debtors of your actions.

"What if all fails and a collections agency must be called? Some companies use a network of private detectives who knock on doors and ask embarrassing questions. This is where the information on the credit application comes in very handy. It's embarrassing to a debtor to have references called and asked if they've had trouble collecting from this company. So a detective will call the debtor and say he's in town to do an asset and liability search on the debtor's company, and will be calling his vendors, suppliers, bank, etc. 'The idea is to exert pressure on the debtor, using the people he does business with.'" (July 2000)

MATERIALS OWNERSHIP 29

Intellectual property becomes more visible and easier to access in a digital world. Protecting it and profiting from it can be a major focus for many businesses. Because disputes arise over ownership (perhaps not as often as financial disputes, but serious nonetheless), having concrete definitions of property is necessary.

On the Internet, where copying is becoming like an expectation—where free is cool—protecting intellectual property makes the owner look like a bad guy. But bad guys they must be, for if the creators of the work can find no compensation for it, then they will have to do something else for income. Only *some* art can be subsidized. But the publishing profession can't work by subsidiary and needs to actively partake in solutions for protection—from security to encryption.

The beginning of policing comes with concrete agreements. The industry tends to rely on assumptions and conventions, especially when projects develop quickly. Knowing who owns what and applying the appropriate information to images is the first step to supervising use later. Ownership can vary from project to project and should never be assumed. Rather, ownership for all publication materials should be specified at the outset as part of the agreement terms. Ownership agreements should be in writing and are often negotiated.

OWNERSHIPS SPECIFIED

Digital ownerships become complicated because there is no true original. Working arrangements should specify who owns:

▶ **preparatory or background material**, such as video or slides—often owned by the provider. However, this is often not the case.

▶ **concepts, ideas, and designs**—a proposal can outline exactly what the client is buying one concept.

▶ **comprehensive presentations**—created to demonstrate concepts for selection and approval.

▶ **images**—such as photographs or illustrations. The default is that the creator owns the images unless otherwise specified and ownership transferred. (See chapter 31, page 346.)

▶ **data entered into a computer**—especially when contracted with a third party for this service.

Difficult disputes can arise between the client and the creative group at the end of the project—before printing or launching and before the creative group receives payment. The client may want to obtain digital files for their library or other uses when they may not, in fact, own those rights. It is best to clarify such issues, in writing, at the beginning, usually in a letter of agreement or proposal.

"THERE ARE SIX CHARACTERISTICS OF WORKS IN DIGITAL FORM:

1. Ease of replication
2. Ease of transmission and multiple use
3. Plasticity
4. Equivalence
5. Compactness
6. Nonlinearity."

—Pamela Samuelson
Professor of Law
University of Pittsburgh
School of Law
Pittsburgh, PA

- **computer programs**—software used to create publications, covered by licensing agreements. (See page 357.)
- **original digital files**—unencrypted along with all native support graphics. Computer files are
  - *commissioned by the client*—with usage agreements.
  - *created by the creative group*—and who generally owns all created work.
  - *refined and prepared for printing*—owned by the production group until payment is made.
- **film or paper for reproduction**—used by the printer as preparatory work (see chapter 23, page 266), often owned by the production group.

DIGITAL ETIQUETTE

Once a work is digital, it is easily accessible and needs as much protection as the owner can give it. Much research into ways to mark and police images is keeping the software industry quite busy and image-creators anxious for further protections. General rules for not violating others' ownership:

- **a digital file may not be copied without permission** from the owner—unless it is in the public domain. (See chapter 14, page 174.)
- **copying digital files constitutes theft** of intellectual property—no one may copy, adapt, or alter a work without the permission of the copyright owner.
- **the creator of the file may protect property with a clear copyright notice** alongside the image or an embedded identifying mark, such as company name or logo. Watch for good encryption programs.
- **place the work in an "encapsulated" form** (such as Encapsulated PostScript) that cannot be altered—this protects native image files from falling into the wrong hands.
- **professional ethics and courtesy must reign**—because a file is not physically protected.

Regardless of the agreement, ownership is not confirmed until payment is made. (See chapter 31, page 350.)

Policing only really becomes an issue when there are profits involved. High-visibility infringements will gain more attention as more of them reach the media and consequences become public, especially as reported on the Internet where recent cases are easy to follow. The government is also slow to relegislate to match technological levels—being concerned with security, regulations, and helping the industry to grow without restrictions.

Although the Internet is notorious for viewers copying images and code, and stealing content, authors can protect themselves by offering only partial works or keeping all files in very low-resolution.

Clients assume they own everything unless they are told otherwise—sometimes taking educational responsibility may seem contentious to prospects and clients. It is a skill to turn this to advantage in the light of good business practices. To differentiate against competition, you can make it good for the company that you are so conscientious.

Many clients rely on their creative and production groups to inform them about ownership. Often, a client organization is not in the publishing industry and may be unaware of the industry's practices. They must know what they both own and what they are buying to avoid potential problems, due to inaccurate assumptions. To avoid any misunderstanding, ownership should be established before the project begins.

CONCRETE AGREEMENTS

Under typical agreements (and everything is negotiable), the client owns:

► **all material they provide** to the creative and production groups and should investigate the copyright of any found materials. (See chapter 31, page 346.) Anything provided to the creative or production groups is returned to the client after the project is completed.

► **design concepts** only for the purpose specified in the proposal. Unchosen concepts remain the property of the creative group. Any other uses must be negotiated. (See chapter 31, page 351.)

► **usage rights** as granted through licensing agreements. (See chapter 30, page 341.) Under these agreements, the client may be granted the right to use:

▷ *concepts and ideas*—for marketing and publication goals as defined in written parameters.

▷ *copies of computer files:*

> generally, the client does not own the digital files that designers create unless the designer is an employee. (See work-for-hire, chapter 31, page 347.)

> if the client hires the creative group to create a publication for a specific use, the creative group owns and retains the computer files, unless other arrangements are made. (See chapter 29, page 330, and chapter 30, page 340.)

> if the client purchases unlimited rights usage (see chapter 28, page 322), the creative group owns and retains the files. Unlimited use is implied for corporate identity and some publication design.

▷ *original artwork*—such as illustrations or photographs and ownership if granted copyright from the creator.

▷ *final printer's proofs*—but not the film unless the whole project is digital.

Agreements should specify who owns copyright and final output, such as reproduction-quality paper or film. For example, the client may own the output but the creator may own the copyright. Copyright applies to the tangible form of a concept, not the concept itself. Computer files are covered by the copyright requirement of tangibility.

If the client does not own the copyright, the creator owns all digital files. The creator is usually a member of the creative group. Clients who wish to obtain a copy of the file for their archives should specify this in advance. Payment for this purpose needs to be negotiated between the client and the creative group. Technically, if the client wants to alter any file in their possession but of which they do not own the copyright, permission needs to be obtained from the creative group. But a lot depends on the nature of the relationship between the client and the creative group.

If more applications are needed, the client receives more benefit from that design and needs to compensate the creative group appropriately. When the creative group knows all the uses a design must address in the beginning of the project, they can design to meet that level of flexibility and do it most cost-effectively.

When competing for projects with other firms, many creative groups may be reluctant to bring up the issue of ownership to their prospective clients. Ownership is a negotiating point and fees can be lower if the creative group retains ownership. When presenting responsibly to prospects, these concerns can be impressive—especially when the competition is neglectful.

With digital files being so fluid, all clients want their publications digitally. Few projects don't have multi-purposing as part of a marketing plan. Turning over files is usually understood and reflected in the initial fees. Designers need to be cognizant of the repurposing of components and images, and protect those that they don't wish to share.

Some corporations have policies that require they own all material created for them and that the creator cannot use any of the artwork for self-promotion, particularly where the client name would be apparent. This is usually covered by a contract.

▶ **the digital files from the creative group**—only when the client purchases the copyright can they then own, maintain, and alter the work. Corporate identity assumes that a client buys all rights to the image.

▶ **final output**—whether reproduction-quality paper or film (see chapter 23, page 268), after they have paid all invoices, even if that output is stored by the production group.

▶ **final printed published work**—whether print or interactive—after invoices are paid.

OWNERSHIP COMPENSATION

Ownership and usage will impact creative fees. It is important that the client inform the creative group of

▶ **all specific uses** for materials developed (see page 335) as part of the project.

▶ **all parameters** (such as quantities, sizes, and scope), as within the project agreement (see chapter 4, page 38).

▶ **any needed revisions** or such as quantity changes, still for the same original project purpose.

▶ **new applications** for additional purposes, content, or quantity—such as taking publication design and using it promotionally.

▶ **changes in usage** by expanding campaigns—such as translating the content of a print publication online or taking current information and recasting it for cross-media applications.

CREATIVE GROUP

Protection of documents is a critical component in the idea business. It begins with an understanding of originality and who owns what.

KNOW TERRITORY

While keeping in mind that other arrangements may be made, the creative group typically owns

▶ **all unused or unchosen concepts** presented to the client.

▷ *When showing more than one idea* in the concept presentation, it is hoped that one idea will be chosen. The unchosen ideas, which still pertain to the parameters, might also have potential with the client.

▷ *Fees may need to be renegotiated*—often inexperienced clients may assume that they are paying for all the ideas and keep the presentation materials. But if more than one concept is chosen to be developed from a presentation, fees will change. Generally, the production portion of the proposal is set up to produce only one selected design.

► **all comprehensive presentation materials**—including the presentation materials of the approved designs. (See chapter 7, page 62.)

   ▷ *The client owns the application, not the presentation*—many clients keep copies of presented designs, so it is important that they realize these concepts remain in the property of the creative group. Make sure that a © notice is on these presentations along with contact information.

   ▷ *Keep track of copies*—it is very difficult not to release copies of concepts, especially when working with committees. So having protection and policing methods in place will help to protect intellectual property. Unused concepts are an important resource for creative groups and often are the seeds of future projects. Not keeping track of these ideas can be dangerous. In rare cases, some clients have hired other firms to execute concepts they do not own rather than go back to the creative group that created them. Even styles and approaches can be copied, so it is best for comps to remain the purview of the creative group.

► **preliminary work** needed to compose the publication:

   ▷ *original artwork*—such as illustrations, photographs, distinctive concepts, individualistic styles.

   ▷ *original computer files*—native documents in the software that each was created in.

   ▷ *preliminary outputs* and working proofs. (See chapter 24, page 280.)

► **computer files** prepared for final output (see chapter 23, page 268), even when giving copies to the client.

► **the right to use printed pieces as samples** for self-promotion unless the client stipulates otherwise.

CONSTRUCTING FEES

The creative group is compensated by a fee to create the work (see chapter 5, page 50) and cover usage. However, book authors usually receive royalties based on the number of books sold. Each situation is going to be different. But, always, the creative group should secure a written agreement with the client on budgets as part of the proposal. (See chapter 4, page 38.) Several factors need to be established:

If a project meets with great enthusiasm from the client's organization or market, the client may wish to continue to use the design in other materials. There are additional fees from the creative group whenever the usage is expanded. These fees should be communicated to the client before the theme is developed or used further. In maintaining good client relations, there are skillful ways to handle usage changes.

► **stress the art director's role** to communicate ownership:

  ▷ *specify usages and payments as part of the marketing plans*—handling this diplomatically is a challenge, as it is often not requested and offered as an additional factor that a client may not wish to think about. But usage is essential to convey. Provide source material as a good way to bring up the subject in the spirit of the client's own protection.

  ▷ *the client needs to know what they are buying and what they own*—though it is their responsibility to know, they may not know that they need to know this. An inexperienced client depends on the creative group to let them know what is needed. When in doubt, the information needs to be provided, even if it is not what the client wants to hear.

► **appropriate releases**—such as model releases when people appear in a photograph. (See sidebar, page 165.) Often the photographer will handle this, but the art director needs to be sure these forms are part of the file.

► **usage rights.** (See also chapter 31, page 346.) Become an expert in uses and protect original intellectual property. Be able to explain:

  ▷ *first rights*: The design or image is only used for one specific purpose, such as a brochure cover or poster. The client may not make additional uses of the image without negotiating arrangements with the owner.

  ▷ *exclusive or specified rights*: The design or image may be used for additional specified purposes, and the creative group is compensated accordingly. This can cover a series or a campaign.

  ▷ *additional rights*: The design or image is used for more purposes beyond the original agreement, requiring additional compensation to the owner. This might be determined after the images are already used for specified rights. It may result in more work for the creative group, as the files cannot be altered without permission from the owners, usually the creators. Common is the repurposing of print concepts and images to the Web.

  ▷ *unlimited rights*: The client may use the design or image for any purpose without needing permission or paying additional compensation. The fee for unlimited rights is usually based on the first rights creative fee plus 50 percent to 100 percent. Some projects, such as a corporate, product, or Web site identity, automatically imply unlimited use. But most others do have specified limits. Usually, unlimited rights includes purchase of the copyright. The new owner can then not only use the image unlimited, but also make derivative works.

► **when the client does not have unlimited rights** to use a design or image, the creative group may negotiate fees for developing

▷ *a series*—such as adding pieces based on the same design, or using templates for additional publications.

▷ *revisions*—such as updates, alterations, or changes. Another artist mimicking the image or style can cause trouble. (See chapter 13, page 162.) The only person who can alter a creative work is the owner of the copyright.

▷ *derivative works* (used but adapting content and design to additional projects or media). When altering to fit another format, include copying files or templates.

▷ *variations of the design for other uses*—such as other forms of promotional pieces or online presentations. It is most economical for the client to develop a good team to develop a publication and then continue with the team on regular issues.

PRINT GROUP

Generally, the production group is not part of the copyrighting aspects of a project. However, the production group owns key elements in the project evolution and are instrumental in series and updatings.

MATERIAL RESPONSIBILITY

The production group obtains materials owned by both the client group and creative group and maintains them in their possession for use in project preparation. The production group

► **handles supplied materials** that include

▷ *digital files.*

▷ *slides and transparencies.*

▷ *camera-ready art and mechanicals.*

▷ *databases.*

▷ *image libraries.*

▷ *final paper or film output.*

▷ *tapes, diskettes, and other digital media.*

► **is responsible for receiving, storing, and preserving material condition** but is not liable for loss and damage from causes beyond their control. (See chapter 28, page 326.)

WORK-FOR-HIRE

Normally, the creator of a work holds the copyright and grants rights. There are two conditions where the creator does not hold the copyright:

> if the creator works for an organization, the organization holds the copyright.

> if it is a work-for-hire, the client can purchase the copyright if in writing and if it fits one of nine categories. (See chapter 31, page 347.)

- ▶ **is liable for buyer-suppled disks** and other media if the media is damaged or lost. (This responsibility is limited to replacement of blank media.) In a digital world, the loss of originals is less of an issue if there is an adequate backup system in place. Liability for property does not exceed the amount recoverable from insurance held by the production group. (See chapter 28, pages 323 and 326.)
- ▶ **is responsible for return of buyer-furnished materials** after the project is finished.
- ▶ **is not responsible for the accuracy of buyer-supplied materials**. (See chapter 31, page 346.)
- ▶ **owns intellectual property:**
  - ▷ *blank and recorded digital materials*—the blank media need to house digital files.
  - ▷ *intermediate hard copy, make-ready, and film*—which may be disposed of without notice (usually within thirty days after completion of work, but kept as a courtesy for one year. (See chapter 27, page 319.)
  - ▷ *files created to enable the project to output properly*— these include
    - > special "print files" (see chapter 23, pages 268 and 271)—including data, text, formulas, and codes that they prepare to support final output.
    - > codes that are developed by the tech group and could be proprietary to systems developed as a unique approach.
    - > computer code and scripting for online delivery.
  - ▷ *computer programs, systems analysis, and related documentation developed* for a project—no use in whole or part can be made of these without permission and mutually agreed upon payment to the production group.
- ▶ **does not own or is not responsible for**
  - ▷ *digital files provided by the client or the creative group*— computer documents and their component parts remain the property of the buyer who is responsible for their usability.
  - ▷ *final output before printing*—in the form of film or prepared final digital files (prior to imposition). This depends on the collaboration of imaging center and the printer. These materials are owned by the client.
  - ▷ *verifying copyrighted material provided* to them by a buyer. May check for © notice on pieces and remind the buyer to include if not there. It should appear on all printed pieces.

"Licensing fees are analogous to renting a car. When you rent a car, it doesn't mean you *own* the car, it means you have purchased the right to use the car. The longer you retain the car, the more money you pay."

—Jo Ann Calfee
The Stock Market
New York, NY

"For most projects, creating your own is the way to go—original content frees you from licensing negotiations and fees, opens up the potentially lucrative position of licensing your content to others, and gives you the possibility of profitable spin-offs. However, producing content may be more expensive and time-consuming than you like. You must be extremely careful to have work-for-hire contracts, so your helpers don't wind up owning a slice of your pie."

—William Rodarmor
Managing Editor
California Monthly
University of California
Berkeley, CA

"Under the copyright law, ownership of a copyright is distinct and separate from ownership of any material object in which the copyright is embodied; accordingly, ordinarily when a physical painting is sold and the copyright remains with the artist."

—Caryn R. Leland
Licensing Art and Design
Intellectual Property Attorney
New York, NY
www.lelandlaw@erols.com

"Good business is good ethics. The technology is rapidly running away with a lot of peoples' work. There is a difference between ownership and copyright. What should you do?
1. Ask for permission.
2. Get it in writing.
3. Establish records.
"Know fair use versus infringement. Part of what business is about is explaining, giving out information and material. For example, people don't *own* photographs of themselves. The biggest challenge is checking all permissions when under deadline pressure. Everyone wants to deal with companies that have good business practices because your jobs will run more smoothly."

—Elaine M. Sarao
Assistant Professor
Northern Virginia
Community College
Alexandria, VA

— Richard Haukom
Multimedia Developer
San Francisco, CA

"When all the rights are open to negotiation, the deal is everything."

—Charles D. Ossola
Attorney
Hunton & Williams
Washington, DC

"Copying is particularly inviting. How do you police it? In fact, the difficulties in detection are such that it is an incentive to not disseminate new works so they don't get out into general circulation. That is at odds with the very purpose of the copyright law. The use of art in commercial settings demands distribution. It makes no sense to hoard images. Yet, if the image is distributed it will be digitized and copied often. As a creator, you just can't worry about it. *But,* when money is involved, then we all get our backs up."

—Jonathan Band
Attorney
Morrison & Foerster
Washington, DC

"Ownership issues are particularly complex because Congress has amended the Copyright Act numerous times, and which provisions apply depend on when a work was first published."

—Jerry Skapof
KanImage
Long Island City, NY

"We've considered giving our clients a policy statement on how we treat information, copyrights, and handle confidentiality. We could have them sign a form of rights. This would inspire them to do business with a firm like ours because of our ethical concerns."

—Michael Fitzgerald
Senior Writer
Computerworld
Framingham, MA

"Corporate software copying amounts to illegal manufacture and distribution of someone else's product. The Software Publishers Association (SPA) is not indiscriminate in its raiding. It gathers solid evidence, generally, from employees (a temporary worker often has tipped them off to a company's piracy) who may be more ethical than the management or may just be unhappy with the company and want to get back at it. Then the SPA gets a search-and-seizure award from a judge and conducts its raid.

"The average information services manager seems to see the SPA as an ally in the fight to get their bosses and fellow employees to understand that copying software is a serious offense."

SOFTWARE OWNERSHIP 30

Software compatibility and fluidity between programs vexes users. Compatibilities are improving, but so are the amount of choices and needed training. Software can be expensive, and it's easy to copy for "trying out" before purchasing (and then not getting around to purchase, but using it, anyway). Of course, the user receives no documentation this way, but there is help on the Internet. Consequences of stealing software:

> ► **guilty conscience**—only those who know it is illegal and understand how it undermines the industry will suffer this consequence. Unfortunately, others don't understand that it has negative repercussions and feel that if it is easy to do, then there's really no reason not to do it.

> ► **hurts development**—software fees pay for research and development that turns technological wonders into tools. If there is no money to be made, then there will be no money to improve the industry.

> ► **not eligible for upgrades**—at some point, to stay current, either an upgrade needs to be purchased or another theft needs to occur.

Legitimate users need to permeate a code of software ethics into all staff members. The main reason that software is stolen is because the infringer is ignorant of the facts. If employees do not learn about software piracy in school, it becomes the employers' responsibility to convey the habits of following software licenses.

DEFINITIONS OF OWNERSHIP

Software applications include any programs purchased from software developers that are loaded into or accessed by a computer's operating system to create documents. These include:

> ► **word processing programs**—such as Microsoft Word® or Word Perfect®

> ► **drawing programs**—such as Corel Draw,® Macromedia FreeHand,® or Adobe Illustrator®

> ► **painting programs**—such as MacPaint® or SuperPaint®

> ► **image manipulation programs**—such as Adobe Photoshop,® Aldus PhotoStyler,® or ColorStudio®

Software licenses are location-specific, tied to workstations. As networking capabilities expand, documents will travel easily from workstation to workstation, following the user. This may require a new structure to licensing agreements.

Piracy means copying software and giving it to someone else for use. Software piracy denies vendors income needed for further research and development of software applications.

Some software companies are recognizing that a growing number of individual users own and use more than one computer. They have one at home, one at work, and a portable for in between. Licensing commonly allows a user to install on two of their own computers.

► **page composition programs**—such as FrameMaker,® Adobe PageMaker,® Adobe InDesign,® QuarkXPress,® or Ventura Publisher®

► **authoring programs**—such as Macromedia Dreamweaver,® Microsoft FrontPage,® NetObjects Fusion,® PageMill®

► **special effects programs**, plug-ins, filters, and extensions (too numerous to list and changing rapidly).

► **graphic and image resource collections**—such as clip art or photo libraries. There are a great proliferation of them with the Web as a tool for dissemination. (See chapter 15, page 184.)

► **creativity tools**—such as outlining programs, indexing tools, or brainstorming.

► **fonts**—including typeface families and image fonts

► **planning and management software**—such as job tracking systems or workgroup tools (packages proliferate).

► **database software**—to handle large amounts of data and allow cross-media publishing

► **other business software**—such as accounting packages or sector-specific packages

► **utility software**—including calendar programs and organizational tools

► **communication software**—such as e-mail providers and Internet hosting services.

VENDORS' RIGHTS AND RESPONSIBILITIES

The purchase of an off-the-shelf software application program is actually the purchase of a licensing agreement, rather than ownership of a product. Although efforts are being made to standardize licensing agreements, vendor agreements do vary.

► Most **licensing agreements specify** that:

▷ *the software may not be copied*, except for one archival backup.

▷ *the program may only be used on one computer* at a given time (versus being tied to an individual user) unless a network version is purchased.

▷ *violators are subject to legal ramifications* and withdrawal of support.

▷ the vendor has limited liability for software performance.

▷ the user has the right to *run software on one machine at a time* and provides manuals, help functions, Internet, and phone support.

▷ *they can serve networks with site packages*—some licenses cover multiple copies, as for corporate settings. Additional licenses are discounted, but are generally limited.

▷ *title transfers to the upgrade*—the user may not use earlier versions on other computers, sell them, or give them away without permission of the software vendor. (There is leniency when donating software.)

► Developers have the responsibility to **inform registered users of verified bugs** in their software and of solutions or work-arounds, if any.

SOFTWARE USER PRACTICES

Assign a staff individual the responsibility of ensuring proper software use. This individual

► **reads, reviews, and understands** all purchased software agreements.

► **registers for software licenses** by sending in all forms.

► **informs all users** about licensing and legalities.

► **keeps track of version upgrades**.

CLIENT GROUP

Most clients offer project materials to the creative group digitally—from database content to charts to photographs to text, the compatibility between softwares may be an issue. Some clients have proprietary (closed) systems that require translation. The client and the creative group need to resolve translation differences.

SPECIAL NEEDS

If the client has requirements for how the project is to be prepared, such as what software to use or how to prepare files, the client

► **communicates the need to the creative group** when requesting proposals. (See chapter 4, page 38.) If the project

▷ *might be part of a larger publication system*, it has to be flexible for specific uses.

▷ *applies to intranet or Internet*—requirements may influence file construction to conform to standards or templates.

► **purchases software** if not part of the creative group's capabilities, and provides any training needed. Special tools will be part of the design parameters.

► **never loans software** (including fonts) to another party. (See following on page 344.)

"When evaluating a software package for a large installation, there are several questions to ask that buyers don't often think of: How important is the software package to the company who provides it? How much money do they have to add enhancements? Do they listen to you, the customer?"

—Sandra Kinsler
PhotoLibrary Management Service
Ventura, CA

COLLABORATIVE PROJECTS

Most typically, clients and creative groups exchange files back and forth, so compatibility needs to be established. In the most collaborative projects, templates may be designed to house variable and updatable content. If the client purchases templates from the creative group to be used in-house, they must

► **allow experimental time**—integrate its operating systems and applications with the creative groups' systems. Hardware configurations and software versions may vary, even within the same platform. It is best if the client assigns a technical person to enable efficient integration with the creative group.

► **purchase their own copies** of the software application used for the digital files (or vice versa, as above).

► **not be responsible for the legal use of software** by the creative or production groups.

CREATIVE GROUP

As client and creative groups work cooperatively, having compatible systems for file interchange becomes necessary. Some projects require a fluidity of uses, such as in corporate identity or cross-media publishing. How smoothly the technology works is a critical factor in project success.

SYSTEMS INTEGRATION

A member of the creative group may need to serve as a systems integrator to coordinate with the client's technology. The creative group will most often purchase, at their own cost, the programs, fonts, and other software needed for the project (unless the client requires a specialized application—see above). The creative group should

► **adhere to licensing agreements**—stay current with version releases, documentation, and upgrades.

► **be adept at using the software** agreed upon, and determine whether the software will work as intended. If working with new or experimental methods, allow time for testing before being under a deadline crunch. Also, test any new tasks, processes, and integrations before committing to a project schedule and results, if possible.

► **take precautions against software piracy**

▷ *prevent illegal copying of software*—explain studio policy to all staff members and make part of orientation for new members.

▷ *prevent the borrowing of software*—unless it must accompany a specific project for output.

Digital publishing growth is hampered by proprietary systems, both in hardware and in software. Though some packages are multi-platform, and can be linked together, for some multitasking, the future of digital publishing will eventually become more open. Seamlessness between packages will allow document files to utilize features from many packages. It will also require the user to learn more capabilities. Processing speed and memory capacity improvements propel more capabilities.

- **notify the software developer** when bugs are detected that affect the efficient operating of the software according to the described capabilities. The software user (not developer) is responsible for results of the software performance, including results from the interaction between different applications.

- **protect project components** by developing practices to

  - *back up all software*—keep additional off-site storage for archiving which is especially important for document archives and libraries.

  - *keep track of processes relating to versions* so that variations can be made if needed.

PRODUCTION GROUP

The production group must adhere to the same practices as the creative group, and is in the business of supporting all major software packages.

SERVICE INDUSTRY

For prepress and printing (see chapter 23, page 266, and chapter 24, page 280), the software packages are at war with open and closed systems. Compatibility problems and lack of standards inhibit the potential of a continuous and seamless digital work flow. But no matter how seamless it may or may not be digitally, the production group has responsibilities to purchase, maintain, and learn applications. They should also

- **adhere to licensing agreements**—purchase their own copy of the software if the same application is used by both the creative and production groups (except for proprietary software provided by the client).

- **share software** only when the production group does not have the package required for the project—the creative group may provide the software (including fonts) along with the project files for use *only* on that project. The software continues to be owned by the creative group. However, some licensing agreements may prohibit this practice, though efforts are being made to standardize. When in doubt, check specific package agreements.

- **report any detected software bugs to the vendor**, and to buyers, if appropriate.

- **coordinate systems with the creative group**—communicate capabilities of software and iron out compatibilities between software versions before there is deadline pressure. Ideally, digital systems can work with seamless efficiency, given the right infrastructure and experience.

"The variety of software licensing agreements available makes it difficult for a manager to control and keep track of what type of software is under what type of agreement. Software can be licensed per machine, per user, or per server, and on a concurrent-use basis, as well as by site/entity. Often, a manager can have one of each type of licensing agreement in the office, making it difficult to keep users educated.

"Our committee recommends licensing standards for a concurrent-use basis. The vendor licenses a certain number of software copies and puts them on the network. Metering software prevents users in excess of the preset number from accessing the software. The software could then be metered as part of the networking operating system.

"We also request that licensing agreements be written in a more standard language that is clearly understandable to both parties. Documentation should be separated from licenses. PC managers must educate their users on licensing issues and software piracy, as well as establish a standard policy."

—D. Keith Heron
Cochairman
Micro Managers Association

"The new generation thinks nothing of copying tapes, online services, etc. There is no public education to teach the importance of intellectual property. There is a disrespect and even trivialization. Greed can be an overwhelming drive. There needs to be an economic incentive to inspire people to behave."

—Jerry Skapof
KanImage
Long Island City, NY

"Do we complain about technology? No. We revel in it. We find that we have what we've always been missing: control. We control a project from beginning to end. We can produce design in record time. What is the price for this control? Losing design time due to learning software, debugging software, and updating software. More time is spent learning to typeset, scan photos, color correct, composite, and trap than on design. On the up side, the more we learn to do well, the more we can do in-house and bill for it. On the down side, the more we do in a shorter time, the more that is expected."

—Eric White
Binary Arts
Winter Springs, FL

COPYRIGHTS 31

Ownership of digital material becomes more crucial the easier it is to share. Copying, especially through the Internet, is endemic. Everyone needs to be aware of its dangers because everyone owns intellectual property to protect. In a publishing environment, the sharing of information is fluid, but usually defined to a project or use. Tracking down ownerships when a project uses a lot of images can be challenging. Where images come from is not an indication of what rights may apply, so images should not be used unless ownership is apparent.

THE COPYRIGHT LAW

Everyone in publishing needs to have a basic understanding of the copyright law, and know instantly if there is a violation in any materials about to become public media. Description of the copyright law outlines:

▶ **tangibility**—only works of authorship fixed in a concrete medium of expression are protected. *Ideas cannot be copyrighted*, only their expression in a tangible form, which includes a digital file. This covers many different forms of authorship, such as:

 ▷ *literary works*—writing which includes any printed or online text.

 ▷ *pictorial or graphic works*—design, illustration, photography.

 ▷ *audiovisual works*—videos, Web sites, television.

 ▷ computer programs—software of all kinds.

 ▷ *sculptural works*—3-dimensional work, packaging, point-of-purchase display.

 ▷ *musical works and sound recordings*—CDs, tapes, downloads, sound clips.

 ▷ *dramatic and motion picture works*—movies, QuickTime videos, theatrical performance.

 ▷ *choreography*—dance, motion design.

 ▷ *architecture*—residential, commercial, experimental.

▶ **length of ownership**—currently, the duration of a copyright is for the lifetime of the creator plus 70 years. Works owned by corporations are protected for 95 years.

▶ **public domain**—After the copyright expires, the work of the creator becomes "public domain," which means that it

---

"Congress has the power to promote the progress of science and useful arts, but securing for limited times to authors and inventors the exlusive rights to their respective writings and discoveries."

—Constitution of the United States Article 8

ORGANIZATIONS
www.wipo.int
World Intellectual Property Organization

WEB RESOURCES
www.lcweb.loc.gov/copyright
Government Copyright Office

www.publaw.com/new.html
The Publishing Law Center

WEBSTER'S DICTIONARY DEFINITIONS:

> COPYRIGHT—the exclusive right to the publication, production, or sale of the rights to a literary, dramatic, musical, or artistic work, or to the use of a manufacturing or merchandising label, granted by law for a definite period of years to an author, composer, artist, distributor, etc. [Covers artistic creations.]

> TRADEMARK—a symbol, design, word, letter, etc. used by a manufacturer or dealer to distinguish his products from those of competitors, and usually registered and protected by law. [Protects names that have tangible form.]

> PATENT—a document open to public examination and granting a certain right or privilege; letter patent; especially, a document granting the monopoly right to produce, use, sell, or get profit from an invention, process, etc. for a certain number of years. [Covers product ideas and inventions.]

SEND REGISTRATIONS TO:

Register of Copyrights
Copyright Office
Library of Congress
Washington, DC 20559
Applications and forms available:
202/707-3000
General information:
202/707-9100
www.loc.gov/copyright

Many computer databases would be covered under collective works or compilations.

COPYRIGHT REGISTRATION CHECKLIST

> Artwork is registered on copyright Form VA, available from the U.S. Copyright Office by calling the Copyrights Forms number at 202-707-9100 or by visiting http://lcweb.loc.gov/copyright/.

> The cost is $30 (payable by check) to register each work you've created.

> The form comes with instructions and you can make as many copies of the form as you need.

> When you file, don't forget to keep copies.

> It can take as long as 6 months to receive copyright registration.

— Jean S. Perwin
Law Offices of Jean S. Perwin
Miami, FL

---

is no longer protected by copyright and may be used by anyone without permission. However, foundations can continue to hold copyrights for longer, so this should be verified per instance.

▶ **ownership factors**:

▷ *The creator of a work is the owner of the copyright*

> it is wise to attach the copyright symbol, date, and name to the work to ensure protection. The work is protected with or without the © symbol, but creation may need to be proven.

> more protection can be obtained by registering the copyright with the copyright office (see sidebar). This entitles greater damages to be awarded the victims, including court costs and lost revenue. Without this protection, a victim will generally only receive a settlement.

> unless signing rights to another party in writing—best to register this transfer at the copyright office.

▷ *employers holds copyright*—if a work is created by the employee of a business within the scope of employment.

▷ *work-for-hire*—if a client commissions a work from an independent contractor, the independent contractor holds the copyright—unless both parties agree in writing that it is a "work-for-hire." If so, the client owns the copyright (for 95 years) if the work comes within one of nine categories identified in the copyright law. These are broad categories and the law is intentionally vague: (An attorney should be consulted if there is a question.)

> a contribution to a collective work (such as a magazine, newspaper, encyclopedia, or anthology).

> a contribution used as part of a motion picture or other audiovisual work.

> a supplementary work, which may include pictorial illustrations, maps, charts, etc., done to supplement a work by another author.

> a compilation (new arrangement of preexisting works).

> a translation.

> an atlas.

> a test.

> answer materials for tests.

> an instruction text (defined as a literary, pictorial, or graphic work for use in systematic instruction activities).

OWNERSHIP RIGHTS

The copyright owner usually maintains the original artwork, and owns it even if it is in someone else's possession. The owner has the right to

▶ **make derivative works or modifications**—this includes using different media to execute an idea, combining images, or using special effects.

▶ **publicly distribute copies, perform, or display the work**—this includes using it for promotional purposes such as in a portfolio, shown in advertising, or on the Web.

▶ **control reproduction of the work**—setting up licenses (see page 348), which includes:

▷ *First rights* and *specified rights*—for purposes outlined in the proposal. Often, each subsequent license, such as when adding media usage, carries a fee which may include usage rights for defined markets, media, or frequencies.

▷ *Exclusive rights*—prohibit the creator from selling rights to anyone else to use. This can be limited to all markets or to a specific market. It can also be limited to a certain time period.

▷ *Additional rights* or *consecutive use* allows the client the right to use a creative work in several reproductions or in several media. This is useful for campaigns and cases where a series of publications is required. Often, a bundle of rights is transferred to the client, as defined in the proposal. Web sites fit this category, especially because they often use content from print marketing and translate it into further application.

▷ *Unlimited rights* mean that the client purchases the right to use the creative work in any manner, for any market, and for either a specified length of time or an unlimited length of time. The creator still maintains the copyright and can control alterations, variations, or derivations. Most Web sites should use unlimited rights because a successful image on the Web should be used throughout the organization's marketing materials. It only makes economic sense to have this integration, and those organizations who don't build upon an integrated identity are spending investment dollars in a fragmented way. The exception is event marketing—campaigns that carry a theme through a series that is used to promote a finite activity.

"If you plan to post your work on the Web, you should also consider digital watermarking, which 'fingerprints' your files. Should you then find unlicensed copies of your watermarked creations elsewhere, there's a good chance that you can track them to the purchaser who made the illegal copies. Two digital watermarking companies are Digimarc (800/344-4627, http://www.digimarc.com) and Liquid Audio (http://www.liquidaudio.com)."

—Susan P. Butler
Attorney
San Francisco, CA
www.macworld.com/2000/07/create/copyright.html

Know your rights by registering your copyright. Although the default goes to the creator whether registered or not, there are many benefits and legal strengths only available to those who formally registered.

"Practically every country is a party to international copyright treaties; that is, national governments agree to protect copyrights for citizens in other nations. Usually the law of the country in which someone rips you off applies. Under some circumstances, you can sue the thief in a court in your own country.

"Thanks to these treaties, copyright laws around the world are basically similar."

—Susan P. Butler
Attorney
San Francisco, CA
www.macworld.com/2000/07/create/copyright.html

A photographer may grant the use of a photograph for a magazine within the real estate industry to be used only once. The photographer may then grant the use of the same photograph to another client for use in ads placed in travel magazines which may extend for the life of the campaign. And after both contracts are over, the image can be available in stock. As photographers control their images, they can maintain a lucrative practice with good library and distribution systems.

TAKE ACTION

All images used in publishing must be checked for copyright prior to publication. Because the client has their name on the product and makes use of it for marketing goals, any potential liabilities are theirs. But the more the entire team understands copyrights, the greater the protection that proprietary content will not be abused, either by using others' work without permission or by having assets copied and used. Either way is illegal theft of intellectual property. If there is a great potential profit, then the client must be especially careful to have all permissions secured before publishing. It is the responsibility of every publishing organization to perpetuate a culture of respecting and honoring copyrights. The law is explicit that:

▶ **permission should be obtained** for use of an image

▷ *in any form of copying or publication* using whole or part of the image; for example:

> photocopied—out of books or magazines and then redistributing to an audience.

> scanned—found sources like from other printed pieces or original images collections.

> downloaded or copied from the Internet—such as portions of other sites. Free copying and then reusing is a dangerous practice that can have unpleasant legal consequences. It is much cheaper to pay permission fees if wanting to use images that belong to someone else than to pay legal fees if caught using.

> photographed of a person or a nonpublic location.

▷ *as a reference* by an artist or illustrator. If the derivative work demonstrates an obvious relationship to the original, as judged by a layperson, then there is a violation.

▷ *used in presentation materials*—such as concept comprehensive presentation materials or slide presentations—particularly applicable to demonstrate design, sales materials or promotional projects. Whenever there is money to be made by use of the concept or work of a creative originator, compensation needs to be discussed.

▷ *used in reproduction*—such as in printing, on CDs, or broadcast on the Web. When an image has increased exposure, someone benefits. The most appropriate to benefit is the creator of the compelling image. The more images reproduced, the larger the reach. The Web, with high visibility, has an audience of global potential.

To obtain full benefit of the copyright law protection when in conflict, be sure to have the tangible work registered. Protection such as a notarized copy of a presentation drawing, sending the work to oneself via registered mail and keeping it unopened, disclosure agreements, watermarking, or online tracking systems all will help to protect ownership, but it is less effective if not registered.

Many who copy files or scan images are unaware of engaging in a legal violation. Because it is so easy to do, it seems okay to do. But the proliferation of legal disputes indicates that this is a risky practice.

There are several foundations and organizations that are ruthless about protecting their copyright on older images. Two examples are Disney's ruthless protection of its characters and the Einstein Foundation that polices some of the most famous images of the scientist.

A seminar presentation for educational purposes, where examples of printed materials created by others are used to illustrate the speaker's content, would probably be considered fair use.

- **permission does not need to be obtained** when under the "fair use" clause which *allows uses that are not copyright infringements*—(the law is vague; permission should be sought or use should be checked with an attorney):
  - ▷ *includes criticism, comment, news reporting, teaching* (including multiple copies for classroom use), *scholarship, and research.*
  - ▷ *accounts for the purpose and character of the use*
    - > commercial or for nonprofit educational purposes.
    - > the nature of the copyrighted work—whether the work is published or unpublished.
    - > the amount and substantiality of the portion used in relation to the copyrighted work as a whole.

CLIENT GROUP

The client, if providing elements for content, should be aware of who owns those elements. It will ultimately be the client who bears responsibility for ownership once images or statistics are published. For commissioned work, it is best to know ownership and accuracy of data before work begins.

RIGHTS AND RELATIONSHIPS

There are many combinations for who owns what and who is responsible for what. The relationship between the client and creative group, based on rapport and trust, can leave much undefined. It is to everyone's benefit to understand

- **transferring of ownership**—if the client wants to hold the copyright on commissioned elements, the client should
  - ▷ *understand the elements* of the design and contents. Rights (including promotional) should be outlined by the creative group, and the client should be aware that fees are partially determined by these rights.
  - ▷ *obtain a written agreement* with outside creative sources. Understand what can or can't be changed in the digital files.
  - ▷ *apply for copyright* (see sidebar, page 347)—if the creator had already done this, then the registration needs to be transferred.
- **confidentiality**—when the client and creative groups work together, there is an automatic agreement of confidentiality, just like a corporate version of the patient-doctor privilege. Any original information shared within the team, whether it's product information, client lists, membership lists, statistics, or photographs, are kept private.

THE RULES THAT APPLY TO WEB CASTING OF MUSIC AND SOUND RECORDINGS

"Music is one of the most complicated areas of copyright law, so it is not surprising that the rules that apply to Web sites using sound recordings are also complex.

WEBCAST DEFINITION. People mean different things when they use words like streaming or Webcasting. For our purposes, we mean a one-way, 'non-interactive' offering of music over the Internet. A Webcast, like a broadcast, means the transmitting organization, not the listener or viewer, determines the content of the program or the playlist. The rules that apply to Webcasting of music and sound recordings are substantially different from those that apply to traditional broadcasting.

FOUR SEPARATE RIGHTS are involved in Webcasting: the right to Webcast (publicly perform) (1) musical compositions and (2) sound recordings (by means of an audio digital transmission); and the right to copy on computers or servers to enable the Webcast; (3) musical compositions and (4) sound recordings.

PERFORMING RIGHTS TO MUSICAL COMPOSITIONS for Webcasts are voluntarily negotiated between the Web site operator and ASCAP, BMI and SESAC. All three performing rights societies now offer blanket and per-use licenses for Webcasts of music in their repertoires. Terms and conditions of those licenses can be found on their respective Web sites:

(www.ascap.com; www.bmi.com; www.sesac.com).

"A blanket license permits Webcasting of all music in a society's repertory for an annual fee, often a percent of revenues with a minimum payment. Under a per-use license the Webcaster pays fees based on the number of musical works used and the kind of use made of the music (theme, background, length)."

—Paula Jameson
Vice President and Chief
Administrative Officer
Gibson Guitar Corp.
Nashville, TN

With many projects, such as informational brochures or publications, who owns the copyright is not an issue. If the client does not have in-house capabilities or the need to own electronic files, they would work with the creative group on all revisions and updates, anyway. An efficient system and rapport can be established which best serves the client in an ongoing relationship with their creative support.

Although it has always been the responsibility of the creative group to provide copyright information to their clients, traditional processes made who owned what more clear-cut than digital processes. For example, the final artwork was always a keylined board whose *usage* was granted to the client, but copyright and ownership resided with the creator. In digital publishing, processes are blurred, and the client owns different materials depending on the project (see chapter 29, page 332). There are more ways to define rights. However, uses and ethics remain the same.

"COPYRIGHT PITFALLS FOR GRAPHIC DESIGNERS

> combining the images of various copyright sources without permission

> altering an original image without permission

> copying an image exactly

> copying an image by having a photographer or illustrator make a 'similar' picture

>changing the medium but still copying the image

>using an image as a reference without permission."

—Nancy E. Wolff, Esq.
Intellectual Property
Attorney
New York, NY

A copyright does not need to be visible on a work for it to be protected. If the notice is visible, any disputes are more clear-cut, so it is to the creative group's benefit to place a © on all their intellectual property.

► **dangerous assumptions**—copyright infringers are often innocents using technology for how it was designed. But an infringer, when using content from someone else, runs a risk of getting caught and suffering legal and financial consequences. It also undermines creative motivation if allowed to go unchecked.

CREATIVE GROUP

All creative groups need to become intellectual property experts: being in the idea-generation business, design is a product to protect. Because design is as service-oriented as it is product-based, finding the right balance between the two is a challenge. But without protecting the work itself, designers have no creative product.

THE DEFAULT

The creator (or author) automatically holds the copyright on all work created—unless other arrangements are made or unless created by an in-house department. The creative group

► **bases fees partially on the uses** of the design and images outlined in the proposal. (See chapter 4, page 38, and chapter 5, page 50, for fee information.) Pricing considers:

  ▷ *the reach of distribution, quantities, or frequencies* in media—how many people will see it.

  ▷ *application or size of the image used*—how large or in what position, such as a magazine cover or Web homepage.

  ▷ *form of the concept or image used*—such as special effects or including it in a collage with other images.

  ▷ *length of time* the concept or the image is used.

  ▷ *sector of industry or territory* where image is used.

► **takes the initiative** to define rights and explain usages to the client. Do not assume this is understood.

► **negotiates terms with the client**—if done diplomatically, they will appreciate how copyrights are in their best interests.

► **knows employees can be personally liable** when violating copyright as well as their employers.

PRODUCTION GROUP

Copyrights generally do not apply to the production group who

► **is not responsible for content**—but hired to supply technical support.

► **may have to suspend production** during a copyright infringement violation, which doesn't happen often.

"The test for copyright infringement is twofold: (1) Proof of access by the infringer to the work alleged to be infringed must be shown (or, if the similarity between the two works is great enough, this access can be assumed); and (2) The jury must conclude that an ordinary observer would believe one work is indeed copied from another. On the other hand, fair use has a fourfold test: (1) the purpose and character of the use, including whether or not it is for profit; (2) the character of the copyrighted work (use of an informational work is more likely to be a fair use than use of a creative work); (3) how much of the total work is used in the course of the use; and (4) what effect the use will have on the market or for value of the copyrighted work."

—Tad Crawford
Writer and Publisher
Allworth Press
New York, NY

"Thousands of copyright violations happen in corporate America every day without anyone giving it much thought: Many people mistakenly believe that copyright issues don't apply if the application isn't a commercial product. It's only that the smaller your audience and the less commercial your intent, the lesser your chance of being caught. But the risks are real, and they're growing, as content owners pay attention to the practice of casual multimedia sampling."

—William Rodarmor
Managing Editor
California Monthly
University of California
Berkeley, CA

"Copyright is separate and distinct from the work itself: the book, the painting, the photograph, the song, etc. You can buy a photographic print or painting, but you will not own any of the rights under copyright unless you acquire them specifically. Copyright owners can permit many people to use the same work by creating different licenses. The same image can be used as a book cover, a billboard advertisement, or on a Web site. If you want to be the only one who can use a certain photograph for a period of time, known as an exclusive use, that agreement must be in writing. It is recommended that all uses—even nonexclusive uses—are expressed contractually in writing to avoid dispute."

—Nancy E. Wolff, Esq.
Intellectual Property
Attorney
New York, NY

"Not every photograph is copyrightable. To create original works of authorship, photographers must add their vision. This photo's unique lighting, posing, and expression make it an original."

—Susan P. Butler
Attorney
San Francisco, CA
www.macworld.com/2000/07/
create/copyright.html

"Fair use is a tricky doctrine and a trap for the unwary. People who want to use copyrighted material have devised enough rules of thumb to hold a national thumb-wrestling content. You'll hear things like, 'It's OK to take up to 30 seconds of music' and 'Two lines out of an eight-line poem is all right.' These shorthand rules are more dangerous than no rules at all. If you're producing a commercial CD-ROM and it doesn't have an explicit educational, critical, or scholarly message, this small-scale borrowing could be copyright infringement."

—William Rodarmor
Managing Editor
California Monthly
University of California
Berkeley, CA

"For all works created or published for the first time after March 1, 1989, no copyright symbol is necessary for protection. That's the date the United States officially became a member of the International Copyright Treaty. By its very existence, the photo is protected by the copyright law. (For works created before March 1, 1989, and even those reprinted after that date, a copyright notice must be legible on all copies.)"

—Philip Bishop
Writer
San Francisco, CA

"You've sweated over your creations, investing time and energy in them. Then someone comes along and rips them off. Your effort is disregarded, and your reputation as an artist–and your income–may even be threatened. You can defend yourself if you're aware of your rights. You need to understand copyright law.

"Copyright law is complex, but knowing a few basics can help you protect your creations. It's important to understand the legal rights of a copyright owner, the types of work copyright law protects, and how to protect your work."

—Susan P. Butler
Attorney
San Francisco, CA
www.macworld.com/2000/07/
create/copyright.html

"It is much easier to negotiate a reasonable license fee for a use than to pay a settlement after you have been caught. If you use an image in a comp, it is likely that the client will want the same image in the final project. If an image is not available, do not have another photographer create a similar one. This violates the exclusive right to distribute and make copies. It is difficult to create a wholly original work if you have an example of another artist's interpretation in front of you

"If you admire the concept of an image and want to create a variation, request permission and pay an art reference fee. This will avoid a claim of infringement by the owner of the original piece."

—Nancy E. Wolff, Esq.
Intellectual Property
Attorney
New York, NY

BUSINESS, COPYRIGHT, AND THE INTERNET

"Making copies of software, images, and text on the Internet is simple. Most software for using the Internet has built-in features allowing files to be copied with a single command or mouse click. The ease of copying can lead people to believe that the copying is legal. Wrong. Although the ease of copying can create legal arguments about what copying is 'authorized,' copyright law has never said that easy copying destroys an author's rights.

"When an employee creates a copyrightable work within the scope of her employment, the copyright belongs to the employer. But when an independent contractor or consultant develops a work, the copyright will generally belong to the independent contractor unless there is an express assignment of the copyright to the purchaser. In other words, that fancy graphic or the description of your business will be subject to a copyright held by the developer. Therefore, if a business is using an outside Web page developer, the business should be sure to obtain an assignment of all copyrights and a representation that the developer has not infringed on anyone else's rights."

—R. Timothy Muth
Reinghart, Boerner, Van
Deuren, Norris &
Rieselback
Milwaukee, WI

"If you don't act to protect your business materials on the Internet, your business may lose its uniqueness, some of your potential customers, and, ultimately, even your claims to the copyright itself. Make up your mind to pursue this vigorously and relentlessly. A periodic search might just help protect your business; I've found it necessary to protect mine."

—Dr. Ralph F. Wilson
Wilson Internet Services
Rocklin, CA

"Much [copyright infringement on the Internet] could be called incidental. It isn't committed by bad people, but people who can't find the owner and assume that the reproducing rights are publicly available or the owner doesn't care."

—Bruce Davis
CEO
Digimarc Corp.
Portland, OR

"Digital watermarking makes it easy for a potential acquirer to find the owner and license the image. A copyright symbol is visible in the tile bar of the image when it is opened in an image processing tool such as Adobe Photoshop, CorelDRAW or Micrografx Picture Publisher. More extensive information about the image's creator can then be found by reading the watermark.

"The watermark lets that artist be found easily so his work can be licensed or sold. It removes arguments of innocent infringement. You have the option of joining up with Digimarc's MarcSpider, a patrolling cybercop of sorts. For a minimum annual fee of $50, the service periodically searches the Internet for sites that are using your watermark. If Digimarc doesn't have a detected site registration to use your image, you're notified via e-mail. The rest is up to you—and, possibly, your lawyer."

—Russell Shaw
Journalist
Portland, OR

"Few readers know that Congress examined copyright issues in a 1998 law called the Digital Millennium Copyright Act. In one narrow area—copyright infringement—this act provides a 'safe harbor' for administrators. But here's the catch: To take advantage of the act's provisions, you must file with the U.S. Copyright Office. If you follow the law and make it easy for copyright owners to contact you, take material off your site when content owners challenge it properly, and follow rules that protect your users from false challenges, you are protected from lawsuits."

—Andy Oram
Editor
O'Reilly & Associates
Cambridge, MA

"Infringers are those who violate the owners' rights and they can be subject to fines, destruction of the infringing work and in some cases even criminal penalties and imprisonment."

—Nancy E. Wolff, Esq.
Intellectual Property Attorney
New York, NY

"Have you found graphics or content from your Web site popping up on other sites? Or has someone 'borrowed' graphics that you created for print publications and used that material on a Web site without your permission?

"Thanks to Congress, you now have a new remedy for this problem. You can contact the Internet Service Provider for the offending Web site and request that the ISP remove the material. According to the new Online Copyright Infringement Liability Limitation Act, an ISP will not be liable for copyright infringement damages for infringing material placed on the ISP's system by a system user if the ISP, upon receiving notice of the claimed infringement, removes the material claimed to be infringing. This law obviously gives ISPs a powerful incentive for complying with a copyright owner's request that infringing material in a Web site on the ISP's system be removed from the system."

—Dianne Brinson
Attorney and Author
Portola Valley, CA
Internet Law and Business Handbook
www.laderapress.com

"All stock houses, whether on the Web or not, require contracts to be signed before authorizing image use.

"Most designers don't bother studying the terms of a Web site use agreement. But you're bound by it anyway. It's like the shrink-wrap licenses that come with software. Although legal problems arise from binding people to agreements they've never read, your use of materials on a Web site will be governed by the Web site use agreement—whether you read it or not."

—Jean S. Perwin
Law Offices of Jean S. Perwin
Miami, FL

"Section 201 (c) of the Copyright Act governs contributions of text or art to magazines, newspapers, and other collective works. Under this section, a publisher is entitled to hold the copyright in the collective work as whole but the individual contributor, absent a specific writing otherwise, retains the copyright in the particular contribution. The publisher/owner of the copyright in a collective work is presumed to have acquired only the priviledge to reproduce and distribute the individual contribution as part of the collective work in its present form, in a later revision of the collective work, or in other collective works in the same series."

—Caryn R. Leland
Licensing Art and Design
Intellectual Property Attorney
New York, NY
www.lelandlaw@erols.com

KNOWING WHAT'S YOURS

"It's bad enough when the public abuses the property rights of creators, but at least their ignorance is understandable. When art directors, designers, and other 'creatives' routinely swipe images, using them for commercial purposes, it's worse.

"The solution to this problem lies in the rigorous enforcement of copyrights.

"Now, more than ever, creators of visual images need to understand the underlying value of the work they create and never underestimate the leverage they have when negotiating the transfer of rights."

— Paul Basista
Former Executive Director
Graphic Artists Guild
New York, NY

"Because the Internet so easily crosses national boundaries, it would be almost impossible to have a site that could be limited by territory. On the other hand, of course, rights can be restricted with respect to language (so that exclusive Internet rights might be for the English language only, leaving the licensor able to sell French or Chinese rights)."

—Tad Crawford
Writer and Publisher
Allworth Press
New York, NY

"If you register your copyright within three months of when you make your image public or before the infringement, you'll be entitled to attorney's fees and statutory damages. But if you file your registration after the initial three-month period or after the infringement, you'll be entitled only to the actual damages you can prove.

"The benefits of registration become clear when you realize that the actual, provable damages suffered by a creative whose work was copies are usually very small. But attorney's fees and costs in copyright litigation can be as high as $50,000 or more. Without the ability to recover attorney's fees from the infringer and collect statutory damages, many creatives with legitimate copyright claims could never afford justice.

"Not every piece of your work must be registered, but you should file copyright registration on any work that will be seen by a lot of people or that's used frequently."

—Jean S. Perwin
Law Offices of Jean S. Perwin
Miami, FL

FOLLOW-UP 32

Projects don't end when printing is delivered or Web sites go online. Success is measured by the response of the constituents. Do sales increase? Do prospective members inquire? Do the tickets sell? Do readers subscribe? These factors have to be the measure of good (i.e. effective and appropriate) design, but fulfilling those elements is not just the job of the designer. It's more the job of the client who must have the right positioning, timing, and who has chosen appropriate resources. With those prerequisites in place, good design propels an organization's momentum—attracts the right prospects or communicates the right message. Design has a job to do. Creativity wears overalls. Even the great campaigns and brands were noticed because they were placed in front of the right people and conveyed value.

CLIENT GROUP

Observe constituents, responses and investigate the profiles of prospects. Follow up on more than the dollar return for publishing efforts and gain the feedback:

► **build on the successful project**—find out what constituents liked in its message.

► **reevaluate and refine** if results were mixed—adjust to constructive feedback. Use information in planning a follow-up approach.

► **change direction** if results were unsatisfactory—discover from investigation why the project didn't reach its goals and learn from these lessons. Begin experimenting and then plan another approach.

► **evaluate learning experience** at the end of every project—go through various aspects and learn from any difficulties. If it was a large challenge, many of the lessons may be very clear. If not, review the chapter that pertains to that challenge.

► **celebrate and conclude**—be glad the project is over and reward those responsible. Pause before jumping into the next project to enjoy the completion through an appropriate activity.

To show tangible and measurable results is required in business. Although the creativity—the power, strength, and emotion of ideas—is elusive, if concepts are well-focused and speak

How can a client determine if the communication project is effective? The best clients have the vision to identify the right solution for their challenge. They know their judgment is sound when sales pick up, membership booms, their company becomes an industry leader, or they gain market share. And they know that how they manage the project process translates into results, though often intangible to measure.

(An industry joke: 50 percent of advertising attracts buyers. We just don't know which 50 percent.)

Success is judged by what happens *after* a project is finished. The client of vision knows that strong design is a factor, but they also know that timing, positioning, frequency, consistency, and follow-through are also controlled by their judgment and work in concert to attain the best results.

to an audience, they will be considered successful. Herein lies the most important judgment of the client which must be made from a gut-level knowledge of the audience. The best clients are passionate about knowing and understanding their constituents.

CREATIVE GROUP

When the creatives meet the accountants, there can be friction. Designers can't measure levels of good—design is poor, anonymous, effective, or brilliant. Beyond that, the results are dependent on more forces than a good idea and good project execution. The project must further business goals to be considered successful.

Find out as much as possible about constituent reactions to feed the depth and effectiveness of the design. Like the client, the best designers have an intuitive understanding of what will appeal to a given market. Back up ideas with strong market-based reasons during development. But these can quickly fade in the wake of new feedback. Seek out this feedback to grow practices:

► **contact the client**—wait for the appropriate time after a project is finished so constituents have had time to react.

► **site visit where the design performs**—retail centers, events, meetings, etc., to watch how people interact with the design and how it functions in its setting.

► **compare to competition**—having done this as part of the initial research, time has passed and new materials may be published that now compete.

► **interview constituents**—any one-on-one contacts give a more personal attachment to the subject, allowing more emotion that can be used in future work.

► **stay close to the client**—watch for continuing projects so greater market understanding can be utilized to the benefit of both.

HARD LESSONS

If the project did not end well, there are several options for follow-up:

► **point of view**—make sure to be able to define the client's unhappiness in *their* terms. Articulate, from their shoes, so that honest information is considered.

► **discover what areas were skipped or neglected**—this is the most important part of the learning experience because it builds preventative practices.

► **make amends where possible**—sometimes this is a

forced situation if relationships need to be preserved. Sometimes it is best to walk away.

► **move on**—be upset for a while if money and goodwill were lost, but then learn and refocus. Even the best designers come upon difficult, unscrupulous, and closed-minded clients. There can be no good design without a good client.

Strength in design comes from finding solid client collaborations. Designers are generally flexible, working in a variety of industries or causes. Matching the client group to the creative group is something so individual, it can only be nurtured within each situation. Practice knowing, understanding, and servicing clients means forming a partnership where the concerns of the client become the concerns of the designer.

PRODUCTION GROUP

With the client and creative groups between the production group and the audience, finding out the success of a project is mostly a communication challenge. Because the production aspects are the most tangible, the production group's profitability is obvious after a project finish. Measuring buyer success, unfortunately, is usually measured by return business. Instead, measure success through building relationships:

► **call immediately after delivery**—check on buyer satisfaction to be sure entire project is satisfactory.

► **call buyer to discuss**—wait an appropriate time for the audience to respond. Ask about strengths and weaknesses of processes and ask for constructive feedback to use to improve services.

► **prospect the buyer for more projects**—if the first one went well. Suggest ways to build upon it.

► **stay in touch with buyers and trends**—expand services for new opportunities to fuel expansion.

FOUNDATIONAL FOLLOW-UP

A conscientious team will have post-project discussions—especially in preparation for the next direction. Balancing marketing consistency versus marketing change is always a challenge. Know when to build or when to explore a new approach—both depend on a strong foundation in business practices. To have a successful design practice means balancing skills in business, technology, and creativity. To be a successful client of design means enthusiasm, involvement, and understanding.

"Make a project follow-up call at about the time you think your work has been in use by the client long enough for them to answer the question you are asking. When you call, you ask, 'How is the design (or illustration or photo) working for you on that project?' The idea comes from the fact that our definition of 'concept to completion' has changed. Completion is no longer just the delivery of the work. At that point, everyone loves it! Rather, it is the performance of the creative work. Today, completion is the successful use of your work in a specific project. Not only will this technique keep clients coming back for more, you also can find out about upcoming new work."

—Maria Piscopo
Creative Services Consultant
Costa Mesa, CA
http://MPiscopo.com

"People are drawn to entertainment media that has been created to attract them, hook them, stimulate them, and reward them. Our institutions are learning that it is no longer enough to simply communicate. Educational institutions, religions, political parties, and businesses must compete for their audiences against powerfully appealing competitors.

"The communication of identity in this arena presents new challenges. Identities must have relevance to the audience. They must be less rigid, singular and structured and more interesting and thematic, with variations in imagery and qualities conveyed. Entertain them and they will respond."

—Philip Marshall Durbrow
Vice Chairman
Frankfurt Balkind Partners
San Francisco, CA

"Graphic designers have an obligation to present messages in clear and understandable ways. It is not our job to judge a client or company's message. But it is our job to be true to our personal beliefs and private ethical standards. I am obligated to decline a job that violates my personal standards of taste and morality. We may not be lawfully responsible for the content of a message, but we still have to sleep with ourselves at night."

—Leslee E. Paquette
Renton Technical College
Renton, WA

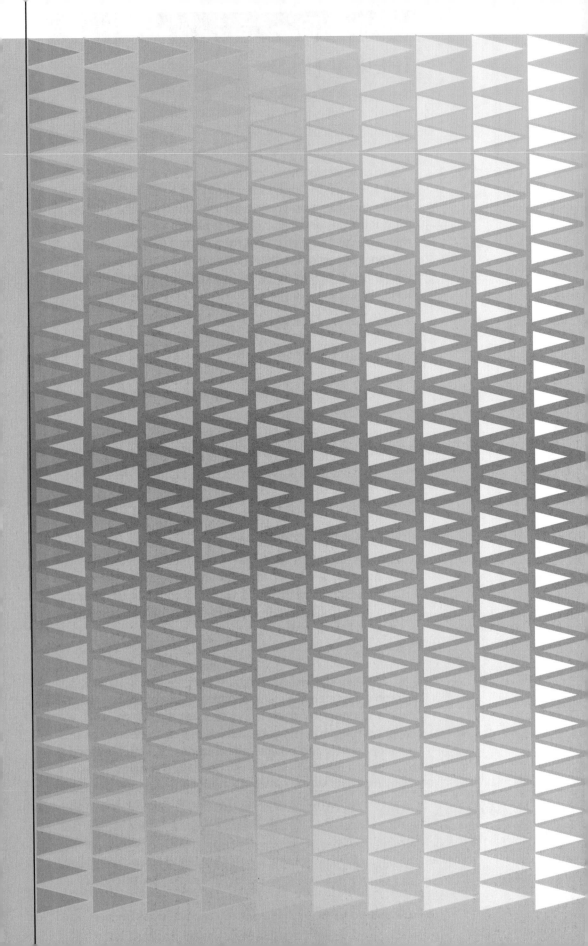

INDEX
GLOSSARY
RESOURCES

GLOSSARY

additive color—projected light in RGB (red, green, blue phosphors) to mix color, as in video monitors.

Adobe Systems—software developers of the PostScript® page description language that made desktop publishing possible. Based in Mountain View, CA., they offer PostScript compatible products, such as Adobe Illustrator®, Adobe PhotoShop®, Adobe PageMaker®, PostScript fonts and other software packages.

alterations—a change in the scope of a project's parameters and specifications. Alterations made at the client's request will likely add to the client's costs. Changes that must be made because of the creative or production group's errors, called corrections, do not add to the client's costs. *See corrections.*

animation—still images strung together to create illusion of movement.

applications—see software application.

Apple Computers, Inc.—personal computer company who revolutionized publishing with the release of the Macintosh computer and then the LaserWriter® in 1985. based in Cupertino, CA, they manufacture and market computers and other hardware and software.

Approval®—high-end digital proofs by Kodak.

artwork—materials created to develop a publication—the ingredients needed to make film and then printing plates. Artwork includes the creation of documents and finished pages, keylines, illustration, photographs, and combinations of all these elements.

asset management—method for organizing intellectual property into accessible components.

audit, design—*see design audit.*

B

background information—previous organizational information, history, articles on a specified industry, competitors' literature, market research data, slides and videos, and other appropriate documentation. This is supplied by the client to the creative group as part of their research.

backups—an extra copy of a digital file in case the original gets damaged or lost. Good backup habits, with methods of saving and naming files, ensure dependability in case of system failure. It is best to maintain an off-site backup copy as further insurance.

ball-park—a quick guess of how much a project will cost. Often given without research, it is dangerous to be too hasty with estimates. Although clients say that they will not hold a creative group to a ball-park, it sets up expectations and then often held to.

bandwidth—the amount of data that can be sent through a connection.

bid—an estimate, quote, or proposal submitted to a buyer. Often a bid is compared to other bids submitted by competitors for the same project.

bitmap—rasterized shape represented by and stored as an array of pixels. *See pixel.*

blank formats—*see dummy.*

bleed—printing color off the edge of the page. Generally, the page composer leaves 1/8th inch of extended image outside the trim edges in documents for bleed. This provides a tolerance area for trimming.

blended media—themes that can be carried from one media form to another, utilizing aspects of each. An audience will remember messages delivered once in five different ways than five times the same way.

blue line proofs—one-color proofs made on photosensitive paper exposed to final film. These are generated for one-color or two-color projects (the second color usually appears as a lighter image), and in addition to color proofs to show format. A Dylux® proof shows if all the images are in register, and indicates of the way pages will fold and bind.

brainstorm—group thinking session where each team member contributes ideas. Meant to create ideas for a specific challenge and is best when the agenda is clear and the participants are open to new approaches.

brand—unique visual identity associated with a product or company, centered around logo or distinguishing theme.

browser—application that enables access to the Internet or the World Wide Web. Most provide the ability to view pages, navigate , download files, copy and print materials from the Web.

bugs—programming errors that undermine the efficient operating of software.

built elements—assembling and creating type, graphics, and images placed within a page. Some visuals are for position only, to be stripped in later by the print group. Others are created within the digital document, often in layers to designate different colors.

calibration—see color calibration.

camera-ready art—reflective artwork or keylines created with traditional techniques and placed in a copy camera for the print group to shoot film. Contains instructions, text, line art, crop marks, and in position images for reflective art.

campaign—series of related projects that support a marketing strategy. Typical campaigns may include brochures, advertisements, direct mail , displays, presentation materials, press kits, proposals, and Web sites.

CD-ROM—read-only documents burned onto a CD.

ChromaCheck®—an overlay proof method by DuPont.

claims—requests for price adjustments, generally by the buyer to their supplier.

clip-art—designs and illustrations that come from old books or collections of previously produced work. Many are public domain images that can be used without permission.

closed system—proprietary hardware or software platform that is not compatible with other systems.

collating—bindery operation of assembling pages in the correct sequence before finishing.

color calibration—a way to measure color created on the designer's monitor, based on RGB to convert to CMYK for printing.

color compression—condensing color files, especially photographs and images that require large amounts of memory, into smaller more portable files. After transfer they can be uncompressed. This system is especially useful for archival storage and back-ups.

ColorKey®—an overlay color proof method by 3M.

color management system (CMS)—combination of hardware and software geared to produce accurate color results.

color registration—*see registration*.

color separation—full color printing with four process ink colors: cyan, magenta, yellow, and black (CMYK). Other colors are closely approximated by halftone screen combinations of these four colors (for example, green is a screen of blue combined with a screen of yellow). To prepare artwork in color, it must be translated into these four process colors, with one piece of film for each color. Photographs and reflective artwork need to be converted into the four process colors, which is called color separation.

comp—abbreviation for composite or comprehensive. Produced during the design process, a presentation layout provides an expression of a concept, approximates the finished product. *See comprehensive presentations.*

composite proofs—proofs that show how colors fit together for multi-colored projects. These proofs are used as guides for the print group to see how the colors register. *See final proofs.*

comprehensive presentations (or comps)—mock-ups that represent a concept tangibly to a client. Many take the form of sketches, magic marker drawings, black-and-white laser mock-ups. Color photocopies, low resolution color laser proofs, or "super comps" (that look almost printed).

compression—reduces the transmission time of media and application files across the Web.

concepts—written or visual elements that convey a statement, theme, direction, or campaign. These ideas are represented initially by presentation materials or comps, and ultimately by finished artwork.

continuous tone—color shadings where there are no halftone dot screens to define the color saturation and which offer very smooth blends of one color to another in a high resolution reflective or transparency format. Continuous tone is inherent in original photographic images, and appears in several digital proofing methods, particularly Iris®, Rainbow®, and Digital Matchprint® proofs. Offset lithography printing requires dot screens to reproduce color shadings. There are a few uncommon printing methods that will print continuous tone color.

contract proof—*see final proof.*

conversion of data—changing an image from one format to another.

copy camera—high-quality reproduction camera (similar to a photostat camera) where art boards are held under pressure as they are optically photographed onto film. This film is developed, stripped, and used to expose printing plates. Camera-generated film is giving way to computer-generated film.

corrected client proof—made by the page composer for the client's review, and then refined into the final document, a corrected client proof incorporates the changes from the client, who might check it again for accuracy.

corrections—changes undertaken to fix errors generated by the creative or production groups, not by the client. Changes to correct client errors are called alterations.

CorelDraw®—a drawing software application published by Corel Corporation.

CricketDraw®—a drawing software application published by Computer Associates.

critical path—key activities in the project process that can hold up the entire process if not fulfilled. These must be identified and managed with priority to keep processes moving forward.

Cromalin®—an integral proofing system developed and sold by DuPont.

crop marks—lines made by the page composer to indicate the trim edges of a document. These are placed in the margins outside the edges of the publication. Fold marks serve a similar function, but are dashed lines that indicate where folds are to be made.

cross-media publishing—concepts that can be applied to a variety of delivery methods, usually in a planned campaign. These might include print, a Web site, and a presentation graphics, all of which can carry a unifying theme to give market support through several media. People absorb one message through five different delivery media more than five times through the same media.

cross-purposing—finding additional delivery opportunities for marketing messages to further utilize themes, campaigns, branding, or exposure for identity. Utilizing asset management system to access intellectual property and apply to a variety of delivery media.

cyan, magenta, yellow, and black (CMYK)—the four process colors. *See color separation.*

D

damage 323, 326
damaged artwork 336
damaged images 327
data, computer 330
data conversion 46
data, submitted 29
data transfer 189
data translation 80
databank companies 184
database construction 80
database publishing skills 67
database software 341
databases 208, 249, 260
deadlines 24, 27, 35, 39, 219, 262,
    263, 265, 267, 295, 300
deal 339
decentralized work methods 14
decision-makers 24
decision-making meetings 25
decision-making plan 42
decisiveness 65, 151
define rights 351
defined responsibilities 230
delays 264
delivery 300, 360
delivery hours 301
delivery instructions 291
delivery schedule 296
demand letter 329
department, creative 43
departments, in-house 51
derivative works 336, 348
description of content 74
design audit 26, 28
design capabilities 132
design concepts 332
design development 105
design guidelines 130
design levels 14, 15
design prices 54
design priorities 132
design process 119
design proofs 280
design team 43
designers 4, 5, 196, 210, 214, 280
designers, computer 157
designer's contribution 102
designers, experienced 157
designs, ownership 330
desktop publisher 5, 152

data—numbers that make up content in digital files and images.

database—organized collection of information. Common databases include mailing lists, membership lists, image libraries, resources.

database publishing—linking databases to page composition systems, allowing targeted time-sensitive marketing and easily-produced directories.

design—creation of visual themes, elements, concepts, arrangements, and layouts.

design audit—review and analysis of an organization or product's image, brand, and corporate identity. Usually requires the gathering of all published materials and an assessment of weaknesses, strengths, and forming of strategies for furthering, altering, or redoing graphic approach.

desktop publishing—*see digital publishing*

die-cutting—special bindery operation for unusual cuts, such as windows in covers of publications, pockets for folders, shaped labels, unusual envelope sizes, or inserts. Dies are made from metal razor-edged rules set into a wooden holder. A special press stamps much like a cookie-cutter.

digital—numbers, images, or line data translated into numerical values for manipulation and reproduction by computers.

digital color printing—images are formed on substrate by a computer controlled printing device, including ink jet, laser, or thermal transfer processes.

digital high-end proofs—electronically generated proofs such as Digital Matchprint® or Approval.® They are not from halftoned film but imitate the halftone dot structure, and do not reflect the actual screen patterns that will appear in a printed piece.

digital imaging—process of image capture, manipulation, and final form via computers.

Digital Matchprint®—digital high-end proofs by 3M.

digital presses—printing presses that accept digital information directly from disk to make plates. See direct-to-press printing.

digital publishing—publishing done on computers. Originally called desktop publishing, but now includes midrange and high-end systems. Digital methods streamline and combine many traditional processes. Writers work on word processors, so text is entered only once. The word processing file is imported into the page composition system, where the page composer creates the documents. Film or plates are generated digitally.

digital scanning—capture of photographs or line art into digital form that can be used in a computer, stored on magnetic media, etc. The scanner is a device that "reads" images and converts them into digital information. These files may then be incorporated into electronic pages or output separately.

direct-to-press printing—presses that accept information directly from computer disk to create plates on press (without film and strip-ping).

directory—compilation of location, information, grouping, and arranging files so that can easily be found.

disputes—disagreements over payment amounts, which require negotiation, generally between a buyer and a vendor.

direct billing—when a service business has a vendor bill a client directly. Compensation to the service group for managing processes comes out of a supervision fee, delineated in the project planning.

disclaimer—document that sites limitations of responsibilities and defining of ownerships.

disclosure—*see nondisclosure agreement*

distribution—delivery of messages, services, and products to audiences, customers, or constituents. Common forms include mailing to prospects, circulating informational materials, advertising, Web sites, or broadcast.

document—*see software documents.*

domain—part of naming hierarchy on the Internet address name that represents a Web site. It consists of a sequence of parts separated by dots. In the United States, most commercial sites end in .com, educational sites in .edu, and organizational nonprofit sites in .org. Most names begin with www with the unique url followed by the domain ending. And example is www.allworth.com.

dots—small circular units that make up halftones in color separations for offset printing. Densities are fixed, but sizes are variable.

dot gain—effect in printing when halftone dots expand as the ink is absorbed by a substrate. This is most prevalent on an uncoated, porous paper, and can happen to a lesser degree on coated paper. Because the dots actually print larger than shown on the film, the printing comes out darker than intended—unless dot gain is compensated for in prepress preparation.

dots per inch (dpi)—resolution of image measured by how many pixels are defined in the boundary of an inch.

download—transferring information from one hardware device to another. Typically, a file is downloaded from one computer to another via a modem.

drawing programs—illustration software applications that utilize vector graphics (unlike paint programs which are based on bit mapped graphics), such as FreeHand® or CorelDraw.®

DreamWeaver®—authoring program for Web site design and construction by Macromedia.

drum scanner—optical scanner where reflective or transmissive art is mounted on a rotating drum, allowing light from the image to enter a lens and recorded in a series of fine lines.

dtp—acronym for desktop publishing.

dummy—a blank format or mock-up to show paper stock and weight, size, and bindery of a printed piece. It is created by the print group to provide the buyer with a "feel" of the project, to function as a production guide, and to form a printing budget. Sometimes used by the creative group to sketch in the arrangement of content and components.

duotone—monochromatically printed image created by two overlapping halftone screens of different colors. Tonal range is greater than if image is printed in only one color.

dye-sublimation proofs—continuous tone proofs, such as Iris® or Rainbow®, that uses a colorant deposited on a substrate and treated with heat and pressure, made directly from a digital file. These are often used as preliminary color proofs. They contain no dot structure and show a smoother blend than a halftone dot screen. They may look better than a printed piece, for halftone dots coarsen images, and dyes are more vivid than printing inks.

Dylux® proofs—see blue line proofs.

E

electronic books—fiction or nonfiction set up interactively with CD-ROM or Internet delivery.

embossing—bindery operation similar to die-cutting. A special press holds a custom die to impress images onto paper to achieve a raised surface. Common uses of embossing are raised logos, lettering, or illustrations on stationery, book covers, plastic credit cards and badges, greeting cards, etc., and other specialty uses.

e-mail—electronic mail as a message sent over a network from one computer to another. This is the most popular service on the Internet.

emulsion—the light-sensitive coating on one side of a piece of film or paper.

encapsulated postscript—see EPS.

encrypted files—process of scrambling a signal or file so that special hardware or software is needed to receive and reconstitute for reading or printing.

enhancement—improvement of an image either through color or density changes.

enterprise solutions—software and hardware that work in tandem to carry integrated packaged throughout a LAN for shared usership.

EPS (Encapsulated PostScript)—communication file format that contains all the relevant text and graphics into a readable form for transferring and exporting into other software packages. Generally, the file cannot be altered. *See TIFF and PICT.*

extensions—abbreviated code at the end of a filename that indicates format. It is read by browsers for identification. For example, JPEGs would have filenames ending with .jpg.

fatal error—when a digital document will not open for the imaging center.

file formats—unique arrangement of digital information saved from an application program. Most applications can import and export graphic and text from other applications.

file server—computer which feeds data to other computers, workstations, or devices.

film—transparent material that is coated with photosensitive material, or emulsion. Exposed film becomes black after development, while unexposed areas become transparent. Generally, the transparent areas will print; this form is called film negatives. Film positives are when the black areas will print.

final low resolution proof—document prepared by the page composer on desktop printer as the last step before outputting the file for printing. The client needs to check all elements and proofread. Beyond this point, changes may entail extra time and fees. The final low resolution proof accompanies the digital document to the production group. Low resolution proofs may check color specifications, but the color can be inaccurate. See low resolution proofs.

final proofs—made from the film that will be exposed to the printing plate. Composite proofs incorporate all page elements together as high resolution (or contract) proofs. Used for final client approval. If photographs or illustrations are involved, color corrections have generally been made to these elements on random (or loose) proofs.

finished artwork—*see artwork.*

Flash—a plug-in file format and authoring tool from Macromedia.

flat—in offset printing, paper or film holding segment films required to expose a printing plate.

flatbed scanner—images are placed in a flat glass platen and scanned with linear arrays.

fluorescence—pigments that contain reflective metal powders that

G

absorb light and re-radiate as an extra light or brilliance.

folding—bindery operation that folds press sheets into pages as specified in the prepress format instructions. Some projects need to be scored before they are folded.

foil—colorant of thin metal or plastic that rolls over print area and is adhered on substrate by combination of heat and pressure.

foil stamping—similar to die-cutting or embossing processes. A special press holds a wide ribbon of foil, which is applied to paper when a heated die strikes. The foil that was not heated and pressed onto the paper falls away. Foils are generally shiny, like silver or gold, but are also available in a variety of colors and textures.

font—a single typeface style (which includes all letterforms in the alphabet, numbers, and punctuation). In traditional typesetting, each font was a single size, and had to be formed individually. With PostScript fonts, a typeface drawn in one-size can be digitally scaled to any size.

formats, blank—see dummy.

format, typographical—a group of type specifications such as size, typeface, style, and leading (line space). Often style sheets are set up so that text can be converted or formatted efficiently. *See style sheets.*

four-color process printing—printing with the combination of the four color process inks: cyan, magenta, yellow, and black. *See color separations.*

FPO (for position only)—low resolution representation of an image intended to be replaced by the production group with a high resolution image before final output. It indicates position on a page, but is not meant to represent the output quality. *See build elements.*

FrameMaker®—a page composition software application published by Frame Technology, Palo Alto, CA.

FreeHand®—a PostScript drawing software application published by Macromedia.

FTP (File Transfer Protocol)—in Internet protocol that enables users to remotely access and download files from other computers.

gamut—a viewable or printable color range that can be captured by a device. If a chosen color is outside the gamut of reproduction, the closest approximate color is used.

giclée—French word meaning the "spraying of ink." Fine-art digitally produced prints.

gif—a bitmapped, color graphic file format. GIF is often used on the Web because it compresses files efficiently, but is generally too low resolution for print applications.

gluing—bindery process using adhesive to create the spines of the flat-edged books as the cover is placed on. The pockets in the back of folders are usually glued. Some gluing, as in the latter, is accomplished manually.

gradation—transition between two colors, also a gradient.

gravure—intaglio printing process for large quantities using durable engraved cylinders that retain ink and can print on most substrates.

hand-outs—printed materials to give an audience that parallel presentation.

halftone dots—breaking up an image or color into a series of dots that repeat in a regular pattern, creating the illusion of continuous tone. Halftone screens reproduce color shadings to simulate photographs or illustrations. Screens are available in many resolutions, where the size of the dots and the number of dots-per-inch vary. The finer the dot grid the higher the quality of the reproduction. Digitally, halftone dots are composed of pixels. *Also see screens*.

hard copy—print outs onto paper from digital files, used as reference for checking documents, and as tangible backup.

Hexachrome®—color matching system for the combination of six colors to create larger gamut of reproducible color.

high-fidelity color—specialized color printing process that utilizes balanced dot frequency to extend the color gamut of tradition 4-color offset printing and reproduce near-continuous tone quality images.

high-resolution—generally over 1,000 dots-per-inch, high enough for most output to appear perfect. For halftones to appear perfect, they need to generally be imaged over 2,000 dpi.

html (Hyper Text Transfer Protocol)—is the programming language that browsers and Web servers read to compose pages so that they are formatted appropriately and linked to other documents for use on the Web.

hypertext—text that is linked to documents, extending the page through additional definitions or background information.

IBM (International Business Machines)—hardware company located in Armonk, NY that manufactures computers.

Illustrator®—a drawing software application published by Adobe Systems.

imagesetter—an output device that produces the high resolution output from digital documents onto film or paper.

imposition—the arrangement of pages prepared for a press sheet so that after printing, the pages fold to be sequential. "Printer's spreads" are this arrangement of pages that fold to signatures. "Reader's spreads" are used in presentations where the pages are viewed in the order of reading versus printing.

ink-jet proofs—made directly from the digital file with colored ink sprayed onto paper. These are often used as preliminary color proofs to predict how the color will look.

integral proofs—predict full-page approximations of the way printed pages will look reflecting the halftone dot structure of the film that will be used to expose the plate. Usually made on coated white paper, they cannot show how the halftone dots will behave on the paper. Made from ink-matched powders and laminated material, they have slight color limitations. However, they provide high resolution, registration accuracy, and are reasonably priced.

integrated publishing—the combination of interactive media used with print. Many integrated publishing projects have a disk included in a pocket of a printed document.

integrator—professional who consults, designs, purchases equipment, installs, writes customized software, handles training, and maintains digital systems.

intellectual property—concept, ideas, approaches that are original and made tangible in visual form. It is best if artwork is registered for copyright.

interactive media—a non-printed form of publishing where the file is opened on the computer of the recipient. The creators of interactive documents program the possible choices; the recipient can immediately access the information of interest. Interactive media responds to choices and input made by the recipient.

interface—communication between a system's hardware and software components. Also an interconnection between devices.

intermediate proofs—single-image proofs made from high resolution film. More colloquially called "random" (or "loose") proofs, these show the client photographic or illustrative elements separately from the pages. Prepared prior to full page separations, they are used to adjust color balance, later to be incorporated into full page separations.

Internet—hardware and software that support the interconnection of computer networks, allowing a computer located anywhere in the world to communicate with any other computer also connected to the Internet. The Web supports a variety of services including the World Wide Web.

interplatform exchange—a method where a document can be created in one software application and then opened in another, to utilize the second application's features.

Intranet—local area network system that operates like a private Internet.

Iris® proofs—ink-jet continuous tone color proofs by Scitex.

ISDN (Integrated Services Digital Network)—technology to support both voice and data services over telephone lines.

Java—cross-platform programming language developed by Sun Microsystems, supported by Web browsers.

JavaScript—scripting language that extends html developed by Netscape.

jpeg (Joint Photographic Experts Group)—standardized compression format to reduce the size of graphics used commonly on the Internet.

kerning—adjusting inter-character spacing between letters, usually specific pairs that need special treatment to make more legible and aesthetic.

keyline—manual composition of elements resulting in a pasteup. This artwork is often called a keyline, but the term keyline is a coding method of guides and marks carrying printing instructions for color or images. In digital publishing, such keylines are handled by the computer and the term keyliner (the person who creates keylined artwork) becomes the page composer.

keyword—specially chosen words that represent topic to be researched. Used commonly in Web search engines and online portals. All Web sites are coded with keywords that can be read by various search devices.

kill fee—cancellation fee charged by illustrator or photographer if project is discontinued after begun. Usually 50% of creative fee plus expenses.

laminating—special heat bindery process of bonding clear plastic adhesive material onto printed pieces for protection or appearance. It gives a highly glossy appearance to a printed piece, and is often used as protection on menus or placemats.

LAN (Local Area Network)—used to link specific workstations in one location and requires a dedicated computer as a server. A LAN is a closed system to outsiders; most businesses utilize a LAN system for operations. Collaborative computing as "groupware" is becoming more prevalent: Documents can be worked on by more than one user at a time and resources shared.

laptop computer—a battery-powered portable personal computer with a flat screen monitor that folds for travel.

large format—printer, media, or print that is larger than 24" in width.

licensing agreement—document to define intellectual property ownership and usage, usually specifies size, frequency, and audience coverage.

lien—the right to hold a buyer's property until payment is made.

line art—artwork that has no halftone dot or shading. It is black or white and requires no replacement screens in prepress. Usually a single color in diagrams or drawings, possessing sharp edges and high contrast between areas of the image that have ink and those areas that do not. The higher the resolution, the sharper the image.

M

link—words or graphics that act as pointers to other Web objects or locations. Links as text are usually underlined and may appear in a different color. The connection sends information from one computer to another, for greatest use in e-mail, resources lists, and references to url addresses.

Linotronic® output—a high resolution imagesetter manufactured by Linotype-Hell. The term is often used generically—and incorrectly—to refer to high resolution output from any imagesetter.

Listserv—compiled list of interested audience members, often prospective customers or clients. Communications can be automated, yet customized and personalized, according to responses generated from interactive relationships.

lithography—method of printing from a plane surface, using an ink-receptive carrier. Non-printing areas repel ink when applied to a substrate. The most common kind of printing used today ithe form of offset.

live file—a document file in its original software application to make a "print file."

local area network—*see LAN.*

loose proofs—*see intermediate proofs.*

low resolution proofs—generally, proofs under 1,000 dots-per-inch. Used for checking elements during various phases of a project.

lpi (lines-per-inch)—a measure of halftone screen resolution. *See dpi.*

Macintosh computer—hardware workstation manufactured by Apple Computer, based in Cupertino, CA. The Macintosh with the Apple LaserWriter and early software such as Adobe (formerly Aldus) PageMaker, pioneered desktop publishing.

MacDraw—drawing software program for low resolution images, published by Claris Corp.

Macromedia—San Francisco-based software company that produces Director®, Dreamweaver®, FreeHand®, and Flash®, Shockwave®, among others.

make-ready—preparing a press for printing, using test sheets to check color, density, and parameters, before full project is run.

MatchPrint®—an integral proof product by 3M.

mark-up policies—a handling fee or percentage added onto the cost of outside purchases. This is to cover the financing, accounting, debt carrying, and management of projects.

mechanicals—camera-ready artwork that possess all the assembled elements and instructions for the printer to shoot film. Uses conventional means such as fpo's for images and line art to set up boards for shooting in a stat camera. See keyline and camera-ready art. *See fpo.*

Microsoft Corporation—a Redmond, WA based software company that develops both operating systems (such as MS-DOS® and Windows® for the IBM) and applications (such as Word®, Excel®, Project®, etc.)

Microsoft Windows—*see Windows.*

mock-ups—approximation of concept execution, shown as comprehensive presentation, prototype, or maquette. Usually created by the designer, the purpose is to express how an idea would be carried out and applied.

modem—a device for using the telephone to link two computers. The computers can be different kinds and still exchange files and other information. Telecommunications software is needed to make a modem usable. Modems are the basis for calling up online information services.

monitor calibration—process used to regulate monitor's display of saturation and brightness, synchronizing the visual effect to a standard, helping to approximate the color of final output.

multimedia—the use of more than one media for a communication message, such as the combination of video, sound, animation, and still images for a presentation. Interactive media is a subset of multimedia.

Multiple Masters™—font system developed by Adobe to allow the user to vary characteristics such as width and thickness. (This book is set in Multiple Masters Myriad.)

native files—original document files in the software version in which it was created. It is editable, as opposed to encapsulated export format.

navigation—route through various segments of a document or Web site. Good navigation helps the reader find information easily, is consistent from segment to segment, and sets up a visual language that codes information through color, symbols, buttons, or typographic treatments.

negative—film made in an imagesetter from digital file where image areas expose as open or clear and non image areas are opaque. Open areas will allow light to expose a plate and hold ink, where the areas that are opaque will keep light from plates, allowing ink to be repelled.

network—computers that are connected for the sharing of information and resources. Often involves an extra computer, called a file server, to store and distribute files to all the other computers on the network. See LAN and WAN.

nondisclosure agreement—a document generated by the client that specifies information given to the creative group and the production group as confidential and cannot be shared with anyone outside their organizations. The agreement binds the entire creative and production groups to confidentiality concerning any elements of the project.

notebook computer—*see laptop computer.*

numbering—special printing process where sequential numbers are imprinted on each impression. For example, many forms need individual numbers. This process is often used in specialty direct mail campaigns.

off-the-shelf software—application software packages as purchased from a vendor that only require installation and training, not needing additional programming or system integration.

offset printing—lithographic process that transfer ink from a plate to a rotating blanket that contacts paper, reproducing text and images in quantity.

one-color printing—a single ink color applied to substrate in quantity. It can be black or a special ink color. The color of the paper is not considered a color in printing.

online—activity that occurs while on the Internet or Web.

open architecture—design of a computer system that allows non-proprietary hardware or software development to be used without factor installation.

operating system—the foundational computer software that supports all installed software applications. The operating system makes the computer usable, ruling the user interfaces and basic functions such as saving, copying, printing, etc.

original concept—an idea that is inherently different from other concepts or ideas—one that has not been seen elsewhere by the creator (although all ideas are the combination of pre-existing elements put together in new ways). It is considered original if it contains no recognizable pre-existing copyrighted work. *See concept.*

out-of-pocket expenses—materials, services, purchased outside an organization. Common out-of-pocket expenses include messenger, travel, mailing costs, paper, film, stats, cabs, telephone calls, etc.

outside purchases—services or materials that an organization does not handle in-house. For a design firm, outside purchases could include photography, illustration, writing, prepress, etc. If the origination is billed for services that are then billed to the client, the service provider is a "third-party" firm. The purchaser becomes a liaison between the client and the service provider.

overlay proofs—show each color on a separate overlay, such as Color Key® or ChromaCheck® proofs. Such proofs are an inexpensive way to check spot color, registration, and trapping. They only come in a limited number of colors. This proof technique was developed for four-color process printing, but now is used more often to check two- and three-color projects.

overruns—variations in the printing quantity delivered from quantity ordered. Over- and under-runs are generally not to exceed 10% on print orders for up to 10,000 copies. For orders over 10,000, the acceptable percentage should be established in advance.

page composition—creation of a digital document that contains all the text and graphic elements assembled into publishing layouts. A number of intermediate proofs can be created that provide the client and the creative group flexibility in development. Two of the most common page composition systems are Adobe PageMaker® and QuarkXPress.®

page layout—process and software that prepare artwork, text, and graphic elements compiled and prepared for printing or interactive publishing.

PageMaker®—a page composition software application published by Adobe Systems for the Macintosh and IBM PC. This application helped to pioneer desktop publishing in the early 1980s when introduced by Aldus (later purchased by Adobe).

pagination—formatting of multiple pages in a sequence for output.

paint programs—bit mapped software application to create pixel-based images, such as MacPaint® or PixelPaint®.

Pantone—company name that produces a widely used color system by the same name to control ink colors. Many page composition systems incorporate the Pantone color system. When using process printing, Pantone colors can only be approximated.

parameters—project specifications that include goals, audience, scope, size, deadline, etc.

pasteup—putting adhesive on traditional typeset galleys and photostats, then applying them to keylined boards. *See keylines.*

payment—compensation for work performed.

pen-based input—method of entering data and interacting with a computer using a writing-like device. The user can handwrite and select options on many hand-held computer devices.

personalization—the ability to merge databases with the delivery of a customized messages to allow each delivery to bear customized content for each recipient.

Photoshop®—an image scanning and retouching software application published by Adobe Systems. It pioneered photo manipulation on the desktop.

piracy—*see software piracy.*

PICT—Macintosh software format for storing and transferring visual images from drawing or painting programs.

pixel—short for picture element, a single dot on a computer screen, which is the smallest unit. See bitmap.

plate—a printing plate made of a photosensitive coating applied to think sheet metal. Stripped film is used to expose ("burn") the plate to light. *See negative.* After developing the exposed plate, the areas where there is an image attract ink, while areas that should not be printed resist ink. When the plate is mounted on press, the inked areas transfer to the paper. *See offset lithography.*

platform—hardware or software design and setup. For example, a Macintosh computer and an IBM computer are different hardware platforms. Adobe PageMaker® and QuarkXPress® are different software platforms.

point—unit of measure where twelve points make a pica, which is approximately 1/72th of an inch. Most commonly used in typography and page composition.

point-of-purchase display (POP)—sign or presentation of products in retail environment.

Polaroid®—photographs on an instantly developing substrate.

portals—Web site that compiles information on particular subject, usually providing comprehensive listing of related sites and resources. Presented in a hierarchy based on user choice, links are instant to selected sites.

portfolio—samples of a creative group's work, usually as finished printed pieces done for other clients. These are shown to demonstrate the group's experience and approach to design. A client can expect similar quality from future projects.

PostScript®—a proportionally-based programming language for describing images in terms for printing. Developed by Adobe Systems in the early 1980s, it has become the industry standard device-independent page description language.

pre-flighting—process of checking a project for possible problems (such as missing fonts, Postscript errors, or color inconsistencies) before sending for final output.

preliminary client proof—a low resolution proof showing the rough formatted text and graphic elements in position. The creative group submits this to the client for content and style refinements. It is the best time for the client to make changes.

prepress—preparation of digital files for printing by adding all the necessary technical specifications for outputting. After proofing and approval in low resolution, documents go to an imaging center for prepress functions. Film is prepared, then placed imposition) and used to "burn" the plates that are then mounted onto the printing press. *See negatives, plate, and offset lithography.*

prepress proofs—created by the print group, high resolution approximations are approved by the client. These most closely approximate printing. Final prepress proofs are made from the film that will be exposed to the printing plate. Composite proofs incorporate all page elements together. They are contract proofs, used for final client review. If photographs or illustrations are involved, color corrections have generally been made in these elements on random proofs, then combined with other elements as part of the prepress process.

presentation materials and comps—*see comprehensive presentations.*

press proofs—setting up a printing press to do a preliminary small run. This test-printing the project is the most accurate and expensive proofing method. This is the only way to proof specialty inks such as metallic or varnishes. It can predict the exact way halftone dots will behave on the chosen paper. See progressive proofs.

press sheet—a large sheet of paper for a printing press that may accommodate many pages of a document. Most printing projects, up to a quantity of 50,000 impressions, are printed "sheet fed" which means that images are printed on large sheets, assembled, folded, and then trimmed into a finished publication. Larger printed quantity projects are generally printed "web fed" or on printing presses that handle large rolls of paper, which are then folded and cut.

print file—a file of PostScript code saved to disk and ready to be sent straight to the output system. Advantages of using print files rather than printing the document from the application (see live files) include a cleaner trace if anything goes wrong; inclusion of fonts in the document; ability to link files together for continuous imaging, etc.

print-on-demand—allows copies of printed materials to be ordered in small and exact quantities, rather than in larger quantities to be stored and used as needed.

pro bono—doing work without fee, usually as a donation for a worthy cause, for a relative, or as a favor. Care needs to be taken to know the current laws of the IRS. Trading services is not the same as donation and is taxed differently. Some donations can be written off of taxes. Always check rather than assume.

proofs—approximation of visual intentions at various stages before output occurs. The earliest proofs are typically low resolution and element placement. The later proofs are typically first composite and then made from the same film that will expose the printing plate(s). See final proofs and prepress proofs.

proofread—checking over text and graphic elements for any errors.

process color—color printing using the four ink colors of cyan, magenta, yellow, and black (CMYK). Most colors can be achieved through combinations of halftone screens of these four colors, which produce the widest range of colors with the fewest numbers of inks. Color ranges are limited by the color gamut (see gamut).

profile—database or file of values that apply to a device to allow color specifications. Used to match other devices in color or image capabilities and ensure successful output.

progressive proofs—proofing each color of a print project separately and in various combinations with the other ink colors. For example, the blue image is proofed alone, then in combination with the yellow, then with the yellow and magenta, etc. until all fifteen combinations have been printed. This is generally done for critical color projects to precheck how one color reacts with another before engaging the full press run. See press proofs.

proofs—prints used to evaluate particular phase of production process prior to final output, reproduction, or delivery. See prepress, preliminary proofs, final proofs.

purchase order—document that gives brief requested product or service description with fee agreements, securing order or commission to proceed with work. Used most often for purchasing materials or services in advance of receiving an invoice for payment.

QuarkXPress®—a page composition software application published by Quark, Inc. based in Denver, CO. Modular additions are called QuarkXTentions.®

QuickDraw®—a screen representation model within the Macintosh operating system. It is also used for dot-matrix printing and other non-PostScript printers, by duplicating (perhaps in higher resolution) what shows on the screen.

QuikTime®—desktop device-independent video presentation software developed by Apple Computer.

Rainbow®—dye-sublimation continuous tone desktop color proffer from 3M.

RAM (Random Access Memory)—high-speed portion of a computer's memory that is allotted for use in running applications or performing procedures.

random proofs—*see intermediate proofs.*

readers' spreads—pages laid out in the order that the recipient will view them, usually as two-pages side-by-side as in book form. *See signatures and printers' spreads.*

reflective art—photographs, illustrations, or other two-dimensional visuals that are created with traditional means such as prints, paintings, drawings, etc. These need to be mounted into a copy camera for color separation or digital conversion. With the introduction of drum scanners, it became more efficient to shoot a photographic transparency of reflective art, to mount around the scanner's cylinder for color separation. With digital publishing, more artwork is created electronically and color separations are part of the image manipulation or prepress software.

register marks—a symbol (usually a cross and circle with the same center) placed outside the trim of a publication, to guide color alignment. The page composer places some on digital documents, others are used in prepress, and still others appear on press sheets. Occasionally the same register marks are used throughout all three phases.

registration—how one color image fits with another color image when reproduced. Because image of different colors are on different plates in the printing process, how they print in relation to one another is critical to quality. Generally, out-of-register printing leaves white gaps between colors and seems to perceptually blurs halftones.

remote printing—publications set up so that the output can be on the receivers' end. Quality depends on the printing device of the recipient.

repurposing content—adapting messages, information, and a variety of content from one medium to another.

research—seeking answers and background on a desired subject which includes direct experience, experiments, reading publications, investigating sources for information, gathering data, collecting facts, and interviews with appropriate people. In graphic projects, this may include interview with client management staff, customers, and suppliers. The Internet has opened many access doors for research and development, as the Web was initially set up for this purpose.

Researcher—person who finds all relevant fact, handles any market studies, and accesses information sources.

resolution—coarseness or fineness of images based on dots per inch (dpi) or pixels per inch (ppi). The higher the number of dots the higher the resolution, the greater the detail of an image.

retainer—spreading out project payments into a regular schedule to smooth out cashflow.

retouching—removing imperfections from an image. Traditionally done on prints or transparencies by an airbrush craftsman, digitally done in image manipulations software.

RIP (Raster Image Processor)—interpret digital publications into dots for high resolution output. Operates with mathematical algorithms that manipulate and print text and images.

rollover—effect in Web site design to make navigation buttons change when the user's cursor goes over it. Generally signifies a selection to be made and a popup window with more information.

ROM (Read Only Memory)—allows images to be read, but not changed or rewritten.

royalties—percentage of unit price sold.

rush charges—charges resulting from overtime or priority work with the production group. The group who requests changes normally are responsible for related charges. Generally, the creative group does not have rush charges but is compensated for client changes through direct estimates and additional billing.

saddle-stitching—*see stapling.*

scanning—digitally capturing a reflective image. Refers to traditional color separation process where images are placed in a camera or on a drum or bed. *See digital scanning.*

scrambling—mixing up or coding information so that it cannot be read without translation.

screens—color or shading created by halftone dots. Screens are measured by the shape of dots and number of dots-per-inch. The dots have space between them which gives the effect of a lighter color. The lower the screen percentage, the lighter the color. *See dpi.*

scripting—programming sequences of actions and animations for presentations. Often uses programs such as Macromedia Director® or JavaScript©.

search categories—using key words to find information through

S

browsers or portals.

separated proof pages—final low resolution proofs that have one page (printed in black) for each color layer. This shows the production group how the color separations are behaving.

separations—splitting or converting images into four-color process colors for offset printing.

service bureau—company that offers custom output services.

sheet-fed printing—common printing method that uses sheets, rather than rolls of paper as substrate. *See Web printing.*

shelf life—length of time that a product is usable after it is produced.

ShockWave®—software plug in produced by Macromedia to make files Internet-ready. Handles animation and Quiktime® movies.

signature—printed sheet containing several pages that are arranged to fold, trim, and bind to create a multi-page finished piece.

signature proofs—printed pages made on a small specialty press to show inks on the actual paper to be used for printing.

site management—keeping track of elements and content of a Web site, maintaining functionality, updating, and interactive communications. There are software programs that to help author and manage sites.

soft proof—image reviewed on a computer monitor. Some creative groups show images to the client on screen for preliminary approval. This can be useful for stock photographs and illustrations, or to show initial rough layouts of elements. It is also good for presentations over the Internet to decision-makers in remote locations. If used skillfully, it can save significantly on travel costs.

software application—programs purchased from software developers that are loaded into or accessed by a computer's operating system to create documents.

software document—a prepared digital file, as opposed to the software application used to create it. A document is generally commissioned by the originating client, created by the creative group, and printed or distributed by the production group.

software piracy—the borrowing or copying of software. Because this is so easy to do, many users do not realize that this practice is illegal. It is also very dangerous for the development of the software industry, as income needed for research and development of future products is lost when users don't pay for the product. If a user wishes to try out a software package, try a demonstration version which the Web makes easy to download.

spot color—refers to how color is handled without color separations in many two- or three-color projects. The printing inks can be specialized such as metallics or varnishes. The second or third color is handled in selected areas such as text or line elements. Many page compositions are quite sophisticated in their handling of spot color separations.

spread—two pages of a printed document as they will appear across from each other in the printed piece. *See readers' spreads.*

stapling—bindery operation of fastening a publication together along the spine. Usually for booklets, brochures, and magazines. This process is often called saddle-stitching.

stat camera—a large copy camera that is configured to shoot line art to size or fpo low resolution images. For keylining. Stat is short for photostat. *See fpo.*

stationary—letterhead, envelop, and business card design as essential business papers.

stochastic screening—random-placement of printing dots to render high resolution and large format images. Unlike halftones, dots are uniform in size, and darker areas have more dots closer together.

stock images—images that already exist, either as outtakes from previous creative projects, or made specifically for sale as stock. Stock images may exist in traditional or digital form. Ownership needs to be checked before using.

stock photographs—existing images that can be purchased for use instead of commissioning a photographer to shoot a specific photograph.

strip-ins—last minute changes made on high resolution paper or film output. The text or images that change are pasted into the output.

stripping—assembly of high resolution film into the configuration, or imposition, necessary for exposing plates. In digital publishing, pages are often composed in position with low resolution fpo. *See fpo.*

style sheets—*see formats, typographical.*

substrate—material that receives a printed image, most commonly paper, cloth, or vinyl.

subtractive color—reflective color space usually prepressed by CMYK in traditional printing.

supervision fee—payment for overseeing third-party activities, such as when an art director charges for printing supervision, but the printer bills the client directly.

system configurations—computer hardware and software.

systems integrator—professional or company that combines various hardware and software products into a single operating system. Often encompasses programming, maintenance, and updating.

templates—electronic page grid and style sheets for design and typography of publications. Templates are set up for recurring publications such as newsletters, magazines, brochure series, reports, or proposals. Many clients hire design firms to create templates which are then carried out by production staff from issue to issue.

thermal transfer—printing process that uses heat to apply colored dye, wax, or resin onto paper.

third-party services—*see outside purchases.*

thumbnail—small, low resolution representation of a larger, high resolution image.

TIFF (Tagged Image File Format)—standard for gray-scale or color images, usually scanned, for transfer, storage, or placement into a digital document. Especially for graphics that will be used in many applications or platforms.

trade secrets—proprietary organizational information that gives competitive advantage. Usually protected by a confidentiality agreement. *See confidential agreements.*

traditional publishing—begins the same way as the digital process: with a client-inspired project. The manuscript is typed on a typewriter, then re-keystroked by a typesetter. Rough galleys (repro proofs) are generated for proofreading, followed by corrected final galleys. The production artist (keyliner) trims final galleys, applies adhesive, and keylines them onto boards. Illustration and photography as handled by fpo "position stats" (black and white low resolution representations), also applied to the boards. The printer receives the keylines, with the illustrations and photographs separately, to place into a copy camera to separate and shoot film. Color images are separated on a copy camera using filters. The resulting pieces of film are stripped together into flats (larger sheets that match the way the pages will print on the press sheet). The film is contacted (exposed by light in a vacuum frame) to a photo-sensitive, lithographic printing plate. The plate is mounted onto the printing press where it accepts the ink, transfers it onto a substrate, usually paper. *See keyline and fpo.*

tree format—figuring the various possibilities of choices in interactive presentations, predicting various user paths to make sure essential information is conveyed.

trimming—a bindery operation which takes assembled press sheets and cuts the publication to the specified size. *See signatures and bleed.*

trapping—when one color overlaps another so there is no white space showing when they print slightly out of register. This overlap is traditionally handled by the print group, but with the advances of digital publishing, more and more page composers are handling it in page composition. *See screens and register.*

TrueType™—a typeface system created jointly by Microsoft and Apple.

turnkey—computer software or hardware system that is operable right out of the box without additional programming or purchases.

typography—the craft of putting letters together to form words that are easy to read and aesthetic. Covers entire range of type design and usage.

under-runs—*see over-runs.*

unit price—total project costs divided by the quantity of final pieces.

UNIX®—a multi-tasking operating system used in Apollo, Sun MicroSystems, Silicon Graphics, Hewlett-Packard, and other work stations, developed by AT&T.

upgrade—to improve a computer system by replacing versions with newer ones. Usually upgrades are denoted by version numbers.

url (Universal Resource Locator)—specification identifying any file on the Internet. It contains the name of the protocol by which the file should be accessed, the domain name of the site, and the path name of the file on the server. *See domain.*

vaporware—"not-yet-released" but much talked-about software. Not a good investment.

VAR (value-added retailer)—dealers who cross-sell with other brands, touting connectivity, or offer additional services (such as training or free upgrades) when selling products.

vector graphics—graphics (such as PostScript) characterized by direction and scale pixels rather than bits. It is based on start and end points, and relate to line segments, type, and tints, using mathematical descriptions of paths and files to define images. See bitmaps.

Velox® **proofs**—reproduction-quality proofs made directly from film negatives. They are high-resolution black images on white paper.

Ventura Publisher®—a page composition software application published by Xerox Corporation.

vignette—transition from one color or intensity to another. *See gradations.*

virtual—having the appearance or representations of existence rather than actual reality.

virtual reality—interactive 3D graphics that simulate realism to users.

W

WAN (Wide Area Networking)—a hard-wired system similar to a LAN but designed to operate efficiently over greater distances. The network can link several LANs together. It also most commonly is linked to the Internet. *See LAN.*

watermarking—distinctive identification in a file to protect ownership. Originally watermarking is an embedded background image into the formulating of paper, such as a logo or an emblem.

web printing—printing method that uses rolls (webs) rather than sheets of paper to achieve high quantity, quality, and speed. See sheet-fed printing.

Web site—online document with html tags to specify formatting and links, with its own unique domain name. *See domain.*

windows—rectangular panels as interface devices on the computer screen to show choices, processes, messages, documents, options, etc. The Macintosh was the first mainstream computer to use windows, which were then mirrored for the IBM PC by Microsoft in their operating system called Microsoft Windows.

work-arounds—alternative method or short-cut for achieving a result (usually with software) that can't be directly achieved due to software bugs or system failure.

workflow—progression of project functions and activities from one process to another, sometimes facilitated by management software.

workgroup—interactive team collaborating on project, utilizing integrated systems and software.

working proofs—digitally generated intermediate tests, generally with no halftone dot-structure. They are mainly used during the development phases of a project and are not intended to be seen by the client. *See proofs.*

www (World Wide Web)—Internet service that provides files linked by html incorporating formatted text and graphics to be viewed by computers interactively anywhere in the world. The Web is the most popular Internet service, and e-mail the most popular application.

WYSIWYG (What You See Is What You Get)—monitor configured to represent a displayed image to approximate output result.

XENIX®—Microsoft's version of the Unix operation system.

QUOTED PROFESSIONALS

INDEX

ORGANIZATIONS  RESOURCES

MAGAZINES

www.3dartist.com
3D Artist 126

www.adbusters.org
Adbusters 101

www.bigpicture.net
Big Picture 208

www.commarts.com
Communication Arts 50, 101

http://cgw.pennnet.com/home.cfm
Computer Graphics World 208

www.adage.com
Creativity 101

www.dsphere.net
Design Sphere 126

http://lexx.nbm.com/index.php?page=dggeneral
Digital Graphics 288

www.digitalout.com/
Digital Output 266

http://ep.pennnet.com/home.cfm
Electronic Publishing 208

www.gammag.com
Graphic Arts Monthly 288

www.dgusa.com
Graphic Design USA 101

www.graphis.com
Graphis 101

www.howdesign.com
HOW Magazine 50, 101

www.idnworld.com
IdN Magazine 101

www.modrepro.com
Modern Reprographics 288

www.internet.com/newmedia
Newmedia Magazine 208

www.newmedia.com
NewMedia Magazine 126

www.pdn-pix.com
Photo District News 158

www.printmag.com
Print Magazine 101

www.publish.com
Publish Magazine 196

www.productpress.co.uk/sdphome.html
Screen & Digital Printer 288

www.media.sbexpos.com
Seybold Reports 208

www.transformmag.com
Transform Magazine 184

www.webtechniques.com
Web Techniques 208

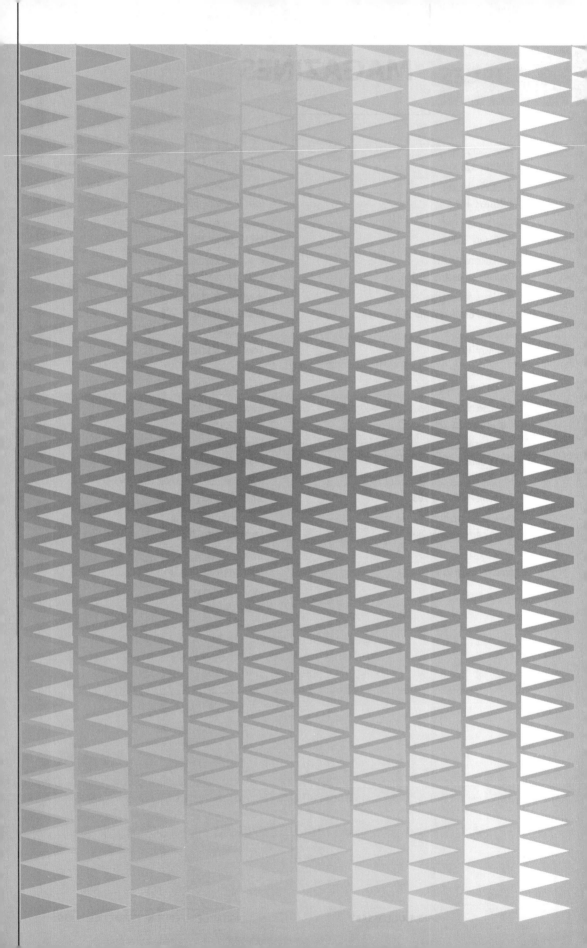

ABOUT THE AUTHOR

Liane Sebastian is a publishing artist: art director, graphic designer, writer, illustrator, and painter. With a career that spans every facet of graphic visual communication, Liane has served clients as a freelancer, design firm president, corporate manager, and a specialized boutique. With a BFA from Northern Illinois University and post-graduate studies at the School of the Art Institute of Chicago, Liane brings a strong artistic and printing background to her work. This is her ninth book: Previous titles include two prior editions of this book, two books on the IDCA design conference, several commissioned corporate profiles, and *Ways of Wisdom,* containing her illustrations with quotations (see **www.wofw.com**). Liane has served national and international clients with award winning work for the American Bar Association, Borg-Warner Corporation, the Economic Development Commission of Chicago, Helene Curtis Incorporated, Herman Miller Corporation, Hyatt International Corporation, the Great Books Foundation, the Illinois Arts Council, and World Book Encyclopedia, among other organizations (see **www.michaelight.com**). As president of ADEPT (Association for the Development of Electronic Publishing Technique) from 1990–1994, Liane's skills have grown up with the technology and all the issues it raises in running a design business. Writing about design and digital publishing, she has had articles in *HOW* magazine, *Publish!, Step-by-Step Publishing,* and *Crain's Chicago Business.* She has exhibited drawings, paintings, and prints in Chicago, Santa Fe, St. Louis, and Detroit galleries. Speaking engagements include Seybold, Graphx, AIGA, PIA, Folio, and CONCEPPTS. Liane lives and works in Evanston, IL.

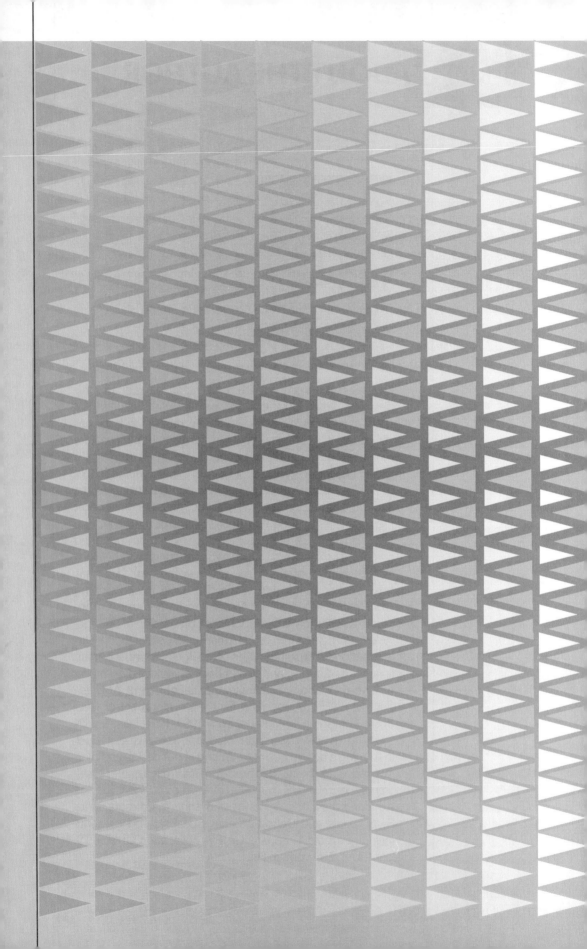

𝔄 Books from Allworth Press